Lecture Notes in Computer Science   9464

Commenced Publication in 1973
Founding and Former Series Editors:
Gerhard Goos, Juris Hartmanis, and Jan van Leeuwen

Lina Yao · Xia Xie
Qingchen Zhang · Laurence T. Yang
Albert Y. Zomaya · Hai Jin (Eds.)

Advances in Services Computing

9th Asia-Pacific Services Computing Conference, APSCC 2015
Bangkok, Thailand, December 7–9, 2015
Proceedings

 Springer

Editors

Lina Yao
School of Computer Science
University of Adelaide
Adelaide, SA
Australia

Xia Xie
School of Computer Science
and Technology
Huazhong University of Science
and Technology
Wuhan
China

Qingchen Zhang
School of Software Technology
Dalian University of Technology
Dalian
China

Laurence T. Yang
Department of Computer Science
St. Francis Xavier University
Antigonish, NS
Canada

Albert Y. Zomaya
School of Information Technologies
University of Sydney
Sydney, NSW
Australia

Hai Jin
School of Computer Science
and Technology
Huazhong University of Science
and Technology
Wuhan
China

ISSN 0302-9743   ISSN 1611-3349   (electronic)
Lecture Notes in Computer Science
ISBN 978-3-319-26978-8   ISBN 978-3-319-26979-5   (eBook)
DOI 10.1007/978-3-319-26979-5

Library of Congress Control Number: 2015954994

LNCS Sublibrary: SL3 – Information Systems and Applications, incl. Internet/Web, and HCI

Springer Cham Heidelberg New York Dordrecht London

Springer International Publishing AG Switzerland is part of Springer Science+Business Media
(www.springer.com)

Preface

It is our great pleasure to welcome you to the proceedings of the 2015 Asia-Pacific Services Computing Conference (APSCC 2015).

APSCC 2015 was held in Bangkok, Thailand, during December 7–9, 2015. The event was the ninth meeting of this conference series, after APSCC 2006 (GuangZhou, China), APSCC 2007 (Tsukuba Science City, Japan), APSCC 2008 (Taiwan), APSCC 2009 (Singapore), APSCC 2010 (Hangzhou, China), APSCC 2011 (Jeju, Korea), APSCC 2012 (Guilin, China), and APSCC 2014 (Fuzhou, China).

APSCC is recognized as the main regular event of the Asian-Pacific region that covers many dimensions of services computing, Web services, cloud computing, security in services, and social/peer-to-peer/mobile/ubiquitous/pervasive computing. APSCC 2015 intended to play an important role for researchers and industry practitioners to exchange information regarding advancements in the state of the art and practice of IT/telecommunication-driven business services and application services, as well as to identify emerging research topics and define the future directions of services computing.

We received a large number of submissions this year, showing by both their quantity and quality that APSCC is a premier conference on services computing. In the first stage, all papers submitted were screened for their relevance and general submission requirements. These manuscripts then underwent a rigorous peer-review process with at least three reviewers per paper. Finally, 23 papers (17 full papers and 6 short papers) were accepted for presentation at the conference and are included in the main proceedings. To encourage and promote the work presented at APSCC 2015, we are delighted that some of the papers will be accepted in special issues of several international reputable journals. All of these journals have played a prominent role in promoting the development and use of services computing.

An international conference of this scale requires the support of many people. First of all, we would like to thank the steering chairs, Hai Jin and Liangjie Zhang, for nourishing the conference and guiding its course. We appreciate the participation of the invited speakers, Tarek El-Ghazawi and Stephen S. Yau, whose speeches greatly benefited the audience. We are also indebted to the members of the Program Committee, who put in hard work and long hours to review each paper in a professional way. Thanks to them all for their valuable time and effort in reviewing the papers. Without their help, this program would not have been possible. Thanks also go to the entire local arrangements committee for their help in making the conference a wonderful success. We take this opportunity to thank all the authors, participants, and session chairs for their valuable efforts, many of whom traveled long distances to attend this conference and make their valuable contributions. Last but not least, we would like to express our gratitude to all of the organizations that supported our efforts to bring the conference to fruition. We are grateful to Springer for publishing the proceedings.

December 2015

Lina Yao
Xia Xie
Qingchen Zhang
Laurence T. Yang
Albert Y. Zomaya
Hai Jin

Organization

Executive Committee

Steering Committee

Hai Jin   Huazhong University of Science and Technology, China
Liang-jie Zhang   Kingdee International Software Group Company Limited,
                           China

General Chairs

Albert Y. Zoyama   University of Sydney, Australia
Laurence T. Yang   St. Francis Xavier University, Canada

Local Chair

Punpiti Piamsanga   Kasetsart University, Thailand

Program Chairs

Lina Yao   University of Adelaide, Australia
Xia Xie   Huazhong University of Science and Technology, China

Publicity Chair

Wenbin Jiang   Huazhong University of Science and Technology, China

Program Committee

Liang Chen   Zhejiang University, China
Shuiguang Deng   Zhejiang University, China
Zhijun Ding   Tongji University, China
Wei Dong   Zhejiang University, China
Bin Guo   Northwestern Polytechnical University, China
Jinsong Han   Xi'an Jiaotong University, China
Haiwu He   Inria, France
Yuan He   Tsinghua University, China
Chunming Hu   Beihang University, China
Xiapu Luo   The Hong Kong Polytechnic University, China
Yutao Ma   Wuhan University, China
Zhengwei Qi   Shanghai Jiao Tong University, China
Kaijun Ren   National University of Defense Technology, China
Hongbing Wang   Southeast University, China
Jian Yu   Swinburne University of Technology, Australia

Liping Zhao	The University of Manchester, UK
Tadashi Dohi	Hiroshima University, Japan
Nuno Laranjeiro	University of Coimbra, Portugal
Chung-Ming Huang	National Cheng Kung University, Taiwan
Zhenhua Li	Tsinghua University, China
Jin Zhao	Fudan University, China
Hongzhi Wang	Harbin Institute of Technology, China
Lei Wang	Southeast University, China
Zhihui Zhan	Sun Yat-Sen University, China
Jian Wang	Wuhan University, China
Yuhua Qian	Shanxi University, China
Fu Chen	Beijing Foreign Studies University, China
Qi Yu	Rochester Institute of Technology, USA
Shangguang Wang	Beijing University of Posts and Telecommunications, China
Zhuofeng Zhao	North China University of Technology, China
Hao Chen	Hunan University, China
Xianzhi Wang	The University of Adelaide, Australia
Jing Bi	Tsinghua University, China
Minghong Liao	Xiamen University, China
Yepang Liu	Hong Kong University of Science and Technology, China
Hailong Sun	Beihang University, China
Yanghua Xiao	Fudan University, China
Yongluan Zhou	University of Southern Denmark, Denmark
Xumin Liu	Rochester Institute of Technology, USA
Yinliang Yue	Chinese Academy of Sciences, China
Xuanzhe Liu	Peking University, China
Eduard Babulak	Fairleigh Dickinson University, Canada
Chun-Yuan Lin	Chang Gung University, Taiwan
Qing Liu	CSIRO, Australia
Teo Yong-Meng	National University of Singapore, Singapore
Depei Qian	Beihang University, China
Lai Xu	SAP Research, Switzerland
Venky Shankararaman	Singapore Management University, Singapore
Qiang He	Swinburne University of Technology, Australia

Invited Talks (Abstracts)

When HPC, Big Data Science, and Wireless Technology Merge: The World of Endless Opportunities and Challenges

Tarek El-Ghazawi

Abstract. The rapid spread of interest and use of cloud computing as an accessible and expandable, as needed, enabling computing facility on the go, from one side, and the advances and the proliferation of intelligent mobile devices from another side, and the increasing availability of high-performance computing capabilities and parallelism from accelerators on-chips to large scale systems in the cloud are compatible exciting developments. With the ever expanding wealth of big data in the cloud, users can have very powerful enabling tools wherever they go. Together, these technologies have the potential of leaving nobody behind when it comes to computing and data applications whether small and personal or large and organizational, and regardless of geographic boundaries and economic conditions. Furthermore, many applications and services that one felt are simply fiction, will become possible. However, the technical challenges associated with the realization of this dream with the responsiveness and quality needed from the user perspective will be monumental. In this talk we examine some of those possible developments and characterize some of user needs, the associated challenges, and the potential research directions.

Challenges and Future Research Direction of Developing Trustworthy Services Computing Systems

Stephen S. Yau

The rapid advances and growth in deploying services computing systems, including cloud computing systems, in various applications have major impacts on the use of IT technologies, the economy, society, and our daily lives. Trustworthiness becomes a key issue for users to have sufficient confidence in using these systems. In this address, the major challenges and future research directions of developing trustworthy services computing systems will be discussed. The important aspects of trustworthy services computing systems, including quality of services assurance, required system platform support, and sharing resources (including data, infrastructures, and knowledge), and the impact of human factors will be addressed. Possible improvements of relevant higher education curricula to meet these challenges of rapid expansion of IT Technologies and their applications will also be discussed.

Contents

Short Papers

Regular Papers

A Context-Aware Usage Prediction Approach for Smartphone Applications

Jingjing Huangfu, Jian Cao$^{(\boxtimes)}$, and Chenyang Liu

Department of Computer Science and Engineering,
Shanghai Jiao Tong University, Shanghai 200240, China
hfjingjing@gmail.com, cao-jian@sjtu.edu.cn, schumeichel_2003@163.com

Abstract. With the popularity of smartphones, an increasing number of applications (app) are installed on common users' smartphones. As a result, it is becoming difficult to find the right apps to use promptly. Based on the observation from the real data, it can be found the correlative relationship exists between the usage of app and the context, specifically, time and location contextual information. According to this analysis, a context-aware usage prediction model is proposed to predict the probability of launching apps and present this prediction by an ordered list. Furthermore, a dynamic desktop application for android platform is developed to adjust the app icon order on the desktop according to the current time and location information, which facilitates the smartphone users always capable finding their needed ones in the first page. The experiments show that our prediction model outperforms other approaches.

Keywords: Context-aware · Mobile app · Usage prediction model · Dynamic android desktop

1 Introduction

With the increasing functionality provided by smartphone, it has great impacts on common people's lives and becomes an indispensable tool. Diverse mobile apps have been developed to satisfy different requirements at any time or place. For example, users can post their pictures using social apps (e.g. microblog and wechat), plan their driving routs by map application (e.g. Google Map), and search the Internet through an explorer. A survey on 4200 users indicates that 14 % of users downloaded a new app in most recent 30 days [1]. According to statistics, in 2013, 22 apps are installed on each Android smartphones, and 37 for iPhone users respectively on average. With the prosperous of mobile phone application market and upgrading of smartphone storage, the number of applications installed in one phone will definitely increase.

With the increasing number of apps installed, common users have to spend more time finding the right app to user promptly, which reduces user experience. Android and iPhone provide some tools to help users find their app. Users can

© Springer International Publishing Switzerland 2015
L. Yao et al. (Eds.): APSCC 2015, LNCS 9464, pp. 3–16, 2015.
DOI: 10.1007/978-3-319-26979-5_1

organize similar apps into a folder or type in keywords to search for the right app. The former method requires users to maintain their app-list regularly, whereas the latter one requires users to describe the function they need or the features of apps. Both of these two methods are time consuming and can be substituted by more intelligent methods.

The existing widely used methods are Most Recent Use (MRU) and Most Frequently Use (MFU) strategies. These methods are already integrated into smartphone system and help customize user's menu and reach a fair effect in the mass. However, these methods consider none of context information, e.g. user's usage pattern of mobile apps and current time and place. Therefore, they have lower prediction accuracy on mobile app usage.

We propose an app usage history-based and context-aware prediction model to provide a list of apps ranked according to the probability that each app will be launched. Moreover, a dynamic desktop application is developed for Android platform.

The contributions of our work can be summarized as follows:

- We present a usage history-based and context-aware model to predict the probability of an app usage accurately.
- We compare our model with MFU and MRU models in experiment. The result shows our model has evident superiority on prediction accuracy.
- We develop a dynamic desktop application for Android platform to present the prediction result, which can facilitate users finding the right app in the first page.

The rest of this paper is organized as follows. In Sect. 2, we summarize related previous work on usage prediction. In Sect. 3, we introduce a real dataset and evaluation standards. In Sect. 4, we describe our usage history-based and context-aware model mathematically and theoretically. We show the experimental setup and analyze the results in Sect. 5, followed by the introduction to our desktop application in Sect. 6. Finally, Sect. 7 concludes the whole paper.

2   Related Work

Smartphones are carried and widely used in different environments [2]. Therefore, taking advantage of user context information can benefit the analysis of user behavior. e.g. motion pattern recognition [3], health protection [4] and battery management [5]. Many studies revealed the relationship between the context information and user's app usage preference. For example, Eagle et al. revealed how this preference changed with dynamic social environment through Reality Mining [6]. They found user preferred to use voice communication than text message. Froehlich et al. analyzed the relationship between locations and the using habit of SMS (Short Message Service). They found user preferred to use SMS when they were moving [7]. Bohmer et al. observed large amounts of users and found the next probably used app was strongly correlated with the previous app usage sequence [8]. However, these studies only provide a non-quantitative model to reveal how the context information has influences on user's app behavior.

Some previous work has attempted to apply machine learning algorithm in app usage prediction. For example, in [9], it was proposed to cluster apps by time and location, and then next probably used apps were predicted through constructed clusters. It has slight 3% accuracy improvement than MFU. This work treated time and location equally in clustering approach, which ignores the difference contribution of time and location. In [10], it was reported a research based on the data collected from 12 kinds of sensors. The approach was based on 37 features and employed an Information Gain (IG) model. However, for a real-time Android system, collecting so many sensor data has to result in heavy burden on battery consumption. Our work aims to reach relatively high accuracy with less sensor data. In conclusion, the existing work considers app usage prediction as a classification problem and applies machine learning models directly without considering the personal inherent app usage pattern.

3 Dataset and Evaluation Metrics

3.1 The Real Dataset

In this subsection, we introduce the real dataset for this work. We use the data collected by a telecommunication company that records user's app usage information, including *UserID, AppName, RecordTime, NetworkMode, City* and *Gcell ID*. The dataset consists of the app usage records from more than 56000 users on appropriate 7700 apps during 60 days.

Our app usage history-based model relies heavily on abundant historical records. So we processed the data set with following restrictions:

- A user with enough data is the one whose records are more than 280;
- The span of the records for each user should be longer than 10 consecutive days;
- For *Gcell ID*, it must be an active public cell that has at least 100 data records.

Finally, we obtain a dataset of app usage records from 885 users on 1633 different apps. The features of the dataset are summarized in Table 1.

Table 1. The summary of dataset

User Number	App Number	Record Number	Time Span/day	Gcell Number
885	1633	372377	60	674

3.2   Evaluation Metrics

In our experiment, we use several metrics to evaluate the performance of our prediction model. We borrow precision and recall concepts from top-N recommendation [11]. Given a sorted prediction list p containing top N prediction result, as well as an actual usage list q, precision is defined as the percentage of those apps appears in p and q simultaneously among p. Recall is defined as the percentage of those apps appears in p and q simultaneously among q. Both metrics can measure the accuracy of prediction list, from different perspectives. The two metrics are computed by Eq. 1.

$$precision@N = \frac{|p \cap q|}{|p|} \qquad recall@N = \frac{|p \cap q|}{|q|} \qquad (1)$$

Since the prediction list is ordered, the larger the prediction result, the more possible it will be launched by the user. So we introduce an evaluation standard based on sorted list. The prediction list is finally presented in the form of a group of icons on the smart phone screen, in a descending order of its predicted usage probability. Suppose if two apps are displayed on the same screen page. User has same accessibility to them. As a result, our prediction aims to put every app in the right page of screen, rather than the exact place in a page. We use *RankScore*, which is a value between 0 and 1 to describe whether an app is put in its right page. 1 indicates it is in the right page. The lower the *RankScore*, the further it is placed from its right page. The following Eq. 2 defines average *RankScore* for all apps in the prediction list,

$$RankScore = \frac{1}{|I|} \sum_{app_i \in I} \frac{1}{|P(i) - Q(i)| + 1} \qquad (2)$$

where I is the set of all apps in the prediction list, $P(i)$ is the page number that app_i falls in the prediction list, and $Q(i)$ is the page number app_i falls in the actual using list. The following Fig. 1 gives an example of *RankScore* in two screen pages situation. In this example, A is a set of those apps belong to page 1 and fall into page 1. B is a set of those apps belong to page 2 and fall into page

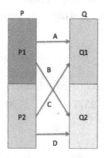

Fig. 1. *RankScore* of two page screen situation

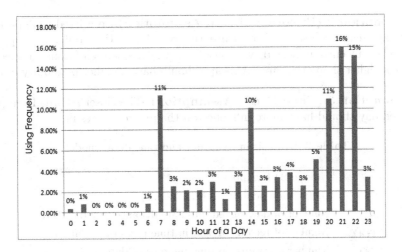

Fig. 2. Using habit varies in a day

1. Similarly, C is a set of apps belong to page 1 and fall into page 2. D is a set of apps belong to page 2 and fall into page 2.

4 A Hybrid Usage Prediction Model

In this section we present our model in detail. It is a hybrid model consisting of three parts. Each part considers different context information. We first introduce each part and then introduce how the three parts are combined together.

4.1 Frequency and Time Decay Model

In our dataset, we segment user's app behavior data hourly and find app usage changes in different periods of a day. For example, Fig. 2 shows a user's usage frequency on app "QQ" in a day varies with time in a day. The user uses "QQ" more frequent in 7:00, 14:00, 20:00, 21:00 and 22:00. Therefore, we suppose most users have fixed app usage habits, but vary with time.

Finally, we decide to divide one day into 8 time-periods, i.e., each period spans 3 hours. The prediction list changes every 3 hours. We believe it is enough to reflect a user's status change. User's historical usage records reflect his app preferences to some extent. Therefore, the first part of our model is based on the following three assumptions:

– **Assumption 1 (Frequent Time Period Assumption)**: An app can be used in any time period of a day. If the frequency of using this app in one period is larger than that of other periods, the app will be used with higher possibility in this time period.

- **Assumption 2 (Preferential App Assumption)**: In a fixed time period, any app can be used. To measure the possibility that different apps be launched in current period, we consider if the frequency of using a specific app is higher than other apps, this app should have a higher possibility in this time period.
- **Assumption 3 (Time Decay Assumption)**: The closer records from predicted day should have more influence on the prediction result.

Table 2. Symbols in frequency and time decay model

i	Unique ID for app
t	ID for time period, $t \in \{1, 2, 3, ..., 8\}$
d	Number of day, casted into date distance to 2015-01-01
$c_{i,d,t}$	Number of records of app_i in time period t of date d
$c_{d,t}$	Number of records in time period t of date d
$p_{i,d}$	Number of records of app_i in date d
$d_{predict}$	Predicted date
$s_{i,d,t}$	Possibility score app_i get in time period t of date d

We calculate every previous day's influence on predicted day, and then use Eq. 3 to compute the possibility score of each app. This Equation is composed of three factors. Each factor corresponds to one of the above assumptions. D in Eq. 3 indicates all previous days that have the records of app_i.

$$s_{i,d,t} = \frac{1}{|D|} \sum_{d < d_{predict}} \frac{c_{i,d,t}}{c_{d,t}} \frac{c_{i,d,t}}{p_{i,d}} e^{-\{1+(d_{predict}-d)\}} \tag{3}$$

4.2 Stickiness and Cycle Model

According to common sense, a user has different usage frequencies for different apps. For example, for social communication apps, like "QQ" and "Wechat", user may use them every day. While other apps like "video", user may use them once several days. Therefore the concept *cycle* is introduced to describe the possible time interval between two consequent times of app use. Moreover, the *stickiness* is a concept of possibility that the user will obey the cycle pattern.

In the dataset, we have a time span of 60 days. In order to get enough observing data from the dataset, we choose a from 1 to 10 as potential cycle. Suppose every time when the user decides whether to use app_i, he is doing a Bernoulli experiment that has binary result of 0 and 1. The parameter of Bernoulli experiment, p, which indicates the possibility of whether to choose app_i, is generated by its prior distribution, Beta distribution as Eq. 4.

$$P(p_{i,a}|s_{i,a}, d_{i,a}) = Be(p_{i,a}; s_{i,a}, d_{i,a}) = \frac{\Gamma(s_{i,a} + d_{i,a})}{\Gamma(s_{i,a})\Gamma(d_{i,a})} p_{i,a}^{s_{i,a}-1} (1 - p_{i,a})^{d_{i,a}-1} \tag{4}$$

Table 3. Symbols in stickiness and cycle model

a	Window size marked by count of days, $a \in \{1,2,3,4,5,6,7,8,9,10\}$
$p_{i,a}$	The possibility of user using app_i with window a
$s_{i,a}$	Latent factor of stickiness. User's tendency of using app_i in cycle a
$d_{i,a}$	Latent factor measures how user's using pattern of app_i distract from cycle a
$v_{i,a}$	Prediction result of wether user will use app_i with window a. $v_{i,a} \in \{-1,1\}$
	1 indicates user will use it and -1 indicates he will not use it
$r_{i,a}$	Observed data of wether user use app_i with window a. $r_{i,a} \in \{-1,1\}$
	1 indicates user use it and -1 indicates he didn't use it

Figure 3 shows the possibility density function (pdf) of $p_{i,a}$ for different combination of $s_{i,a}$ and $d_{i,a}$. It is easy to see that with large $s_{i,a}$ and small $d_{i,a}$, the peak of pdf will move right, i.e., $p_{i,a}$ is close to 1. On the other hand, if $s_{i,a}$ is small and $d_{i,a}$ is large, $p_{i,a}$ is closer to 0. The model fits our common sense and the definitions of two parameters well.

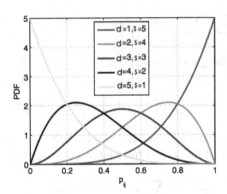

Fig. 3. PDF of Beta distribution

The decision result $v_{i,a}$ can be regarded as a Bernoulli process described in Eq. 5.

$$P(v_{i,a}|s_{i,a}, d_{i,a}) = \int_0^1 (v_{i,a}p_{i,a} + \frac{1 - v_{i,a}}{2})Be(p_{i,a}; s_{i,a}, d_{i,a})dp_i$$

$$= -\frac{d_{i,a}}{s_{i,a} + d_{i,a}} + \frac{1 + v_{i,a}}{2}$$

(5)

Parameters $s_{i,a}$ and $d_{i,a}$ can be evaluated from observed data. Suppose S and D are two parameter matrix we want to evaluate. $\mathscr{D}$ is the set of all observed sample data. We use max-likelihood method to evaluate parameters. The likelihood is computed in Eq. 6.

$$\mathscr{L} = logP(\mathscr{D}|S,D) = \sum_i \sum_{k_i} logP(r_{i,a}|s_{i,a},d_{i,a})$$
$$= \sum_{app_i \in A_u} \sum_{k_i} logP(v_{i,a}|s_{i,a},d_{i,a}) \tag{6}$$

where k_i is all user's record on app_i. We introduce two statistical parameters $n_{i,a}$, $m_{i,a}$ to remove k_i from the Eq. 6. $n_{i,a}$ is defined as the number of records that the user use app_i within the time period a, while $m_{i,a}$ is defined as the number of records that user did not use app_i within a. Their mathematical definition is described as Eq. 7.

$$n_{i,a} = \sum_{k_i} \frac{1+r_{i,a}}{2}$$
$$m_{i,a} = \sum_{k_i} \frac{1-r_{i,a}}{2} \tag{7}$$

After introducing $n_{i,a}$ and $m_{i,a}$, the likelihood in Eq. 6 is transformed into Eq. 8 that contains only known values and parameters to estimate,

$$\mathscr{L} = \sum_{app_i \in A_u} \sum_{k_i} logP(v_{i,a}|s_{i,a},d_{i,a})$$
$$= \sum_i (n_{i,a}logP(1|s_{i,a},d_{i,a}) + m_{i,a}logP(-1|s_{i,a},d_{i,a})) \tag{8}$$
$$= \sum_i (n_{i,a}log\frac{s_{i,a}}{d_{i,a}+s_{i,a}} + m_{i,a}log\frac{d_{i,a}}{d_{i,a}+s_{i,a}})$$

where parameters $d_{i,a}$ and $s_{i,a}$ can be estimated by gradient descent method with partial difference of likelihood function on $d_{i,a}$ and $s_{i,a}$ in Eq. 9:

$$\frac{\partial \mathscr{L}}{\partial s_{i,a}} = \sum_i (\frac{n_{i,a}}{s_{i,a}} - \frac{n_{i,a}+m_{i,a}}{s_{i,a}+d_{i,a}})$$
$$\frac{\partial \mathscr{L}}{\partial d_{i,a}} = \sum_i (\frac{m_{i,a}}{d_{i,a}} - \frac{n_{i,a}+m_{i,a}}{s_{i,a}+d_{i,a}}) \tag{9}$$

The parameters can be iteratively updated by Eq. 10, where α is the learning rate:

$$s_{i,a} := s_{i,a} + \alpha(\frac{n_{i,a}}{s_{i,a}} - \frac{n_{i,a}+m_{i,a}}{s_{i,a}+d_{i,a}})$$
$$d_{i,a} := d_{i,a} + \alpha(\frac{m_{i,a}}{d_{i,a}} - \frac{n_{i,a}+m_{i,a}}{s_{i,a}+d_{i,a}}) \tag{10}$$

For a fixed cycle a, user's using possibility can be described in a periodical function as Eq. 11. Changing window a from 1 to 10 and getting user's stickiness for app_i on 10 different potential cycles will influence final cycle constantly as

Fig. 4. Suppose a user tents to use app in cycle with large stickiness. Then we use stickiness as a weight factor and transform the final periodical function for usage prediction into Eq. 12:

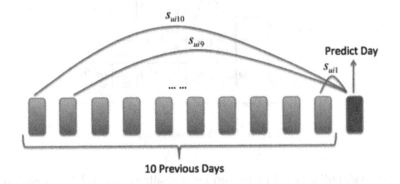

Fig. 4. Stickiness influences cycle

$$\cos\left(\frac{2\pi}{a}\Delta d\right) + 1 \tag{11}$$

$$\sum_{a=1}^{10} w_a \cos\left(\frac{2\pi}{a}\Delta d\right) + 1 \tag{12}$$

$$w_a = \frac{s_a}{\sum_{a=1}^{10} s_a}$$

4.3   Location-based Collaborative Filtering

Some users use different apps in different places. For example, they play games at home and use efficiency apps when they are at work. Moreover, the apps that are always used in the same place may have some connections on function or category. So we introduce location as a factor to cluster apps [12]. The clusters give us more information on which app should be used in a specific location.

First, we build an item-location matrix B like Fig. 5 and defined b_{li} as the location frequency (13):

$$b_{li} = \frac{\text{the number of } app_i \text{ record in location } l}{\text{the total number of } app_i \text{ record}} \tag{13}$$

The location-based similarity of app_i and app_j can be calculated in Eq. 14.

$$s_{ij} = \frac{\sum_{b_{li}\neq 0, b_{lj}\neq 0} b_{li} b_{lj}}{\sqrt{\sum b_{li}^2}\sqrt{\sum b_{lj}^2}} \tag{14}$$

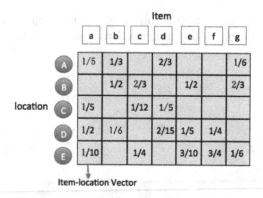

	a	b	c	d	e	f	g
A	1/5	1/3		2/3			1/6
B		1/2	2/3		1/2		2/3
C	1/5		1/12	1/5			
D	1/2	1/6		2/15	1/5	1/4	
E	1/10		1/4		3/10	3/4	1/6

Fig. 5. User-location matrix

By sorting similarity value, we can get a similar app-list $L(i)$ for each app_i. We can estimate the missed frequency in user-location matrix by similar apps in Eq. 15.

$$b_{li} = \sum_{j \in L(i)} s_{ij} b_{lj} \tag{15}$$

4.4   Combination of Three Models

Frequency-Time Decay Model and Stickiness-Cycle Model contain the features in app usage records and time. So we combine these two models by a multiplication function and get a compound function of decay in macroscopic and fluctuate in microscopic, as exemplified in Fig. 6. It can be described in Eq. 16:

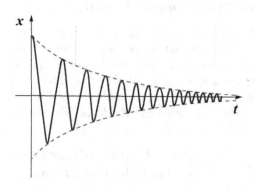

Fig. 6. Compound function

$$s_{i,d,t} = \frac{1}{|D|} \left[\sum_{d<d_{predict}} \frac{c_{i,d,t}}{c_{d,t}} \frac{c_{i,d,t}}{p_{i,d}} e^{-\{1+(d_{predict}-d)\}} \sum_{a=1}^{10} w_a \left(\cos\left(\frac{2\pi}{a}\Delta d\right) + 1 \right) \right]$$

(16)

Location-based CF uses only statistical frequency of all historical records regardless of different time. So we use a linear combination of location-based CF and Eq. 16 to compute our predicted value, as described in Eq. 17, where α is the linear parameter control the importance weight of each model in the final model.

$$s_{i,d,t,l} = \alpha \frac{1}{|D|} \left[\sum_{d<d_{predict}} \frac{c_{i,d,t}}{c_{d,t}} \frac{c_{i,d,t}}{p_{i,d}} e^{-\{1+(d_{predict}-d)\}} \sum_{a=1}^{10} w_a \left(\cos\left(\frac{2\pi}{a}\Delta d\right) + 1 \right) \right]$$
$$+ (1-\alpha) \sum_{j \in L(i)} s_{ij} b_{lj}$$

(17)

5   Experiment and Conclusion

In this section, we present our experimental methods, results and analysis.

5.1   Experiment Methods and Result

In the experiment, we randomly choose a time point in the last quarter of time span, to ensure we have enough historical data for prediction, and then apply Eq. 17 to compute the possibility value for each app, rank them in a descending order and get the prediction list. After that, we choose apps from predicted time period of dataset, and order them by their distance from the chosen time point. Finally we evaluate the prediction list by evaluation metrics introduced in Sect. 3.2. We choose MFU, MRU and Popularity-based model for comparison. In gradient descent method, we choose learning rate $\alpha = 0.005$ and precision $\epsilon = 0.00001$ after several experiments. Table 4 shows the experimental result of all models.

Where N is number of apps that prediction list contains, and n is the number of apps that can be displayed in one page of screen.

5.2   Conclusion

From Table 4, we can draw the following conclusions:

- From the aspect of *Precision* and *Recall*, Frequency and Time Decay Model, Compound Model and Hybrid Model with a large α can get a relatively higher accuracy than MRU, MFU and Popularity-based Model. Because these models consider context information like time and location to increase their accuracy under specific context.

Table 4. Experiment result on Different Models

	Precision		Recall		RankSocre			
	N=5	N=10	N=5	N=10	n=4	n=6	n=8	n=10
MFU	64%	38%	74%	85%	0.76	0.75	0.84	0.90
MRU	67%	42%	72%	83%	0.79	0.84	0.85	0.89
Popularity Model	37%	21%	64%	45%	0.64	0.69	0.72	0.78
Time Decay Model	76%	43%	78%	85%	0.75	**0.84**	0.87	0.91
Location-based CF	34%	35%	33%	62%	0.45	0.47	0.52	0.58
Frequency and Stickiness Compound Model	76%	45%	79%	**88%**	0.77	0.82	0.84	0.93
Hybrid Model $\alpha = 0.2$	45%	38%	44%	65%	0.56	0.59	0.64	0.68
$\alpha = 0.4$	58%	38%	52%	78%	0.72	0.74	0.81	0.88
$\alpha = 0.6$	67%	42%	68%	80%	0.78	0.82	**0.85**	0.89
$\alpha = 0.8$	**78%**	**50%**	**79%**	**88%**	**0.78**	0.82	0.84	**0.94**

- The larger α results in higher accuracy. When α reaches 0.8, *Precision* and *Recall* reach peak. Introducing location information increases *RankScore* over Compound Model.
- Generally, *Precision* is lower than *Recall*. The reason for this is that user's actual app usage list is always smaller than 10 apps. So the increase of prediction number results in the decrease of *Precision*.
- Pure location-based CF does not get a high accuracy. The reason is that the location marked by *cell ID* is a small scope geographically. So the cluster of apps are small and few similar apps can be found.
- It is easy to understand that a larger n results in a higher *RankScore*.

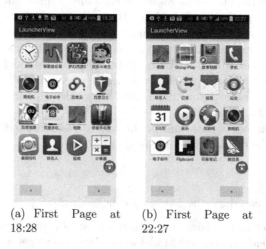

(a) First Page at 18:28

(b) First Page at 22:27

Fig. 7. Launcher change according to time period

6 A Dynamic Desktop for Android Platform

We implemented our Hybrid Model in a real smartphone application, with an Android launcher on a Sumsung Galaxy Note 3. This application can sort the app icons according to time period and location. The historical app usage data is stored in a local SQLite database. The metrics and model are updated every time period. Only simple calculation is needed when time period changes. Therefore it is a lightweight application with a quick response in a smartphone. The user interface is shown in Fig. 7.

7 Conclusion and Future Work

In this paper, we present a hybrid model to predict which app will be used in specific time and location. The hybrid model has 3 components, Frequency and Time Decay model, Stickiness and Cycle model and Location-based CF. Our experiment on the dataset shows this model has evident superiority on prediction accuracy over other approaches. Moreover, we develop a launcher in a real Android smartphone, which tests our model in real environments. However, This model is now not sensitive to new users and new apps in a system. Therefore our future work will be carried out to solve this cold start problem.

Acknowledgement. This work is partially supported by China National Science Foundation (Granted Number 61272438,61472253), Research Funds of Science and Technology Commission of Shanghai Municipality (Granted Number 15411952502, 12511502704).

References

1. http://blog.nielsen.com/nielsenwire/online_mobile/the-state-of-mobile-apps/
2. Blom, J., et al.: Contextual and cultural challenges for user mobility research. CACM **48**(7), 37–41 (2005)
3. Khan, A.M., et al.: Activity recognition on smartphones via sensor-fusion and kda-based svms. Int. J. Distrib. Sens. Netw. **2014**, 11581–11604 (2014)
4. Maitland, J., et al.: Increasing the awareness of daily activity levels with pervasive computing. In: Pervasive Health Conference and Workshops (2006)
5. Ravi, N., et al.: Context-aware battery management for mobile phones. Pervasive Comput. Commun. (PerCom), 224–233 (2008)
6. Eagle, N., Pentland, A.: Reality mining: sensing complex social systems. Pers. Ubiquit. Comput. **10**(4), 255–268 (2006)
7. Froehlich, J.E., et al.: MyExperience: a system for in situ tracing and capturing of user feedback on mobile phones. In: Proceedings of the of Mobisys? vol. 7, pp. 57–70 (2007)
8. Böhmer, M., et al.: Falling asleep with angry birds, facebook and kindle - a large scale study on mobile application usage. In: Proceedings of MobileHCI, pp. 47–56 (2011)

9. Matsumoto, M., et al.: Proposition of the context-aware interface for cellular phone operations. In: Proceedings of the INSS 2005, pp. 233–233 (2005)
10. Choonsung, S., Hong, J.-H., Dey, A.K.: Understanding and prediction of mobile application usage for smart phones. In: Proceedings of the 2012 ACM Conference on Ubiquitous Computing, pp. 51–79. ACM (2012)
11. Paolo, C., Koren, Y., Turrin, R.: Performance of recommender algorithms on top-n recommendation tasks. In: Proceedings of the Fourth ACM Conference on Recommender Systems, pp. 5–12. ACM (2010)
12. Xiang, L.: Recommendation System Practice, p. 6. BeiJing Youdian Publication House, BeiJing (2012)
13. Peifeng, Y., et al.: App recommendation: a contest between satisfaction and temptation. In: Proceedings of the Sixth ACM International Conference on Web Search and Data Mining, pp. 4–11. ACM (2013)
14. Gang, L.: Crazy Android Textbook. Dianzi Gongye Publication, BeiJing (2011)

A Framework for Fast Service Verification and Query Execution for Boolean Service Rules

Soumi Chattopadhyay[1]([⊠]), Saikat Dutta[2], and Ansuman Banerjee[1]

[1] Indian Statistical Institute, Kolkata, India
soumi61@gmail.com
[2] Jadavpur University, Kolkata, India

Abstract. The problem of service rule verification has attracted some attention in recent years. In this paper, we consider service rules in simple Boolean logic and present a new method for business rule verification using simultaneous minimal support set computation. As we show here, the problem is similar in flavor to the problem of prime implicant generation of a given Boolean function which has alluded researchers for several decades and significant efforts in this direction have been reported in literature, with proposals of widely varying algorithms and data structures. In this paper, we revisit this problem in the context of business rules and present a new method that aids in rule verification and also in query execution at runtime. Our method builds on the classical binary decision diagram data structure for representing business rules and generates the test scenarios by a simple traversal algorithm. Experimental results on simulated benchmark rules show the efficacy of our approach.

1 Introduction

In recent years, the services ecosystem has seen a steady emergence of business rule management systems (BRMS) which allow enterprise architects to separate the concerns in an aspect-oriented way and easily define, manage, update and run the decision logic that directs enterprise applications in a Business rule engine (BRE) without needing to write code or change the business processes calling them [1–4]. This enables modern businesses to change their business rules dynamically in order to adapt to a rapidly changing business environment, and they contribute to agility in a service-oriented architecture paradigm by enabling reduced time to automate, easier change and maintenance for business rules.

In this services execution paradigm with business rules, the overall performance of the enterprise services delivered depends critically on the functional correctness and performance of the rule repository used for handling the business use cases. Bugs in rule encoding or implementation may create functional errors in service delivery, which may critically affect business performance. With web services and e-commerce being the order of the day like never before, ensuring correctness of a business rule set before deployment is becoming a critical mandate. In recent years, the problem of business rule verification has emerged as a problem of immense importance, due to failure scenarios of business cases that

© Springer International Publishing Switzerland 2015
L. Yao et al. (Eds.): APSCC 2015, LNCS 9464, pp. 17–32, 2015.
DOI: 10.1007/978-3-319-26979-5_2

have been reported by customers in interacting and working with several web services. This paper addresses this specific problem of business rule verification in a services ecosystem.

Business rules can be described in a declarative style [5,6] (using Declare models), or through a set of logical formulas (in Boolean/temporal logic [7] or their domain-specific variants). In this work, we have adopted a simple intuitive style for modeling business rules, using Boolean logic with predicates over arbitrary variables. Our motivation in this work is to address the problem of business rule verification, by which we can detect violations of business rules from the intended deployments or service level agreements. Indeed, this is of extreme importance, since the business rules form the critical component in a service delivery model. A number of research articles [8,9,14,15] have been reported in recent literature on this problem. These typically model business rules in either the declarative or the procedural style and attempt to use standard web service testing and formal verification methods to check for rule violations. While testing methods have their own limitations in terms of exhaustiveness of the scenarios covered, formal verification methods often fail to scale to the size and complexity of the business rule database that we are looking at. Moreover, most of these methods have restricted modeling styles for modeling business rules. In contrast, we attempt to present in this paper, a foundational framework for business rule verification by delineating the problem from the domain specifics, and instead adopting a simple intuitive classical logical style for expressing business rules. For the sake of simplicity and ease of illustration, we adopt Boolean logic with predicates for expressing business rules, and show that the problem of business rule verification can be cast as a simple instantiation of the classical prime implicant generation problem for Boolean functions. In this paper, we cast the problem of business rule verification as the task of extraction of test scenarios/queries for which the business rule is expected to evaluate to *true* (these are the ones in which a given business rule is supposed to trigger) and to *false* (scenarios where the rule is expected not to be exercised). Indeed, these constitute the scenarios that a business rule logic implementation need to be put against, since any violation of the expected outcomes on these scenarios is undesirable. The ability to extract and test against all the true and false scenarios gives us an added confidence of exhaustiveness akin to formal methods, while at the same time, makes our approach scalable and relevant in practice, since we do not suffer from any computational bottlenecks.

Computation of prime implicants for a Boolean function is one of the fundamental problems of Boolean algebra. Several approaches have been proposed in the literature that deal with this problem [10–13]. For business rule verification, we map the minimal support set generation problem as a simple variant of the classical prime implicant generation problem. Not only are we interested in extracting the scenarios in which a given business rule is expected to be triggered (which maps exactly to the prime implicant generation problem), we are also interested in extracting the use cases where the rule is expected to remain unexercised. This corresponds to another run of the prime implicant generation problem, for the negation of the given rule. In this work, we present a mechanism by which we are able to unify the two tasks and address both the scenario generation problems in one pass. This is indeed an explicit novelty that we add on in this paper, and we contrast our approach through simulations on business rules, to show how much effort we save, in contrast to the two pass approach. Additionally, we address another interesting piece in this work. Not only are we interested in extracting the test scenarios, we also address the problem of computing the minimal cardinality valuations that can make a given rule true or false. This ensures that redundant scenarios are not used in our validation process and we indeed test a given business rule repository against a minimal relevant scenario set, while not compromising on the exhaustiveness of the verification process. The test scenarios synthesized by our approach are expected to be used by a rule logic implementation team to test their deployment models. Additionally, we can also take up these scenarios to check whether any of them leads to violations of the service level agreements. this serves as a key component in the rule verification process in a services deployment ecosystem, and we can use the rules to generate sample queries for the testing task. Another important aspect is the fact that this analysis can expedite the rule execution procedure. If we preprocess the rules and store the minimal support sets of the rule in an efficient manner, the queries can be answered at run time, without even executing the rule set, but using a simple look-up table. This in turn can expedite the query execution procedure as well. This paper has three key contributions, as outlined below:

- We address the problem of business rule verification, with rules expressed in extended Boolean logic and model it as an instantiation of the simultaneous test scenario generation problem.
- We discuss how we can expedite the process of run time query execution using a look-up table.
- We also present an innovative approach for simultaneously computing the exhaustive set of positive and negative test scenarios using a one-pass method, with the help of a novel data structure. This makes our proposition scalable and exhaustive and usable in practice.

2   Motivation for This Work

In this section, we illustrate the motivation of our work using a simple example. We consider an example business decision rule $\mathcal{R}$ in an online shopping framework defined as follows:

Example 1. **Rule:** **If** the brand is ADIDAS (p_1) and *any* of the following conditions is true:

- It is Christmas time (p_2)
- For other times of the year ($\bar{p}_2$), if the customer is a valuable ADIDAS customer (p_3)
- On purchase from the old stock (p_4)
- On purchase above \$ 150 (p_5)

Then announce 10 % discount on every shopping from ADIDAS.

For convenience of notation and simplicity of expression, we introduce the predicates $p_1, p_2, \ldots, p_5$ in the ruleset above that express the different conditions and use them in the following discussion throughout this paper. The triggering condition of the above rule can be expressed as the following Boolean expression.

$$\mathcal{A} = p_1 \cdot (p_2 + \bar{p}_2 \cdot p_3 + p_4 + p_5)$$

A simple analysis of the antecedents of the rules reveals the following scenarios where the rule is always true: {the brand is ADIDAS, this is Christmas time}, {the brand is ADIDAS, the customer is valuable}, {the brand is ADIDAS, purchase is done from the old stock}, {the brand is ADIDAS, purchase is more than $ 150}. If we notice these scenarios carefully, we can observe that all the conditions are not present in every scenario, but still we can decide the rule as *true* and can actuate the corresponding consequent (discount announcement) of the rule. In fact these scenarios are minimum in cardinality, i.e., if we remove any of the conditions from inside any of the scenarios (comma separated list), the rule cannot be decided for a truth value. Similar is the situation for scenarios for which the rule is always false: {the brand is not ADIDAS}, {it is not Christmas time, the customer is not a valuable ADIDAS customer, the purchase is not done from the old stock, the purchase is less than or equal to $ 150}.

If we can pre-process the antecedents of a given rule as discussed above, we can generate the sample queries to guide the functional testing of the rule set. A lot of approaches exist in literature which attempt to find these scenarios, using methods based on, test generation using support sets for Boolean function. Karnaugh map (K-map) [10] is one such popular method which can generate support sets, but it does not scale with the number of variables. Another popular method is the one proposed by Quine McCluskey [11]. The main disadvantage of this approach is, we cannot use this approach for a large number of inputs, because all minterms of a function need to be stored simultaneously in memory and to generate both the scenarios discussed above, we need to execute this method twice. Our proposed method, on the other hand, is able to generate both the scenarios simultaneously.

Another important use case, as already discussed is that our preprocessing proposal helps to expedite the run time service execution against incoming queries as well. When the actual query (valuation of the variables appearing in the rule) comes at run time, without even executing the rule set, we can decide which rules are true by simply using a look-up table, which can be used to store the support set valuations for which the rule evaluates to true or false. This greatly expedites the query evaluation process, since we do not need to explicitly evaluate a query for every input scenario. The support sets can guide us here as well. We explain in the following section, the technical details.

3   Detailed Methodology

In this section, we formally discuss our methodology. Figure 1 shows the different components of our method. The input of our method is the set of service rules.

The rule has two parts an *If* part and the *Then* part. In this paper, we assume that the *If* part is expressed in simple Boolean logic. *Then* part is the consequent part of the rule, it is actuated when the *If* part of the rule is evaluated as true. From here onwards, in our discussions, we consider the *If* part of the rule and we show how we can compute the supporting and refuting scenarios for the *If* part simultaneously that can expedite rule execution. The preprocessor preprocesses the rules and generate exhaustive set of positive and negative test cases which are stored in a lookup table. On one hand, this lookup table is used to verify the services, on the other side it is used for faster query execution at run time. We discuss both of these later in this section.

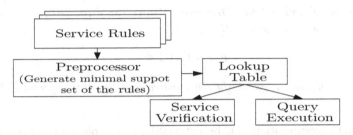

Fig. 1. Component diagram of our method

Before going into the details of the algorithm, we define a few terminologies that are necessary to develop our algorithm. We begin with some background concepts.

Definition 1. *Support Set:* *A set $\mathcal{U} = \{(u_1, x_1), (u_2, x_2), \ldots, (u_l, x_l)\}$ is said to be a support set of a service rule Φ defined over a set of Boolean propositions $\mathcal{P} = \{p_1, p_2, \ldots, p_n\}$, where $x_i \in \{0, 1\}$ and $u_i \in \mathcal{P} \ \forall i = 1, 2, \ldots, l$, if Φ evaluates to either 0 or 1, when $u_1 = x_1, u_2 = x_2 \ldots, u_l = x_l$.* ■

Example 2. Consider a service rule $\Phi = p_1.(p_2 + p_3 + p_4.p_5 + p_6.p_7.p_8)$. $\{(p_1, 1), (p_2, 1), (p_4, 1), (p_5, 1)\}$ is a support set of Φ, since Φ evaluates to 1 on the assignment $p_1 = 1, p_2 = 1, p_4 = 1, p_5 = 1$. ■

A support set of a service rule Φ for which Φ evaluates to true is called a *positive support set* and a support set for which Φ evaluates to false is called a *negative support set*.

Definition 2. *Minimal Support Set:* *For a service rule Φ, a support set $\mathcal{U}$ is said to be a prime, if no proper subset of $\mathcal{U}$ is a support set of Φ.* ■

It is worth noting that we redefine the classical notion of minimal support set by including the minimal support set with its valuation.

Example 3. For the function given in Example 2, $\{(p_1, 1), (p_2, 1)\}$ is a minimal support set of Φ. However, $\{(p_1, 1), (p_2, 1), (p_4, 1), (p_5, 1)\}$ is not a minimal support set of Φ, since it is a superset of $\{(p_1, 1), (p_2, 1)\}$, which is a minimal support set of Φ. ■

We now define the co-factor of a service rule.

Definition 3. *Co-factor:* *The positive (negative) co-factor of a service rule Φ defined over a set of Boolean propositions $\mathcal{P} = \{p_1, p_2, \ldots p_n\}$ with respect to a proposition $p_i \in \mathcal{P}$ is obtained by substituting 1 (true) / 0 (false) in Φ.*   ■

The positive co-factor, denoted by Φ_{p_i} is obtained by substituting the variable p_i with 1 in Φ, i.e., $\Phi_{p_i} = \Phi(p_1, p_2, \ldots, p_i = 1, \ldots, p_n)$. Similarly, the negative co-factor, denoted as $\Phi_{\bar{p}_i}$ is obtained as $\Phi(p_1, p_2, \ldots, p_i = 0, \ldots, p_n)$. The co-factors are independent of the proposition p_i with respect to which they are computed. We now define the concept of decomposition of a service rule. This follows as a straightforward application of Shannon's expansion [16].

Definition 4. *Service Rule Decomposition:* *The decomposition of a service rule Φ with respect to a proposition $p \in \mathcal{P}$ is obtained as:*

$$\Phi = p.\Phi_p + \bar{p}.\Phi_{\bar{p}},$$

where Φ_p and $\Phi_{\bar{p}}$ are respectively the positive and negative cofactors of Φ with respect to p.   ■

Service rule decomposition can be extended to multiple propositions as well. From the definition of a minimal support set, it is trivial to observe that the co-factor of a service rule Φ with respect to its minimal support sets is always constant, i.e., either 0 or 1. Essentially, if $\{(p_1, 1), (p_2, 0)\}$ is a minimal support set of a service rule Φ, then $\Phi_{p_1 \bar{p}_2} = $ constant, i.e. 1 or 0. We now define a few concepts which are important for our methodology and serve as the foundation.

Definition 5. *Strong Proposition:* *A proposition u is said to be a strong proposition with respect to a minimal support set $\mathcal{U}$ of a service rule Φ, where, $\mathcal{U} = \{(u_1, x_1), (u_2, x_2), \ldots, (u_k, x_k)\}$ and $(u, x) \in \mathcal{U}$, if $\Phi_{p=\bar{x}}$ is independent of $\{u_1, u_2, \ldots, u_k\}$.*   ■

Example 4. Consider the service rule $\Phi = p_1.p_2 + p_3.(p_4.p_5 + p_6)$. $\{(p_1, 1), (p_2, 1)\}$ is a positive minimal support set of Φ. p_1 is a *strong* proposition with respect to $\{p_1, p_2\}$, since, $\Phi_{p_1=0}$ (i.e. $\Phi_{\bar{p}_1}$) is independent of p_2. On the other hand, $\{(p_1, 0), (p_3, 0)\}$ is a negative minimal support set of Φ. It is easy to see that $\Phi_{p_1=1}$ (i.e. Φ_{p_1}) is *not* independent of p_3. Therefore, p_1 is not a strong proposition with respect to $\{(p_1, 0), (p_3, 0)\}$.   ■

Definition 6. *Strong Minimal Support Set:* *A minimal support set $\mathcal{U}$ is said to be a strong minimal support set with respect to a service rule Φ, if all the propositions appearing in $\mathcal{U}$ are strong.*   ■

Example 5. Consider $\Phi = p_1.p_2 + p_3.p_4$. Here, $\{(p_1, 1), (p_2, 1)\}$ is a strong minimal support set, since, $\Phi_{\bar{p}_1}$ is independent of p_2 and $\Phi_{\bar{p}_2}$ is independent of p_1.   ■

It is intuitively obvious that a minimal support set with cardinality 1 is trivially a strong minimal support set.

Definition 7. *Derivative:* *The derivative of a service rule Φ with respect to a proposition p_i is defined as the exclusive-or of the positive and negative co-factors of Φ with respect to the proposition p_i, i.e., $\partial\Phi/\partial p_i = \Phi_{p_i} \oplus \Phi_{\bar{p}_i}$.* ∎

In order to determine whether a proposition is strong with respect to a function Φ, we use the concept of the derivative.

Definition 8. *Critical Proposition:* *A proposition p is said to be a critical proposition with respect to a service rule Φ, if the derivative of Φ with respect to the proposition p is 1, i.e., $\partial\Phi/\partial p = 1$.* ∎

3.1   Algorithm for Minimal Support Set Generation

In this section, we discuss our algorithm for simultaneous generation of positive and negative minimal support sets for a given service rule. Consider a service rule Φ defined over a set of Boolean propositions $\mathcal{P} = \{p_1, p_2, \ldots, p_n\}$. A naive approach for computing the minimal support sets of a service rule Φ is as follows: We choose a combination $\mathcal{C} = \{(p_1, x_1), (p_2, x_2), \ldots, (p_k, x_k)\}$, where, $x_i \in \{0, 1\}$, $\forall i \in \{1, 2, \ldots, k\}$. We compute the co-factor of Φ with respect to $\mathcal{C}$ and check whether Φ evaluates to a constant. We start with a combination of cardinality 1 and then gradually increase the cardinality. While doing so, we keep track of the combinations for which Φ evaluates to a constant and do not consider any super set of such a combination. For each proposition p_i in the service rule Φ, there are three possibilities, either $(p_i, 0) \in \mathcal{C}$ or $(p_i, 1) \in \mathcal{C}$ or p_i does not belong to the propositions appearing in $\mathcal{C}$. Therefore, this naive procedure will lead us to explore all $3^n - 1$ combinations, which is very inefficient. We propose below our modified approach that does the same job in a more efficient way. We use the reduced ordered binary decision diagram (ROBDD) [17] data structure representation for Boolean functions as the backbone of our method. Algorithm 1 presents our approach for generating the minimal support sets of a service rule. It is an iterative algorithm. We gradually build up a tree $\mathcal{T}$ to find out the minimal support sets. Each step of the algorithm is discussed below.

We now explain the detail of each step below. We start with a dummy node $\mathcal{S}$ and incrementally iteratively build the complete minimal support set tree.

Simplify: In this step, we simplify the service rule Φ by removing all the critical propositions of the function. We identify all the critical propositions with respect to Φ and substitute them by 0 in Φ. The simplified function is used in the later steps of the algorithm.

Consider the propositions $u_1, u_2, \ldots, u_l$ such that $\partial\Phi/\partial u_j = 1$, for $j = 1, 2, \ldots, l$. Then Φ can be written as, $\Phi = u_1 \oplus u_2 \oplus \ldots \oplus u_l \oplus \Phi_{\bar{u}_1.\bar{u}_2.....\bar{u}_l}$. Consider $\Phi_1 = \Phi_{\bar{u}_1.\bar{u}_2.....\bar{u}_l}$. We find the minimal support sets of Φ_1 and then combine them with the 2^l combinations of $u_1, u_2, \ldots, u_l$ to obtain the minimal support sets of Φ. Lemma 1 expresses the correctness of this step.

Lemma 1. *If $\partial\Phi/\partial p = 1$, every minimal support set of Φ has $(p, 0)$ or $(p, 1)$.* ∎

Algorithm 1. *GenerateMinimalSupportSet*

1: *Input:* A service rule Φ; *Output:* Set of minimal support sets
2: Construct the $ROBDD(\mathcal{B})$ for Φ;
3: **Simplify** Φ; Create a start node $\mathcal{S}$;
4: **while** Φ is not constant **do**
5:     **Generate** partial minimal support set and construct $\mathcal{T}$;
6:     **Substitute and simplify** Φ;
7: **end while**
8: **Back propagate** from leaf node to start node;

Proof. Using Shannon's expansion we have, $\Phi = p.\Phi_p + \bar{p}.\Phi_{\bar{p}}$. Since $\partial\Phi/\partial p = 1$, $\Phi_p = \bar{\Phi}_{\bar{p}}$. $\Phi = p.\bar{\Phi}_{\bar{p}} + \bar{p}.\Phi_{\bar{p}} = p \oplus \Phi_{\bar{p}}$. $\Phi_{\bar{p}}$ is independent of p. The positive minimal support set of Φ = positive minimal support set of $\Phi_{\bar{p}} \cup \{(p,0)\}$, since positive minimal support set of $\Phi_{\bar{p}}$ makes it 1 and if we substitute p by 0 then we get 0 and XOR of 0 and 1 is 1. As a result we get the positive minimal support set of Φ. Similarly positive minimal support set of Φ = negative minimal support set of $\Phi_{\bar{p}} \cup \{(p,1)\}$. On the other hand, negative minimal support set of Φ = positive minimal support set of $\Phi_{\bar{p}} \cup \{(p,1)\}$, or negative minimal support set of $\Phi_{\bar{p}} \cup \{(p,0)\}$. Hence, every minimal support set of Φ contains either $(p,1)$ or $(p,0)$. ∎

The advantage of this step is, if a service rule Φ contains only those propositions, with respect to which the derivative of the function is 1, we do not need to proceed further. Each combination of the propositions gives a minimal support set of Φ in such a situation.

Example 6. Consider $\Phi = p_1 \oplus (p_2 + p_3)$. Since, $\partial\Phi/\partial p_1 = 1$, the minimal support sets of $(p_2 + p_3)$ are $\{\{(p_2 = 1)\}, \{(p_3 = 1)\}, \{(p_2 = 0), (p_3 = 0)\}\}$. The minimal support sets of Φ are $\{\{(p_1 = 1, p_2 = 1)\}, \{(p_1 = 0, p_2 = 1)\}, \{(p_1 = 1), (p_3 = 1)\}, \{(p_1 = 0, p_3 = 1)\}, \{(p_1 = 1, p_2 = 0, p_3 = 0)\}, \{(p_1 = 0, p_2 = 0, p_3 = 0)\}\}$.

Partial Minimal Support Set Generation: In each iteration of the algorithm, we modify the original service rule Φ. In the next step, we discuss the modification of Φ. In this step, we generate the minimal support set of the modified service rule, we call it its partial minimal support set. Later, when we back trace through the minimal support set tree $\mathcal{T}$, we modify the partial ones to get the minimal support set of Φ. Algorithm 2 shows the formal procedure to generate a partial minimal support set. There may be multiple partial minimal support sets at this step. However, we generate one and proceed to the next step.

Once we obtain a strong minimal support set of the modified service rule, we create a leaf node $\mathcal{L}$ and an intermediate node $\mathcal{I}$ of the tree in the same level, i.e., both of them have the same parent. The parent node of the first level is the start node $\mathcal{S}$. The leaf node $\mathcal{L}$ contains two parameters: the strong minimal support set of the modified service rule and the corresponding functional value,

---

Algorithm 2. $GeneratePartialMinimalSupportSet$

1: *Input* : ROBDD for the modified function Φ_1
2: *Output* : A strong minimal support set
3: **for** each proposition p associated with Φ_1 **do**
4:     **loop**
5:         Find a positive minterm $\mathcal{M}_1$ associated with p from $\mathcal{B}$;
6:         Find a minimal support set $\mathcal{P}_1$ from $\mathcal{M}_1$;
7:         **if** p belongs to the proposition set of $\mathcal{P}_1$, **then** break;
8:     **end loop**
9:     **if** $\mathcal{P}_1$ is a strong minimal support set, **then** return $\mathcal{P}_1$;
10:     **loop**
11:         Find a negative minterm $\mathcal{M}_2$ associated with p from $\mathcal{B}$;
12:         Find a minimal support set $\mathcal{P}_2$ from $\mathcal{M}_2$;
13:         **if** p belongs to the proposition set of $\mathcal{P}_2$, **then** break;
14:     **end loop**
15:     **if** $\mathcal{P}_2$ is a strong minimal support set, **then** return $\mathcal{P}_2$;
16: **end for**
17: return NULL;

---

i.e., 1 if it is a positive minimal support set and 0 if it is a negative one. The intermediate node also has two entries.

- A set of Boolean propositions assigned with a value, combined using OR.
- An assigned value, either 0 or 1.

Algorithm 3 shows the formal procedure to create an intermediate node. Initially, we pass $\mathcal{S}$ in Algorithm 3 as the parent node. The intermediate node(s), obtained from the current level, is (are) going to be the parent node(s) in the next level. The interpretation of an intermediate node is as follows:

- If the assigned value of an intermediate node is 0, it implies that the content of the intermediate node is combined with a positive minimal support set generated at any of the levels lower than the one to which the intermediate node belongs to.
- Similarly, if the assigned value of an intermediate node is 1, it implies the content of the intermediate node is combined with a negative minimal support set generated at any of the levels lower than the one to which the intermediate node belongs to.

If we do not obtain any strong minimal support set in this step, instead of creating a single intermediate node, we create two intermediate nodes. The first intermediate node contains $(p, 0)$ in its first field and X (unknown) in its second field while the second intermediate node contains $(p, 1)$ in its first field and X in its second field, where p is a proposition associated with Φ_1. If the assigned value of an intermediate node is X, it implies that whether the content of the intermediate node is combined with a minimal support set generated at any of the levels lower than the one to which the intermediate node belongs to is

---

Algorithm 3. *CreateIntermediateNode*

1: *Input :* Strong minimal support set of Φ_1: $\mathcal{U}$, parent Node $\mathcal{P}$
2: **if** $\mathcal{U}$ is NULL **then**   ▷ No strong minimal support set exists
3:       Create two intermediate nodes $\mathcal{I}_1$ and $\mathcal{I}_2$;
4:       Choose a proposition p associated with Φ_1;   ▷ preferably first node of the
        $ROBDD$ of Φ_1
5:       Assign $(p, 0)$ to the first field of $\mathcal{I}_1$ and X to the second field;
6:       Assign $(p, 1)$ to the first field of $\mathcal{I}_2$ and X to the second field.
7:       Add two edges from $\mathcal{P}$ to $\mathcal{I}_1$ and from $\mathcal{P}$ to $\mathcal{I}_2$;
8: **else**
9:       Create an intermediate node $\mathcal{I}$;
10:      Assign tuples $(u_i, \bar{x}_i)$ corresponding to each tuple $(u_i, x_i) \in \mathcal{U}$ combined using
        OR, to the first entry of the intermediate node.
11:      **if** $\mathcal{U}$ is a positive minimal support set **then**
12:          Assign 1 to the 2nd field of the intermediate node;
13:      **else if** $\mathcal{U}$ is a negative minimal support set **then**
14:          Assign 0 to the 2nd field of the intermediate node;
15:      **end if**
16:      Add an edge from $\mathcal{P}$ to $\mathcal{I}$;
17: **end if**

---

decided after verification, i.e., we have to verify whether the minimal support set generated at any of the levels lower than the one to which the intermediate node belongs to, is a minimal support set of this level or not. This step is justified by the following lemmas.

Lemma 2. *If* $\mathcal{U} = \{(u_1, x_1), (u_2, x_2), \ldots, (u_k, x_k)\}$ *is a positive (negative) minimal support set of* Φ, *where,* $x_i \in \{0, 1\}, i = 1, 2, \ldots, k$, *every negative (positive) minimal support set contains at least one* $(u_i, \bar{x}_i)$, *such that,* $(u_i, x_i) \in \mathcal{U}$. ∎

Proof. To prove this lemma, we need to prove two things: The set of propositions in a negative minimal support set $\mathcal{U}_1$ of Φ has a non empty intersection with the set of propositions in $\mathcal{U}$ and every negative minimal support set contains at least one $(u, \bar{x})$, such that, $(u, x) \in \mathcal{U}$. We prove both the claims by contradiction. Let us first consider a negative minimal support set of Φ, $\mathcal{U}_1 = \{(w_1, y_1), (w_2, y_2), \ldots, (w_l, y_l)\}$, where, $y_i \in \{0, 1\}$ for $i = 1, 2, \ldots, l$ and $w_i \notin \{u_1, u_2, \ldots, u_k\}, \forall i \in \{1, 2, \ldots, k\}$. Therefore the truth table of Φ contains at least one row satisfying $\mathcal{U}$ and $\mathcal{U}_1$ simultaneously, since the intersection of the proposition set in $\mathcal{U}$ and $\mathcal{U}_1$ is empty. This contradicts the fact that $w_i \notin \{u_1, u_2, \ldots, u_k\}, \forall i \in \{1, 2, \ldots, k\}$. So, the proposition set of $\mathcal{U}_1$ contains at-least one $u_i \in \mathcal{U}$, for $i = 1, 2, \ldots, k$. Now we consider the fact $\mathcal{U}_1$ does not contain any $(u, \bar{x})$, such that, $(u, x) \in \mathcal{U}$. Hence, we assume $\mathcal{U}_1$ contains $\{(u_{i_1}, x_{i_1}), (u_{i_2}, x_{i_2}), \ldots, (u_{i_l}, x_{i_l})\} \subseteq \mathcal{U}$. Now if we merge the elements of $\mathcal{U}$ and $\mathcal{U}_1$, it remains a support set, say $\mathcal{W}$. It is easy to see that, $\mathcal{U}$ and $\mathcal{U}_1$ are both subsets of $\mathcal{W}$, but $\mathcal{U}$ is a positive minimal support set of Φ and $\mathcal{U}_1$ is a negative support set of Φ. This contradicts our assumption. ∎

Lemma 3. *If $\mathcal{U} = \{(u_1, x_1), (u_2, x_2), \ldots, (u_k, x_k)\}$ is a strong positive (negative) minimal support set of Φ, where $x_i \in \{1, 0\}, i = 1, 2, \ldots, k$, every negative (positive) minimal support set of Φ contains exactly one $(u_i, \bar{x}_i)$, where $(u_i, x_i) \in \mathcal{U}$. Also no other positive (negative) minimal support set of Φ contains any (u_i, x_i) or $(u_i, \bar{x}_i)$ where $(u_i, x_i) \in \mathcal{U}$.* ∎

Proof. We present the proof for the positive minimal support set case and the proof for the negative minimal support set is similar. The proof is as follows: Since $\mathcal{U} = \{(u_1, x_1), (u_2, x_2), \ldots, (u_k, x_k)\}$ is a strong positive minimal support set of Φ, $\Phi_{u_i = \bar{x}_i}$ is independent of $\{u_1, u_2, \ldots, u_{i-1}, u_{i+1}, \ldots, u_k\}$. If another minimal support set contains $(u_i, \bar{x}_i)$, it cannot contain any proposition from $\{u_1, u_2, \ldots, u_{i-1}, u_{i+1}, \ldots, u_k\}$.

Therefore a minimal support set can contain at most one $(u_i, \bar{x}_i) \in \mathcal{U}$. From Lemma 2, it follows that each negative minimal support set contains at least one element from the positive minimal support set with opposite polarity. Therefore, every negative minimal support set contains exactly one $(u_i, \bar{x}_i)$ for $i = 1, 2, \ldots, k$.

Now we prove the second claim, i.e., no other positive minimal support set of Φ contains any (u_i, x_i)or$(u_i, \bar{x}_i)$ where $(u_i, x_i) \in \mathcal{U}$.

Case 1: Consider a positive minimal support set $\mathcal{W} = (w_1, y_1), (w_2, y_2), \ldots, (w_l, y_l)$ of Φ such that $(u_i, \bar{x}_i) \in \mathcal{U}$ and $(u_i, \bar{x}_i) \in \mathcal{W}$. $w_j \notin \{u_1, u_2, \ldots, u_{i-1}, u_{i+1}, \ldots, u_k\}$ for $j = 1, 2, \ldots, l$, since $\mathcal{U}$ is a strong minimal support set. Consider a tuple $(u_j, x_j) \in \mathcal{U}$ and $u_j \neq u_i$. Clearly, when we substitute u_j by $\bar{x}_j$, the function Φ does not become independent of u_i which contradicts the fact that $\mathcal{U}$ is a strong minimal support set.

Case 2: Consider a positive minimal support set $\mathcal{W} = (w_1, y_1), (w_2, y_2), \ldots, (w_l, y_l)$ of Φ such that $(u_i, x_i) \in \mathcal{U}$ and $(u_i, x_i) \in \mathcal{W}$. Then there exists, at-least one $(u_j, x_j) \in \mathcal{U}$ such that $u_j \neq u_i$ and $(u_j, x_j) \notin \mathcal{W}$, otherwise $\mathcal{W}$ would not be a minimal support set. If $(u_j, \bar{x}_j) \in \mathcal{W}$, then this case would be similar to Case 1. We can conclude that u_j does not belong to the proposition set of $\mathcal{W}$. Therefore, when we substitute any u_j by $\bar{x}_j$, the function Φ does not become independent of u_i, which again contradicts the fact that $\mathcal{U}$ is a strong minimal support set. ∎

Substitution and Simplification of Φ: Once the intermediate node is created, we modify the service rule, which we have now, say Φ_1 for the sub tree rooted at the intermediate node. Consider a strong minimal support set $\{\mathcal{U}\}$ generated in the current iteration. Assume, $\mathcal{U} = \{(u_1, x_1), (u_2, x_2), \ldots, (u_k, x_k)\}$. Let us consider a tuple (u_i, x_i) from $\mathcal{U}$. We substitute $(u_i = \bar{x}_i)$ in Φ_1 and simplify the function to get the modified Φ_1.

Back Tracing Through the Minimal Support Set Tree T: Once the entire minimal support set tree is created using the steps discussed above, we back trace through this tree to compute the minimal support sets. All the minimal support sets of Φ_1 generated in this step, have to be modified in order to get the minimal support set of Φ. For that, we need to back trace through the intermediate nodes till the start node is obtained. If we consider the positive minimal support set of Φ_1, while traversing backward we consider only the intermediate nodes which have assigned value 0. On the other hand, if the minimal support set of Φ_1 is negative, we similarly consider only the intermediate nodes with assigned value 1. If the assigned value of an intermediate node is X, we verify whether the minimal support set is going to be combined with the content of the intermediate node, in order to be a minimal support set of the service rule corresponding to the step to which the intermediate node belongs to.

3.2   A Complete Example

In this subsection, we explain the working of Algorithm 1 using an example. Consider the following rule:

Rule: If the brand is ADIDAS (p_1) and *any* of the following conditions is true:

- It is Christmas time (p_2)
- The customer is a new ADIDAS customer (p_3) and he purchases above \$ 150 (p_4)
- The customer is a frequent ADIDAS customer (p_5) and he purchases from new stock $(\bar{p}_6)$
- The customer is a infrequent ADIDAS customer $(\bar{p}_5)$ and he purchases from old stock (p_6)
- The customer is a frequent ADIDAS customer (p_5) and he purchases above \$ 100 (p_7)

Then announce 10 % discount on every shopping from ADIDAS.
The *If* part of the rule can be expressed as:

$$\Phi = p_1.(p_2 + p_3.p_4 + p_5.\bar{p}_6 + \bar{p}_5.p_6 + p_5.p_7)$$

We wish to find all the minimal support sets (positive and negative) for Φ.

[**Step 1:**] (Simplify Φ): We construct the ROBDD for Φ. We find the derivative of Φ with respect to all the propositions appearing in Φ. We create a start node S of the minimal support set tree T.

[**Step 2:**] The aim of this step is to find a strong minimal support set of Φ_1 and create one/two intermediate node(s) of T as needed. The iteration is started from this step. Let us assume the first proposition we consider here is p_1. It is easy to observe that $(p_1, 0)$ is a strong minimal support set of Φ_1. We find a positive and a negative minterm from the ROBDD of Φ_1 which contain p_1 and from these two minterms, we find a positive and a negative minimal support set containing p_1 as described in Algorithm 2. This has already been substantiated in Lemma 3. Eventually, we get a strong minimal support set $(p_1, 0)$. We create a leaf node $\mathcal{L}$ containing $(p_1, 0)$ in its first field and 0 in its second field. We also create an intermediate node $\mathcal{I}$ containing $(p_1, 1)$ in its first field and 0 in its second field as shown in Fig. 2.

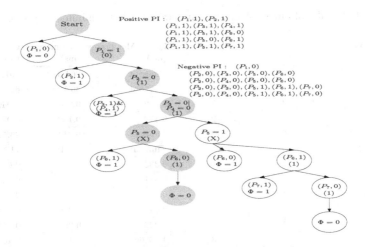

Fig. 2. Minimal support set computation for $\Phi_1 = p_1.(p_2 + p_3.p_4 + p_5.\bar{p}_6 + \bar{p}_5.p_6 + p_5.p_7)$

[**Step 3:**] We substitute p_1 by 1 and modify Φ_1. Now the modified function is $\Phi_1^{(1)} = p_2 + p_3.p_4 + p_5.\bar{p}_6 + \bar{p}_5.p_6 + p_5.p_7$. For the sake of simplicity of illustration, we have used superscripts to differentiate the functions generated at each iteration and differentiate from the ones generated in other iterations. In this way, we create all the intermediate nodes corresponding to a strong minimal support set as shown in Fig. 2. In iteration 4, we have the modified function as $\Phi_1^{(2)} = p_5.\bar{p}_6 + \bar{p}_5.p_6 + p_5.p_7$. As we can see, the algorithm fails to find a strong minimal support set. Therefore, we create two intermediate nodes in this step $\mathcal{I}_1$ and $\mathcal{I}_2$. Assume the proposition which we consider in this step is p_5. Hence, $\mathcal{I}_1$ contains $(p_5, 0)$ in its first field and X in its second field. Similarly, $\mathcal{I}_2$ contains $(p_5, 1)$ in its first field and X in its second field. The modified function for the subtree corresponding to $\mathcal{I}_1$ is $\Phi_1^{(3)} = p_6$ and the modified function for the subtree corresponding to $\mathcal{I}_2$ is $\Phi_1^{(4)} = \bar{p}_6 + p_7$. We again start to find the minimal support sets of $\Phi_1^{(3)}$ and $\Phi_1^{(4)}$ according to our algorithm. Once the tree is constructed fully, we start back tracing in order to find the minimal support sets of $\Phi_1 = p_1.(p_2 + p_3.p_4 + p_5.\bar{p}_6 + \bar{p}_5.p_6 + p_5.p_7)$.

[**Step 4:**] (Back Propagation): Consider the shaded path in Fig. 2. The leaf node indicates that we are going to construct a negative minimal support set, since the leaf node contains $\Phi = 0$. The assigned value of the previous intermediate node is 1, hence we consider its first field $(p_6, 0)$. The next intermediate node contains an assigned value X, therefore we have to verify whether we consider $(p_5, 0)$ as follows. The service rule corresponding to this level is $\Phi_1^{(2)} = p_5.\bar{p}_6 + \bar{p}_5.p_6 + p_5.p_7$. It is easy to see that we have to consider $(p_5, 0)$ in order to get a minimal support set, since $\Phi_1^{(2)}$ does not evaluate to constant, if we substitute $(p_6, 0)$ in $\Phi_1^{(2)}$. The next intermediate node contains 1 in its second field. Therefore we have to consider its first field. Here we get two minimal support sets, one combining $(p_3, 0)$ and another combining $(p_4, 0)$. Similarly we have to consider

Table 1. Results of our implementation (MSS stands for minimal support sets)

Rules	Input variables	Positive MSS	Negative MSS	Quine McCluskey (ms)	Our method (ms)	CUDD (ms)
1	30	9936	13	Timeout	264.319	497.556
2	21	9	1296	Timeout	30.132	82.645
3	4	4	3	0.922	0.226	0.046
4	4	4	2	0.879	0.132	0.039
5	18	9	512	Timeout	6.814	23.527
6	36	64	87205	Timeout	30445	113824
7	3	1	3	0.333	0.111	0.033
8	3	1	3	0.331	0.079	0.029
9	3	1	3	0.33	0.1	0.042
10	2	1	2	0.04	0.066	0.021
11	4	1	4	4.029	0.09	0.041
12	7	5	3	4127.66	0.163	0.131
13	3	3	1	0.33	0.088	0.034
14	3	3	1	0.443	0.083	0.03
15	3	3	1	0.331	0.085	0.032
16	3	2	2	0.147	0.108	0.031
17	29	282	2196	Timeout	1219.72	1257.8
18	36	4434	11209	Timeout	155199	158713
19	35	1905	10228	Timeout	27078	62563.8
20	32	802	4692	Timeout	5333.16	9520.12

the next intermediate node as well. We do not need to consider the intermediate node containing $(p_1, 1)$ since its second field contains 0. The next node is the start node itself and therefore the back propagation terminates. In this step we get two negative minimal support sets: $\{(p_2, 0), (p_3, 0), (p_5, 0), (p_6, 0)\}$ and $\{(p_2, 0), (p_4, 0), (p_5, 0), (p_6, 0)\}$. The final step is to combine either $(p_0, 0)$ or $(p_0, 1)$ with each minimal support set in order to get the minimal support sets of Φ.

4   Implementation and Results

We implemented our methodology in C++. We ran our experiments on manually created random service rules of varying sizes and complexity. On one side, we obtain the positive and negative minimal support sets related to any proposition, as required by Algorithm 2 by iterating over the paths of our data structure. We also implemented the Quine McCluskey algorithm in C++ to provide a contrast of its efficiency against ours, which is shown in Table 1. The CUDD [18] Boolean function manipulation package provides several programmable interfaces for minimal support set generation, and we contrast our approach against the CUDD routine as well in the same table. Table 1 presents the experimental results. Columns 5, 6 and 7 show the time taken by Quine McCluskey, our method and CUDD Implementation respectively. As evident from Table 1, our method takes much less time as compared to Quine McCluskey on all the benchmarks. Quine McCluskey is inherently nonscalable, hence we obtained timeouts on some of the larger cases. We also have comparative performance improvement over the CUDD API as evident on some of the cases. As explained earlier, we employed both the Quine McCluskey and CUDD API on the original function and its negation and recorded the combined times. As it can be seen, the CUDD implementation fails to generate all minimal support sets in many of the cases.

5   Conclusion and Future Work

In this paper, we address the problem of business rule verification and query execution, with rules expressed in extended Boolean logic. We discuss how we can expedite the run time execution using a look-up table and finally we present an innovative approach for simultaneously computing the exhaustive set of positive and negative test scenarios using a one-pass method, with the help of a novel data structure. This makes our proposition scalable and exhaustive. Experimental results on simulated benchmark shows the efficacy of our proposal. As evident from the results, our method efficiently computes the positive and negative scenarios as well. We are currently working on real business decision rules to see how our method works when put into real practice. Also we are experimenting on query evaluation using our method.

References

1. Paschke, A., Teymourian, K., AG Corporate Semantic Web: Rule based business process execution with BPEL+. In: I-SEMANTICS (2009)
2. Rosenberg, F., Dustdar, S.: Business rules integration in bpel-a service-oriented approach. In: E-Commerce Technology, CEC 2005 (2005)
3. Weigand, H., van den Heuvel, W.-J., Hiel, M.: Rule-based service composition and service-oriented business rule management. In: ReMoD (2008)
4. Paschke, A., Kozlenkov, A.: A rule-based middleware for business process execution. In: Multikonferenz Wirtschaftsinformatik (2008)
5. JRULEENGINE. http://jruleengine.sourceforge.net/
6. DROOLS. http://www.drools.org/
7. Pnueli, A.: The temporal logic of programs. In: 18th Annual Symposium on Foundations of Computer Science. IEEE (1977)
8. Deutsch, A., et al.: Automatic verification of data-centric business processes. In: Proceedings of the 12th International Conference on Database Theory. ACM (2009)
9. Shi, Y.-L., et al.: TLA based customization and verification mechanism of business process for SaaS. Jisuanji Xuebao (Chin. J. Comput.) **33**(11), 2055–2067 (2010)
10. Karnaugh, M.: The map method for synthesis of combinational logic circuits. Am. Inst. Electr. Eng. Part I: Trans. Comm. Electron. **72**(5), 593–599 (1953)
11. McCluskey, E.: Minimization of Boolean function. J. Bell Syst. Tech. **35**, 1417–1444 (1956)
12. Coudert, O.: Two-level logic minimization: an overview. Integr. VLSI J. **17**(2), 97–140 (1994)

13. Ron, R.: An SE-tree-based prime implicant generation algorithm. Ann. Math. Artif. Intell. **11**, 351–365 (1994)
14. Foster, H., Uchitel, S., Magee, J., Kramer, J.: Model-based verification of web service compositions. In: Proceedings of the 18th IEEE International Conference on Automated Software Engineering, 6–10 October 2003, pp. 152–161 (2003). doi:10.1109/ASE.2003.1240303
15. Zhu, Y., Gao, H.: A novel approach to generate the property for web service verification from threat-driven model. Appl. Math. **8**(2), 657–664 (2014)
16. Hachtel, G.D., Somenzi, F.: Logic Synthesis and Verification Algorithms. Kluwer Academic Publishers, Dordrecht (2000). ISBN:0792397460
17. Huth, M., Ryan, M.: Binary decision diagrams. In: Logic in Computer Science: Modelling andReasoning About Systems, Chap. VI, pp. 316–374 (2000)
18. CUDD: CU Decision Diagram Package Release 2.5.0. http://vlsi.colorado.edu/~fabio/CUDD/cuddAllDet.html

A Novel Reactive-Predictive Hybrid Resource Provision Method in Cloud Datacenter

Guorui Sun[1,2], ZhiHui Lu[1,2(✉)], Jie Wu[1,2], Xueying Wang[1,2], and Patrick Hung[3]

[1] School of Computer Science, Fudan University, Shanghai, China
{13210240060,lzh,jwu,13210240064}@fudan.edu.cn
[2] Engineering Research Center of Cyber Security Auditing and Monitoring,
Ministry of Education, Shanghai, China
[3] University of Ontario Institute of Technology, Oshawa, Canada
patrick.hung@uoit.ca

Abstract. Dynamic resource provisioning is an important way of ensuring performance and Service Level Agreement (SLA) guarantees for applications under changing workload. However, it is always hard to meet exactly the amount of resources required at every second. Thus, how to optimize the resource provision becomes the key problem. In this paper, we propose a Reactive-Predictive Hybrid Resource Provision Method (RPHRPM), which combines reactive and predictive methods together to benefit from both. We take advantage of ARIMA model to predict the workload and get resources pre-provisioned. Meanwhile, a reactive method is also enabled to deal with the unpredictable situations. More importantly we describe a novel mechanism which will be involved when conflicts between these two methods happen. It can help to keep better performance when encounter could burst. The experiment results show that RPHRPM not only has better performance compared with other provision schemes, but also be energy-efficient.

Keywords: Provision · Reactive method · Prediction model · Cloud datacenter

1 Introduction

In recent years, more and more enterprises and organizations began to take advantage of cloud to provide their services. So a widely concerning problem is approved - how to provision cloud resource not only ensuring application performance, but also minimizing the cost. In the traditional situation, resource allocation is always static. As the workload of an application is usually dynamic, allocating resources based on application peak workload will lead to over-provisioning. On the contrary, if resources are provided based on minimum workload, applications will experience SLA violations because of the insufficiency of resources.

In actual running environment, systems usually take SLA in the first place. So there may be thousands of hosts used to meet application demands, just to satisfy the highest workload. Therefore the data center resources are usually in low utilization. Based on the report of Data Center Efficiency Trends for 2014 from 'Energy Manager Today' [1], in current data center environments, server utilization rates are typically very low,

© Springer International Publishing Switzerland 2015
L. Yao et al. (Eds.): APSCC 2015, LNCS 9464, pp. 33–47, 2015.
DOI: 10.1007/978-3-319-26979-5_3

currently averaging in the 6–12 percent range. A completely idle server still draws 60 percent of its maximum power. In particular, it has been reported that energy-related costs account for approximately 12 % of overall data center expenditures. For large companies like Google, a 3 % reduction in energy cost can translate to over a million dollars in cost savings. As a result, reducing energy consumption has become a primary concern for today's data center operators [2]. Under the request of cost-saving, resource over-provisioning for near-peak performance should be considered by the resource management solutions. One of the most effective approaches for reducing energy costs is to dynamically adjust the data center capacity by launching required VMs and turning off free instances, or to set them to a sleep state. This is supported by the evidence that an idle machine can consume as much as 50–60 % of the power of when the machine is fully utilized [2]. Thus, to achieve the goal of energy-efficient resource management, utilization of cloud resource must be optimized. Dynamically adjusting and minimizing the number of active machines in a data center is a fundamental way to reduce energy consumption while meeting the SLAs of applications.

Dynamic resource provisioning has a lot of potential for increasing resource utilization in data centers. Most dynamic resource provisioning approaches can be categorized into two types: predictive and reactive. However, there are obviously advantages and disadvantages between these two kinds of method. A common assumption in resource provisioning is that workload demand can be predicted [3]. Therefore, prediction-based resource provisioning is required so as to deal with the periodic resource usage pattern. However, some unpredictable load spikes or fluctuations can hardly be found, which suggests that purely predictive approaches might be insufficient for handling data center workloads. As for reactive provisioning, although it can detect any kind of workload change immediately, it still can't ensure SLA guarantees all the time due to the assignable set-up time, especially in public cloud.

Based on this background, the key contribution of this paper is to develop a novel reactive-predictive hybrid resource provision method in cloud datacenters. RPHRPM analyzes the historical data of application workloads and predicts future workloads using ARIMA model, then computes the minimum resources that can ensure the application performance. Additionally, because of the dynamic nature of the workloads, when predictive results are not suitable for current situation, RPHRPM can discover this happening and adjust resource provision adaptively through a reactive method.

The remainder of this paper is organized as follows: In Sect. 2, we discuss related work on reactive and predictive provisioning methods. In Sect. 3, we introduce the architecture of the Reactive-Predictive Hybrid Resource Provision Method in cloud datacenter. Then we present our provision scheme in Sect. 4. In Sect. 5, we carry out experiments and related analysis. In Sect. 6, we make a conclusion of this paper and give a prospect of future work.

2   Related Work

With the rapid development of cloud computing, more and more enterprises are moving their business into cloud. In order to contribute to a high efficiency cloud datacenter, dynamic resource provisioning attracts much attention. In recent years, there have been many research works and studies on resource provisioning schemes and most of these approaches can be categorized into two types: predictive and reactive.

Firstly, we present some basic prediction techniques. Roy et al. [4] develop a model-predictive algorithm for workload forecasting, in order to achieve efficient auto-scaling in the cloud. They use a second order autoregressive moving average method (ARMA) filter to have a single look-ahead prediction. Gong et al. [5] present a novel predictive elastic resource scaling (PRESS) scheme for cloud systems. This scheme leverages Fast Fourier Transform (FFT) and discrete-time Markov chain to forecast future demand. It can handle both cyclic and noncyclic workload. Another kind of resource management scheme adds feedback control to the prediction models. Shen et al. [6] present CloudScale, a system that automates fine-grained elastic resource scaling for multi-tenant cloud computing infrastructures. It employs online resource demand prediction and prediction error handling to achieve adaptive resource allocation. Padala et al. [7] present AutoControl system. In this system, there is a model estimator, which inputs past allocation, past performance and leverage ARMA model to achieve the future performance value. With the predicted performance value, the optimizer module determines the resource allocation. There are some other research works which adopt multiple time series to build prediction models. Khan et al. [8] present a multiple time series approach for workload characterization and prediction in the cloud. Tan et al. [9] present multi-resource prediction models for resource sharing environments, like cloud, data center and so on. Jiang et al. [10] present an online system to model and predict the cloud VM demand. They utilize two–level ensemble method to capture the characteristics of the high transient demand time series.

Besides predictive method, some studies about cloud resource reactive dynamic provisioning technology are also proposed. In the paper [11], Guo et al. describe Seagull, a system designed to facilitate cloud bursting by determining which applications should be transitioned into the cloud and automating the movement process at the proper time. Seagull optimizes the bursting of applications using an optimization algorithm as well as the overhead of deploying applications into the cloud using an intelligent pre-copying mechanism that proactively replicates virtualized applications, lowering the bursting time from hours to minutes. Lo et al. [12] present PEGASUS, a feedback-based controller that significantly improves the energy proportionality of warehouse-scale computer (WSC) systems, as demonstrated by a real implementation in a Google search cluster. PEGASUS uses request latency statistics to dynamically adjust server power management limits in a fine-grain manner, running each server just fast enough to meet global service-level latency objectives. In large cluster experiments, PEGASUS reduces power consumption by up to 20 %. Gandhi [13] introduce a dynamic capacity management policy, AutoScale, that greatly reduces the number of servers needed in data centers driven by unpredictable, time-varying load, while meeting response time SLAs. Auto-Scale scales the data center capacity, adding or removing servers as needed.

Generally speaking, predictive methods can be very suitable when dealing with periodic or seasonal workloads. However, it fails when the workload spikes happen unpredictably. While reactive methods possess good ability dealing with aperiodic workload. But there is always delay when reacting to the workload changing, even when it is seasonal. Therefore, we design a novel hybrid scheme involving both predictive and reactive process, which help system keep better application performance under periodic workload, as well as irregular workload.

3   System Architecture

In this paper, we propose RPHRPM: a reactive-predictive hybrid resource provision method. Figure 1 shows the overall architecture of RPHRPM. It mainly consists of four parts: monitor, predictive process, reactive process and resource controller.

The monitor module will keep track of workload and collect application-level performance metrics, such as response time. Predictive process and reactive process are parallel, and they will not interfere with each other. For predictive process, a set of historical workload traces will be input into the prediction model. Then it will calculate the predicted workload for a certain future time. Next, the resource analyzer will analyze how much resource will be required based on the predicted workload to meet the user-defined application SLA. For reactive process, the first step is to check whether it needs to trigger the reactive dynamic resource provision or not. Observed application performance metric received from the monitor module and user-defined application SLAs are the two inputs of reactive trigger. This trigger will compare these two input statistics and decide whether to send the triggered signal to the resource analyzer. If triggered, the resource analyzer in reactive process will acquire current workload from the monitor module and the application SLA defined by user. Then it will calculate how much resource is actually required.

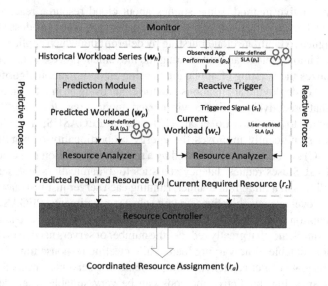

Fig. 1.   RPHRPM architecture

We can consider predictive process as the main process, and reactive process as a supplement of prediction. This is because the reactive method has a natural defect that is the non-negligible delay caused by the set-up time of launching a new instance. But the predictive method can be very close to the ideal solution if the prediction accuracy is very high. So we suppose that the predictive process can always get the appropriate result in most cases. Reactive process is the supplement of it when the unpredictable situation happens. Thus, we set a trigger in front of the reactive process to avoid unnecessary computations, and make the system more efficient.

After we get both predicted required resource (r_p) and current required resource (r_c), the resource controller will be enabled to coordinate these two requests. There are four conditions: (1) $r_p' > 0$ and $r_p' > r_c$, means the system is predicted to be overloaded, but now the reactive part doesn't recognize it or not that much. (2) $r_p' > 0$ and $r_p' < r_c$, means the system is predicted to be overloaded, and the overload situation already appeared, and is even more serious than the predicted level. (3) $r_p' < 0$ and $r_p' < r_c$, means system is predicted to be over-provisioned, but now the reactive part doesn't recognize it or not that much. (4) $r_p' < 0$ and $r_p' > r_c$, means system is predicted to be over-provisioned, and the over-provisioned situation already appeared, and is even more serious than the predicted level. In order to deal with all these possible conflict situations, we design an algorithm to coordinate them and give the final resource assignment decision.

4 RPHRPM Modules and Components

In the following subsections, we will describe the internals of every RPHRPM components. The monitor module is responsible for collecting required raw data. The predictive process consists of the prediction model and resource analyzer, while the reactive process

Table 1. Description of variables

Variables	Description
p_s	User-defined application performance SLA (response time)
p_o	Observed application performance (response time)
w_h	A vector of workload data (number of requests in the system) for a period of past time
w_p	A vector of predicted workload data (number of requests in the system) for a period of future time
w_c	Current workload (number of requests in the system)
r_p	A vector of predicted required resource (number of VMs)
r_p'	Current predicted required resource (number of VMs)
r_c	Current actual required resource (number of VMs)
r_a	Coordinated resource assignment (number of VMs)

contains the reactive trigger and resource analyzer. Results of these two processes will be inverted by resource controller into coordinated resource provision request.

For easy reference, Table 1 summarizes the mathematical symbols that will be used for key parameters and variables in these modules.

4.1 Resource Monitor

In order to keep the system running in a good performance, we need to get in charge of the whole circumstance of the active applications. Therefore, we design the monitor module to keep track with workload changes and collect the corresponding application performance metrics. The monitor module will periodically fetch these statistics and store them. Then the data will be prepared for other modules.

To implement the monitor module, there are many mature tools that can be integrated, including business products and open source projects. Actually most cloud providers also provide their own monitoring service, such as CloudWatch of Amzon EC2. As in private cloud, for example, OpenStack, also has the ceilometer project to provide some monitoring function. However, if the provided service does not satisfy user requirements, third-party tools also can be involved, such as LoadRunner etc. In our experiment, we implement this monitor module on a simulation platform. We select the number of requests in the system (w) to represent workload (we will explain why we choose this workload signal in the resource analyzer section below), and we choose response time (p) as the monitored performance metric, considering the user-defined application SLA is also supposed to be response time (ps). Meanwhile we set the monitor to record w as well as corresponding p every minute, and format them for the following modules.

4.2 Prediction Module

The aim of predictive method is to help the system know the workload change in advance, so that it can take action before the change happens. In other words, this means it can prepare the required resource in advance, thus it can eliminate the response delay caused by the virtual machine's set-up time. So in order to implement this, we need to predict future workload first.

The prediction module will periodically calculate future workload based on the historical workload series. Most classical methods for prediction are based on time series analysis and there also are some advanced models, such as ARMA models, ARIMA models, and state-space models etc. Here we select the autoregressive integrated moving average (ARIMA) model as our prediction model because this model is fitted to time series data either to better understand the data or to predict future points in the series. It is applied in some cases where data show evidence of being non-stationarity, where an initial differencing step can be applied to remove the non-stationarity. The model is generally referred to as an ARIMA (p, d, q) model where parameters p, d, and q are non-negative integers that refer to the order of the autoregressive, integrated, and moving average parts of the model respectively. More research about the ARIMA model and how to apply it in prediction can be found in [7, 14–16], so we do not make a more detailed description here.

In our experiment, we implement the prediction module based on the python module that provides classes and functions for the estimation of many different statistical models called StatsModels. The prediction modeling approach is depicted in Fig. 2. Our prediction module can take the workload data points during a past period time (w_h) as inputs. And then output the workload data points for a period future time (w_p). In order to ensure the accuracy of prediction, and simplify compute process at the same time, it is better to set length(w_h): length(w_p) = 5:1.

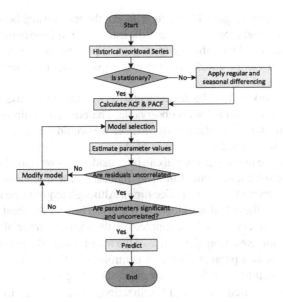

Fig. 2. Prediction modeling approach

4.3 Reactive Trigger

Reactive trigger is a trigger module in the reactive process. It will run periodically to detect whether the predictive method is working well. If the prediction result is accurate, the system should be running in good performance which means the user-defined SLA should not be violated. Under this circumstance, to calculate current required resource is not necessary. So the following process is keeping in a waiting status, only if the active triggered signal is received, the resource analyzer in reactive process will run.

Here we will summarize the logical process applied to the reactive trigger module. As described in the monitor part above, we choose response time as the monitored performance metric as well as user-defined SLA. Therefore, inputs of the reactive trigger module are the current observed app response time p_o pulled from the monitor module and application SLA p_s defined by the user. Then the inputs will be sent to the predicate function. If $p_o > p_s$ is false, this means that the current observed performance meets the requirement of the SLA. Then it will return and wait for the next execution. If $p_o > p_s$

is true, this means the current observed performance has already violated the settled SLA, and also means that the current workload is more heavy than the predicted. Thus it will trigger the resource analyzer to calculate the actual required resource, so that the whole system can be aware of this situation and require more resource to ensure the application performance.

4.4 Resource Analyzer

The resource analyzer is responsible for figuring out the relationship between workload and desired resource, and essentially inverts the estimated model to compute the desired resource allocations in order for the system to meet its performance SLA. So the function of resource analyzer can be concluded into two steps: model building and computing required resource.

As we know, workload is the fundamental factor that will cause fluctuations in application performance. Workload is observable. Therefore, if we find the relationship between workload and application performance, we would be able to infer desired resource to meet user-defined performance SLA.

However, there are many metrics that can be considered as workload signal. Request rate could be the most popular one. But using request rate will ignore one of the most important factors – request size (or service time). Although request rate is the same, for different request sizes the actual workload in system can be very different. Besides, there are also other alternative choices, for example, T_{95} (the 95th percentile of response times for requests that complete during the course of the trace). Using the performance metric as a feedback signal is a popular choice in control-theoretic approaches. However it could be quite difficult to infer desired resources according to T_{95}. Also some system-level metrics are wildly used, such as CPU utilization, memory utilization, network, etc. A major drawback of employing these signals is that it is hard to set the threshold for scaling up since our target is to increase resource utilization as much as possible. So we propose using the number of requests in the system (n_{sys}) as the workload signal (w) for scaling up capacity rather than the request rate. We assert that n_{sys} more faithfully captures the dynamic state of the system than the request rate. If the system is under-provisioned either because the request rate is too high or request size is too big or servers have slowed down, n_{sys} will tend to increase. If the system is over-provisioned, n_{sys} will tend to decrease below some expected level. Further, calculating n_{sys} is fairly straight-forward; many modern systems (including Apache load balancer) already track this value, and it is instantaneously available.

After getting n_{sys}, we can't simply scale up the capacity linearly with an increase in n_{sys}. This is because n_{sys} grows super-linearly during the time that the system is under-provisioned. Actually, we try to infer the amount of work in the system by monitoring n_{sys}. The amount of work in the system is proportional to both the request rate and the request size, and thus, we try to infer the product of the request rate and request size, which we call system load, l_{sys}. Formally,

$$l_{sys} = \text{request rate} \times \text{average request size} \tag{1}$$

Once we have l_{sys}, it is easy to get to the required resource (r), since l_{sys} represents the amount of work in the system and is hence proportional to r. Next we will depict the translation process from n_{sys} to l_{sys}, and then from l_{sys} to r. We refer to this entire translation algorithm as the *resource inference algorithm*. The full translation from n_{sys} to r will be given in Eq. (4) below. A full listing of all the variables used in this section is provided in Table 2 for convenience.

Table 2. Variables of resource inference algorithm

Variables	Description
n_{sys}	Number of requests in the system
l_{sys}	System load
r	Required resource (number of VMs)
n_{srv}	Number of requests at a server
l_{srv}	Load at a server
N	Current number of virtual machines
F	The relationship between the number of requests at a server and the load at a server
n_{max}	The maximum number of requests that a server can serve concurrently and meet its response time SLA
l_{max}	The maximum load in the system that a server can take and meet its response time SLA

In order to understand the relationship between n_{sys} and l_{sys}, we first derive the relationship between the number of requests at a single server, n_{srv}, and the load at a single server, l_{srv}. We expect that l_{srv} should increase with n_{srv}, and we can certificate it through experiment. Suppose that $l_{srv} = F(n_{srv})$, for different system the function can be different, but it only need to be analyzed through experiment once for each system. Now we can estimate system load (l_{sys}) using the relationship between l_{srv} and n_{srv}. To estimate l_{sys}, we first approximate n_{svr} as n_{sys}/N, where N is the current number of virtual machines. We then use n_{srv} to estimate the corresponding l_{srv}. Finally, we have $l_{sys} = N * l_{srv}$. In summary, given the number of requests in the system (n_{sys}), we can derive the system load (l_{sys}) as follows:

$$l_{sys} = N \times F(n_{sys}/N) \tag{2}$$

Fortunately, the relationship between the number of requests at a server and the load at a server does not change when request size changes, as well as server speed changes. This is because a decrease in server speed is the same as an increase in request size for system. Thus, while calculating $l_{sys} = N * l_{srv}$, we do not have to worry about the request size and we can simply use it to estimate l_{sys} from n_{sys} irrespective of the request size. The reason why the F is agnostic to request size is because l_{sys}, by definition (see Eq. (1)), takes the request size into account. If the request size doubles, then the request rate into a server needs to drop by a factor of 2 in order to maintain the same l_{srv}. These changes result in exactly the same amount of work entering the system per unit time, and thus, n_{srv}, does not change.

There is another parameter that needs to be determined. In order to analyze the value of r based on l_{sys}, we need to know the maximum number of requests that a server can serve concurrently and meet its response time SLA. The number satisfied this condition is called n_{max}, the corresponding load on that server is called l_{max}. They are constant for a certain system, thus also only need to be computed once. Since l_{sys} corresponds to the total system load, while l_{max} corresponds to the load that a single server can handle, we deduce that the required capacity is:

$$r = l_{sys}/l_{max} \tag{3}$$

In summary, we have:

$$r = \left\lceil \frac{N \times F(n_{sys}/N)}{l_{max}} \right\rceil \tag{4}$$

The design of the *resource inference algorithm* includes a few key parameters: n_{max}, l_{max}, F. In order to deploy the resource analyzer on a given system, these parameters need to be determined. Fortunately, all of the above parameters only need to be calculated once. This is because these parameters depend on the specifications of the system, such as the server type, the setup time, and the application, which do not change at runtime. Request rate, request size, and server speed, can all change at runtime, but these do not affect the value of the above key parameters. This makes it a very robust algorithm.

In our system architecture, there are two analyzers in predictive process and reactive process respectively. Since the process that deduces the required resource from the workload for both the predictive and reactive methods are similar, the core algorithm is the same. The only difference is for predictive process, input is a vector of workload data for a period of future time. While for reactive process, input is current workload. So the output of predictive resource analyzer is a vector of predicted required resource, corresponding to the predicted workload. While the reactive output is the real-time desired resource, corresponding to the newest sampled workload.

4.5 Resource Controller

The most important goal of the resource controller is to take both predictive and reactive results into consideration, and give the optimized decision. For dynamic resource provisioning, those decisions can be categorized into two actions: scale up and scale down. Now we will describe how it determines when to scale up or down and how large the scale should be.

Firstly, we will make clear about the input value. As depicted before, a vector of predicted required resource (r_p) and a current required resource (r_c) will be sent to this module. Each value in r_p represents the required resource for the corresponding predicted time point. And it begins from the results being calculated. In terms of the set-up time of launching a new instance, we need to prepare these resources in advance. Otherwise it will cause delay. In our experiment, we set the predicted period as 15 min, and suppose set-up time to be 1 min, and the monitoring period is 1 min. Thus actually it needs to predict workload for the future 16 min, and calculate corresponding required resource for the 16 min. But the value of the first 1 min are useless, because of the set-up time, it can never satisfy the requirement in the first 1 min. So we just keep values of the following 15 min. Then these values can be saved in a constant array. And every minute the program will take the first value in this array as the current predicted required resource (r_p') and remove it from the array after processing. The inverting of another input r_c is much more simple. It can be saved as a constant integer variable, initialized with the current VM number. This only updates when the reactive resource analyzer is triggered.

Next, the resource controller will periodically compare r_p' and r_c. If r_p' $>= r_c$, which means the preditve capacity is larger than currently needed. It seems to be over-provisioned. However as we explain before, because of the set-up time, the predictive value is not for current need, but to prepare for the future. So we take r_p' as r_a. If r_p' $< r_c$, which means the actual desired capacity is larger than the predicted one. Here we can't simply set r_c as r_a because we don't know how long the increase will last. If it is just an unpredictable load spike, and only appears for seconds, the new resource we added will be wasted. So we propose a *slow-increasing algorithm*.

First, we will introduce a new parameter R to represent the reliability of the reactive value, which also means it's a real number between 0 and 1. The computational formula of R can be described as follows:

$$R = k/\alpha, (\mathrm{k} \in [0, \alpha]) \tag{5}$$

where k is the number of times that r_p' $< r_c$ happens within the last α times of the comparison. α is a empirical value, which indicates the max number that could be accepted that the predictive result turns to be not accurate. In our experiment we set $\alpha = 5$.

Integrating the previous condition, r_p' $> r_c$, here we give a comprehensive way to compute r_a:

$$r_a = \left\lceil N' + (r_c - N') \times R \right\rceil, (N' = \max\left(r_p', N\right)) \tag{6}$$

Initially, if no r_p' $< r_c$ appears, k would be 0, thus, R is 0, $r_a = N'$. It means the current required resource has been no more than predicted value, so until now we totally rely on prediction. If r_p' $> N$ and r_p' $> r_c$, then the system will scale up, and new instances would be added. If $r_c < r_p$' $< N$, then the system would keep stable. When r_p' $< r_c$ happens, k increases, thus, R increases. It means the reliability of the reactive value has increased, and the adjustment function of the reactive process would be enabled. Part of the distance between r_c and N' would be added, so that the degree of scale-up would be increased. Applying the *slow-increasing algorithm* can help the system deal with the short-time load burst wisely and keep the system stable.

Finally, if $r_a > N$, the system will scale up, and send the request to get more resources. If $r_a < N$, this means the system should scale down. However, here we apply a conservative policy. When a server goes idle, rather than turning off immediately, it sets a timer of duration t_{wait} and sits in the idle state for t_{wait} seconds. If a request arrives at the server during these t_{wait} seconds, then the server goes back to the busy state; otherwise the server is turned off. To do this, a routing scheme needs to be implemented. The scheme tends to concentrate requests onto a small number of servers, so that the remaining servers will naturally time out. Benefiting from this conservative policy logically would be very simple in the scale down cases.

5   Experiment and Analysis

In this section, we will present experimental results that demonstrate the effectiveness of our reactive-predictive hybrid resource provision method (RPHRPM). The results will show that our method can dynamically adjust resource scale to ensure the performance SLO of application, with optimized resource provision compared with fixed server number, pure reactive method and pure predictive method.

5.1   Experimental Setup

We evaluate RPHRPM on an open-source framework for modeling and simulation of cloud computing infrastructures and services called CloudSim, with a set of usable extensions from CloudSimEx project. We simulate a cloud environment on this platform. There is one datacenter with three hosts and managed by a webBroker. The datacenter has been set to be able to contain 100 standard VMs at most. And each of the standard VM has been configured with 1 vCPU, a memory size of 512 M and 1000 Mbps network bandwidth. The set-up time of standard VM is set to 1 min.

Here we employ server provisioning on the stateless application servers only, as they maintain no volatile state. Stateless servers are common used in today's application platforms. We generate workload based on the analysis of some real-world traces. Figure 3 describes the changes of workload. We scale the arrival traces such that the maximum request rate into the system is 5000 req/s. Further, we scale the duration of the traces to 1 h. And considering of prediction, we need to know the history workload. So we also generate workload trace for the last 2 h, supposing it has the similar period with the first 30 min in Fig. 3.

In order to prove the effectiveness of RPHRPM, we make 4 experiments with different resource provision method respectively. The performance SLO, mean response time, is set to be 500 ms. *(1) A group of servers with fixed number.* In this senario, we run 20 standard VMs and don't do any adjustment. *(2) Predictive method.* We initialize 10 VMs at beginning, and applying preditive resource provisioning. It use the workload trace of last 75 min to predict future 15 min every 15 min. *(3) Reactive method.* Also

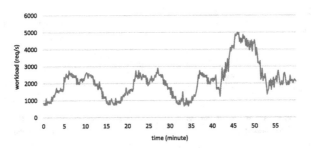

Fig. 3. Workload for experiment

we have 10 VMs first. Then it will trigger the increment of resource only if the mean response time is over 500 ms. *(4) RPHRPM.* There are 10 VMs at first. For predictive process, the policy is same as (2). For reactive process, the policy is same as (3). And as illustrated previously, the key parameter of resource controller, max number that could be accepted that the predictive result turns to be not accurate (α), is set to 5.

The results of these experiments are showed in following figures.

5.2 Results and Analysis

Figure 4 illustrates the mean response time of these four groups of experiments. For the experiment with fixed 20 servers, its mean response time was increasing significantly when every time the workload increased, overstepping performance SLO, 500 ms, a lot. And at around 45 min, because of the rapid growth of request rate, system with fixed number is totally out of service. Comparing with it, experiment applying predicted resource provision method behaves much better. Within the first 45 min, workload goes regularly, which makes the prediction works well. So during this period of time, mean response time is under control. However, since workload begins to be unpredictable after 45 min, the predictive method also becomes invalid. The response time increase sharply. Another experiment run with reactive method has specific advantage and disadvantage. The advantage is that no matter how workload changes, it will finally get the required resource to turn the system back to normal. Just as in our experiment, previous two methods are both failed to deal with the final increased workload, while the reactive one make it. But the cost is delay. So there are inevitable response time spike that will violate SLO. It's not a time-efficient method. Every time when the new resource under preparing, it can cause distinct growth of response time during the set-up process, such as 5 min, 22 min, 35 min and 45 min. Finally, let's observe at the performance of RPHRPM. The red line shows its mean response time during the whole experiment. The results show that it ran very close to SLO, which is set to 500 ms, under the periodical workload. The performance is just as good as predictive method. Meanwhile, for the next stage, it not only decreases the response time below SLO, but also reacts faster than reactive method. In conclusion, RPHRPM seems to have the best performance among these 4 resource provision strategy. It works well under both periodic and irregular workload.

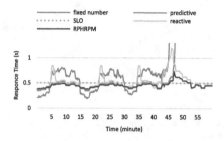
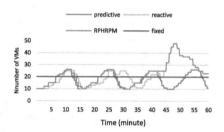

Fig. 4. Mean response time of difference methods

Fig. 5. Number of VMs with difference methods

Furthermore, after analyzing the application performance, we also check the resource cost of each method. We demonstrate the number of running VMs during the 1 h experiment in Fig. 5. It shows that system using predictive method has the lowest resource cost. But it doesn't work well for the last part, as we described before. System with fixed number of servers has the same problem. Then compare the reactive provisioning with RPHRPM. Actually after 45 min, predictive process of RPHRPM had already been inaccurate, which means it can be considered as just reactive method then. Therefore, we can see there is overlap between the two lines of reactive method and RPHRPM at the last section. Thus, we just need to compare the previous part. Since reactive method can't know the future situation, some new resource may only be used for a very short time. It will cause waste of resources, like around 27 min. While RPHRPM benefits from both reactive and predictive method, it can always make the best compromise between performance and resource cost. So that it has good results both in application performance and energy saving.

6   Conclusion and Future Work

In this paper, we have presented RPHRPM, a novel resource provision method which combine reactive and predictive methods together to achieve better performance of application under changing workload with optimized resource cost. We take advantage of ARIMA model to facilitate the predictive process, and applying reactive process as a supplement. Furthermore, we propose a slow-increasing algorithm to deal with the conflict situation, avoiding the instability of resource capacity caused by the short-time cloud burst. Finally we evaluated our method on CloudSim simulation platform. And the results of experiment prove the effectiveness and the efficiency of RPHRPM.

As a next step work, we plan to implement this method in real cloud datacenter environment, such as OpenStack-managed cloud environment. Besides, now we only consider the user cases of stateless application server. In the future, we will continue researching on stateful tier servers scaling, such as database server and the dynamic resource provisioning across multiple cloud datacenters.

Acknowledgements. This paper work is based on the Fudan-Hitachi Innovative Software Technology Joint Laboratory project-cloud virtualized resource management system. We would like to give our sincere thanks to Hitachi for all the support and advice. This work is also supported by 2014–2016 PuJiang Program of Shanghai under Grant No. 14PJ1431100 and 2015–2017 Shanghai Science and Technology Innovation Action Plan Project under Grant No. 15511107000.

References

1. Aaron, R.: Data center efficiency trends for 2014 (30 December 2013). http://www.energymanagertoday.com/data-center-efficiency-trendsfor-2014-097779/. University, June. VMWare. http://www.vmware.com/
2. Qi, Zh.: Efficient resource management for cloud computing environments, PHD thesis in University of Waterloo (2013)
3. Gandhi, A.: Dynamic server provisioning for data center power (2013)
4. Roy, N., Dubey, A., Gokhale, A.: Efficient autoscaling in the cloud using predictive models for workload forecasting. In: 2011 IEEE International Conference on Cloud Computing (CLOUD), pp. 500– 507. IEEE (July 2011)
5. Gong, Z., Gu, X., Wilkes, J.: Press: predictive elastic resource scaling for cloud systems. In: 2010 International Conference on Network and Service Management (CNSM), pp. 9 –16. IEEE (October 2010)
6. Shen, Z., Subbiah, S., Gu, X., Wilkes, J.: CloudScale: elastic resource scaling for multi-tenant cloud systems. In: 2nd ACM Symposium on Cloud Computing (SoCC2011), pp. 1–14. Cascais, Portugal (2011)
7. Padala, P., Hou, K.Y., Shin, K.G., Zhu, X., Uysal, M., Wang, Z., Singhal, S., Merchant, A.: Automated control of multiple virtualized resources. In: Proceedings of the 4th ACM European Conference on Computer Systems, pp. 13–26. ACM (2009)
8. Khan, A., Yan, X., Tao, S., Anerousis, N.: Workload characterization and prediction in the cloud: a multiple time series approach. In: Network Operations and Management Symposium (NOMS), 2012 IEEE, pp. 1287–1294. IEEE (April 2012)
9. Tan, J., Dube, P., Meng, X., Zhang, L.: Exploiting resource usage patterns for better utilization prediction. In: 2011 31st International Conference onDistributed Computing Systems Workshops (ICDCSW), pp. 14–19. IEEE (June 2011)
10. Yexi, J., Chang-shing, P., Tao, L., Rong, C.: Asap: A self-adaptive prediction system for instant cloud resource demand provisioning. In: 2011 IEEE 11th International Conference on Data Mining (ICDM), pp. 1104–1109. IEEE (2011)
11. Guo, T., Sharma, U., Shenoy, P., Wood, T., Sahu, S.: Cost –aware cloud bursting for enterprise applications. ACM Trans. Internet Technol. (TOIT) 13(3), 10 (2014)
12. Lo, D., Cheng, L., Govindaraju, R., Barroso, L.A., Kozyrakis, C.: Towards energy proportionality for large-scale latency-critical workloads. In: Proceeding of the 41st Annual International Symposium on Computer Architecture, pp. 301–312. IEEE Press (2014)
13. Gandhi, A., Harchol-Balter, M., Raghunathan, R., Kozuch, M.A.: Autoscale: dynamic, robust capacity management for multi-tier data centers. ACM Trans. Comput. Syst. (TOCS) 30(4), 14 (2012)
14. Brockwell, P.J., Davis, R.A.: Time Series: Theory and Methods. Springer, Heidelberg (1991)
15. Dashevskiy, M., Luo, Z.: Prediction of long-range dependent time series data with performance guarantee. In: Watanabe, O., Zeugmann, T. (eds.) SAGA 2009. LNCS, vol. 5792, pp. 31–45. Springer, Heidelberg (2009)
16. Chen, G., He, W., Liu, J., Nath, S., Rigas, L., Xiao, L., Zhao, F.: Energy-aware server provisioning and load dispatching for connection intensive internet services. In: NSDI, vol. 8, pp. 337–350 (2008)

A Social Balance Theory-Based Service Recommendation Approach

Lianyong Qi[1,2,3](✉), Xuyun Zhang[4], Yiping Wen[5],
and Yuming Zhou[1]

[1] Nanjing University, Nanjing 210093, Jiangsu, China
lianyongqi@gmail.com
[2] State Key Laboratory of Software Engineering, Wuhan University,
Wuhan 430072, China
[3] Qufu Normal University, Rizhao 276826, China
[4] Machine Learning Research Group, NICTA, Melbourne, VIC 3003, Australia
[5] Hunan University of Science and Technology, Xiangtan 411201, China

Abstract. With the popularity of social network, an increasing number of users attempt to find their interested web services through service recommendation, e.g., Collaborative Filtering (i.e., CF)-based service recommendation. Generally, the traditional CF-based service recommendation approaches work, when the target user owns one or more similar neighbors or friends (Neighbor and friend are interchangeable in the rest of paper) (i.e., user-based CF), or the target user's invoked services own similar services (i.e., item-based CF). However, in certain situations, similar neighbors and similar services are absent from the user-service invocation network, which brings a great challenge for accurate service recommendation. In view of this challenge, a novel recommendation approach *SBT-SR* (Social Balance Theory-based Service Recommendation) is put forward in this paper. Concretely, for the target user, we first determine his/her "enemies" (antonym of "friend", i.e., the users who have opposite preference with target user), and then look for the "potential friends" of target user, based on the "enemy's enemy is friend" rule in Social Balance Theory. Afterwards, the services preferred by "potential friends" are recommended to the target user. Finally, through a case study and a set of experiments, we demonstrate the feasibility of our proposal.

Keywords: Service recommendation · Target user · Similar neighbor · Similar service · Dissimilar enemy · Social balance theory

1 Introduction

Recently, with the adoption of SOA (Service Oriented Architecture) in both academic and industrial areas, more and more business processes are encapsulated into web services, and published in the public service community [1–3]. Therefore, people can easily browse, find and select their interested web services from the service community, so as to further construct various service-oriented complex business applications.

However, as new web services are emerging rapidly, the number of web services registered in service community is becoming larger and larger, which makes it difficult

© Springer International Publishing Switzerland 2015
L. Yao et al. (Eds.): APSCC 2015, LNCS 9464, pp. 48–60, 2015.
DOI: 10.1007/978-3-319-26979-5_4

to find the right web services that a user is really interested in [4, 5]. To solve this problem, many efforts have been made to develop various service recommendation approaches, e.g., Collaborative Filtering (i.e., CF)-based service recommendation [6–8]. Concretely, through the user-service invocation network (including user ratings on services), we can determine the similar neighbors of target user, or the similar services of the services that were invoked by target user, and further develop various recommendation approaches, e.g., user-based CF, item-based CF, or hybrid CF.

However, the above CF-based service recommendation approaches often perform poor, when similar neighbors and similar services are absent from the user-service invocation network (an example is presented in Sect. 2 for illustration), which brings a great challenge for accurate service recommendation. In view of this challenge, a novel service recommendation approach, i.e., *SBT-SR* (Social Balance Theory-based Service Recommendation) is put forward in this paper. Instead of looking for similar neighbors or friends in traditional CF-based recommendation approaches, in *SBT-SR*, we first look for dissimilar "enemy" (antonym of "friend") of the target user, and then further determine the target user's "potential friends", based on the "enemy's enemy is friend" rule in Social Balance Theory [9]. Finally, the services that are preferred by the "potential friends" of target user are recommended to the target user.

The remainder of this paper is organized as follows. In Sect. 2, we formalize the service recommendation problem and demonstrate the motivation of our paper. In Sect. 3, a novel service recommendation approach named *SBT-SR* is brought forth. A case study is introduced in Sect. 4. In Sect. 5, a set of experiments are deployed and the time complexity of *SBT-SR* is analyzed, to validate the feasibility of our proposal. Related works and discussion are presented in Sect. 6. And finally, in Sect. 7, we summarize the paper and point out the future research directions.

2   Formalization and Motivation

In this section, we first formalize the service recommendation problem. And afterwards, an example is presented to demonstrate the motivation of our paper.

2.1   Formalization

Generally, the service recommendation problem in social network could be specified with a four-tuple *Ser-Rec* $(U, WS, \rightarrow, user_{target})$, where

(1) $U = \{user_1, \ldots, user_N\}$ denotes the user set in user-service invocation network, and N is the number of users.
(2) $WS = \{ws_1, \ldots, ws_n\}$ denotes the web service set in user-service invocation network, and n is the number of web services.
(3) $\rightarrow = \{(user_i \xrightarrow{R_{i-j}} ws_j) | 1 \leq i \leq N, 1 \leq j \leq n\}$ denotes the invocation record set in user-service invocation network, i.e., $user_i \xrightarrow{R_{i-j}} ws_j$ means that $user_i$ invoked ws_j in the past and the $user_i$'s rating on ws_j is R_{i-j} after service invocation. Here, for simplicity, we adopt the well-known $\{1^*, 2^*, 3^*, 4^*, 5^*\}$ rating system to depict R_{i-j}.

(4) *user_target* denotes the target user that needs service recommendation, and *user_target*∈*U* holds.

Based on the above formalization, the classic service recommendation problem could be specified as follows: recommend appropriate web services ws_x (∈*WS*) to target user *user_target* (∈*U*), based on the historical user-service invocation record set "→" between *U* and *WS*.

2.2 Motivation

In this subsection, we demonstrate the motivation of this paper by an example (shown in Fig. 1). In Fig. 1, there are three users {*Tom, Alice, Bob*} (*Tom* is the target user) and six web services {ws_1, ws_2, ws_3, ws_4, ws_5, ws_6}; the user-service invocation records as well as their user ratings are also presented in the figure. Then according to the traditional CF approaches, the user similarity (∈[−1, 1]) could be calculated based on PCC (Pearson Correlation Coefficient) [10]. Concretely, $Sim(Tom, Alice) = -0.275$ and $Sim(Tom, Bob) = $ Null; therefore, target user *Tom* has no similar neighbors since no positive user similarity is present. Besides, likewise, the service similarity could also be calculated, i.e., $Sim(ws_1, ws_3) = Sim(ws_1, ws_4) = Sim(ws_2, ws_3) = Sim(ws_2, ws_4) = -1$ and $Sim(ws_1, ws_5) = Sim(ws_1, ws_6) = Sim(ws_2, ws_5) = Sim(ws_2, ws_6) = $ Null; therefore, target user *Tom*'s invoked services (i.e., ws_1 and ws_2) have no similar services, since no positive service similarity is present.

In this situation, the traditional CF-based service recommendation approaches (e.g., user-based CF, item-based CF or hybrid CF) cannot make accurate service recommendation, as the target user has no similar neighbors and the target user's invoked services have no similar services, which brings a great challenge for accurate service recommendation. In view of this challenge, a novel service recommendation approach named *SBT-SR* is put forward in the next section.

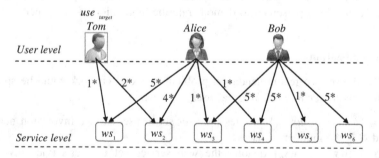

Fig. 1. Service recommendation with no similar neighbors and similar services: an example.

3 *SBT-SR*: A Service Recommendation Approach Based on Social Balance Theory

In this section, we introduce a novel approach *SBT-SR*, to recommend appropriate services to the target user, when similar neighbors and similar services are absent from the user-service invocation network. In general, our proposal is based on the "enemy's enemy is friend" rule in Social Balance Theory. Concretely, *SBT-SR* consists of four steps in Fig. 2.

Step1: Determine target user $user_{target}$'s enemy. Calculate the similarity $Sim(user_{target}, user_i)$ between $user_{target}$ and other users $user_i \in U$. If $Sim(user_{target}, user_i) \cdot P$ (P is similarity threshold), then $user_i$ is an enemy of $user_{target}$.

Step2: Determine target user $user_{target}$'s enemy's enemy. For each enemy (i.e., $user_i$ in Step1) of $user_{target}$, determine his/her enemy $user_k$, through the same judgment process as Step1.

Step3: Determine target user $user_{target}$'s potential friends. According to the "enemy's enemy is friend" rule in Social Balance Theory, determine $user_{target}$'s potential friend set $Po\_friend\ (user_{target})$.

Step4: Service recommendation. Recommend the services preferred by potential friends in set $Po\_friend\ (user_{target})$ (derived in Step3) to the target user $user_{target}$.

Fig. 2. Four steps of service recommendation approach *SBT-SR*.

(1) Step1: Determine target user $user_{target}$'s enemy.

In this step, we first calculate the similarity $Sim(user_{target}, user_i)$ between target user $user_{target}$ and other user $user_i$ ($user_i \in U$); and afterwards, those users with low similarity values are regarded as the "enemies" of $user_{target}$. Concretely, according to the classic PCC, $Sim(user_{target}, user_i)$ could be calculated by (1).

$$Sim\left(user_{target}, user_i\right) = \frac{\sum\limits_{ws_j \in I} \left(R_{target-j} - \overline{R_{target}}\right) * \left(R_{i-j} - \overline{R_i}\right)}{\sqrt{\sum\limits_{ws_j \in I} \left(R_{target-j} - \overline{R_{target}}\right)^2} * \sqrt{\sum\limits_{ws_j \in I} \left(R_{i-j} - \overline{R_i}\right)^2}} \quad (1)$$

Here, set I denotes the common services that were invoked by both $user_{target}$ and $user_i$; $R_{target-j}$ and R_{i-j} denote $user_{target}$'s and $user_i$'s ratings on service ws_j, respectively; while $\overline{R_{target}}$ represents $user_{target}$'s average rating on all his/her invoked web services, and $\overline{R_i}$ represents $user_i$'s average rating on all his/her invoked web services. Then according to (1), we can calculate the similarity $Sim(user_{target}, user_i)$. Specially, if $user_{target}$ and $user_i$ have no common invoked services (i.e., I = null), then their similarity $Sim(user_{target}, user_i)$ = Null.

As can be seen from (1), the user-similarity $Sim(user_{target}, user_i) \in [-1, 1]$. Afterwards, we set a similarity threshold P ($-1 \leq P \leq -0.5$) to judge whether $user_i$ is an enemy of target user $user_{target}$. Concretely, the judgment process is based on the formula in (2). Through (2), we can obtain the enemy set of $user_{target}$, i.e., *Enemy_set* $(user_{target})$.

$$user_i \begin{cases} \in Enemy\_set(user_{target}) & \text{if } Sim(user_{target}, user_i) \leq P \\ \notin Enemy\_set(user_{target}) & \text{if } Sim(user_{target}, user_i) > P \end{cases} \quad (2)$$

(2) Step2: Determine target user $user_{target}$'s enemy's enemy.

In this step, we determine target user $user_{target}$'s enemy (i.e., $user_i$ in Step1)'s enemy (denoted by $user_k$). Concretely, for each enemy $user_i \in Enemy\_set(user_{target})$, we calculate his/her similarity with other users, and further determine his/her enemy $user_k$, through the same judgment process as Step1. Here, the concrete calculation and judgment details are not repeated again.

(3) Step3: Determine target user $user_{target}$'s potential friends.

In Step1, we have obtained target user $user_{target}$'s enemy $user_i$, and in Step2, we have determined $user_i$'s enemy $user_k$. Then according to the "enemy's enemy is friend" rule in Social Balance Theory, we can infer that $user_k$ is a potential friend of $user_{target}$, i.e., $user_k \in Po\_friend(user_{target})$. Here, please note that $user_k$ is just a "possible" friend of $user_{target}$, not a "definite" friend of $user_{target}$. Therefore, to quantify the credibility that $user_k$ is a friend of $user_{target}$, a new criterion $Credibility\_friend(user_{target}, user_k)$ is put forward here, which could calculated by (3).

$$Credibility\_friend(user_{target}, user_k) = Sim(user_{target}, user_i) * Sim(user_i, user_k) \quad (3)$$

According to (2), user similarity $Sim(user_{target}, user_i) \leq P$ and $Sim(user_i, user_k) \leq P$ ($-1 \leq P \leq 0$). Therefore, according to (3), $Credibility\_friend(user_{target}, user_k) \in [P^2, 1]$. For example, if similarity threshold $P = -0.9$, $Sim(user_{target}, user_i) = -0.92$ and $Sim(user_i, user_k) = -0.95$, then $Credibility\_friend(user_{target}, user_k) = (-0.92)*(-0.95) = 0.874$.

(4) Step4: Service recommendation.

After Steps 1–3, we have obtained the potential friends of target user $user_{target}$, i.e., $user_k \in Po\_friend(user_{target})$. Next, we recommend appropriate web services to $user_{target}$, based on the services preferred by potential friends $user_k$. Concretely, only the web services ws_x with high ratings (e.g., 3*, 4*, 5*) from $user_k$ are recommended to $user_{target}$. Here, Rec_Serv_Set is recruited to denote the recommended service set, i.e., $ws_x \in Rec\_Serv\_Set$ holds.

Next, to discriminate and rank all the recommended services $ws_x \in Rec\_Serv\_Set$, we quantify their recommendation-credibility $Rec\_Credibility(ws_x)$ by (4). Here, R_{k-x} (introduced in Sect. 2.1) denotes $user_k$'s rating on service ws_x; while function $weight()$ is utilized to transform user rating R_{k-x} (e.g., 3*, 4*, 5*) into a value in [0, 1]. For simplicity, a naive $weight()$ function is adopted here, which is specified in Table 1 (in fact, only R_{k-x} with 3* ~ 5* are of use here). Then according to (3)–(4), the recommendation-credibility $Rec\_Credibility(ws_x)$ of each service ws_x could be calculated, based on which we can rank all the recommended services $ws_x \in Rec\_Serv\_Set$ in descending order, and recommend them to the target user $user_{target}$. Here, please note

Table 1. A naive rating transformation function $weight()$ recruited in formula (4).

R_{k-x}	1*	2*	3*	4*	5*
$weight(R_{k-x})$	0	0.25	0.5	0.75	1.0

that if ws_x is recommended by multiple potential friends simultaneously, the average $Rec\_Credibility(ws_x)$ value is adopted.

$$Rec\_Credibility(ws_x) = Credibility\_friend\left(user_{target}, user_k\right) * weight(R_{k-x}) \qquad (4)$$

With the above Steps 1–4 of our *SBT-SR* approach, a set of web services, i.e., $ws_x \in Rec\_Serv\_Set$, are recommended to the target user $user_{target}$. More formally, the pseudo-code of our proposal is specified as below. Here, $user_{target}$ denotes the target user in user-service invocation network where no similar neighbors and similar services are present; then through *SBT-SR*, we can recommend appropriate services to $user_{target}$, so as to avoid the cold start problem in service recommendation.

Algorithm: *SBT-SR* (U, WS, →, $user_{target}$)

Inputs: (1)$U = \{user_1, ..., user_N\}$: a set of users in user-service invocation network;

(2)$WS = \{ws_1, ..., ws_n\}$: a set of web services in user-service invocation network;

(3)→$=\{(user_i \xrightarrow{R_{i-j}} ws_j) \mid 1 \le i \le N, 1 \le j \le n\}$:

a set of user-service invocation records;

(4)$user_{target}$: target user that needs service recommendation.

Output: Rec_Serv_Set : web service set recommended to $user_{target}$

1: Set user similarity threshold P

2: **For each** $user_i \in U$ **do**   //determine $user_{target}$'s enemy

3:     calculate $Sim(user_{target}, user_i)$ by (1)

4:     **If** $Sim(user_{target}, user_i) \le P$

5:         **then** put $user_i$ into set $Enemy\_set(user_{target})$

6:     **End if**

7: **End for**

8: **For each** $user_i \in Enemy\_set(user_{target})$ **do**   //determine $user_{target}$'s enemy's enemy

9:     **For each** $user_k \in U$ **do**

10:         calculate $Sim(user_i, user_k)$ by (1)

11:         **If** $Sim(user_i, user_k) \le P$

12:             **then** put $user_k$ into set $Po\_friend(user_{target})$

13:                 calculate $Credibility\_friend(user_{target}, user_k)$ by (3)

14:                 **For each** $ws_x \in WS$ **do**   //determine recommendation-credibility of ws_x

15:                     **If** $(user_k \xrightarrow{R_{k-x}} ws_x)$ exists in set → and $R_{k-x} \in \{3^*, 4^*, 5^*\}$

16:                         **then** put ws_x into set Rec_Serv_Set

17:                             calculate $Rec\_Credibility(ws_x)$ by (4)

18:                     **End if**

19:                 **End for**

20:         **End if**

21:     **End for**

22: **End for**

23: **For each** $ws_x \in Rec\_Serv\_Set$ **do**

24:     rank ws_x based on $Rec\_Credibility(ws_x)$ in descending order

25: **End for**

26: **Return** Rec_Serv_Set to $user_{target}$

---

4   Case Study

In this section, a case study is presented to demonstrate the concrete service recommendation process based on our proposed *SBT-SR* approach. As Fig. 3 shows, there are six users {*Tom*, *user₁*, *user₂*, *user₃*, *user₄*, *user₅*} and six web services {*ws₁*, *ws₂*, *ws₃*, *ws₄*, *ws₅*, *ws₆*} in the user-service invocation network (here, *Tom* is the target user; and for *Tom*, similar neighbors and similar services are both absent from the invocation network).

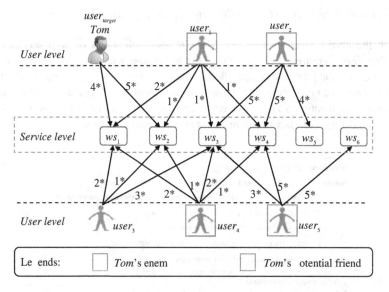

Fig. 3. A user-service invocation network with six users and six web services.

Next, we introduce how to recommend appropriate services to *Tom* based on our proposed *SBT-SR* approach. As Table 2 shows, although *Tom* has no similar neighbors, he has two enemies, i.e., *user₁* and *user₄* (here, we assume the similarity threshold $P = -0.8$). Afterwards, we determine *user₁*'s (or *user₄*'s) enemies, by calculating the similarity between *user₁* (or *user₄*) and other users. The results are listed in Table 3. As *user₁* and *user₄* are both enemies of *Tom*, it is not necessary to calculate their user similarity.

As can be seen from Table 3, *user₂* is an enemy of *user₁*, while *user₅* is an enemy of *user₄*. As *user₁* and *user₄* are both enemies of target user *Tom*, according to the "enemy's enemy is friend" rule in Social Balance Theory, *user₂* and *user₅* are potential friends of *Tom*. Therefore, we can recommend *user₂*'s and *user₅*'s invoked web services with high ratings (≥3*) to *Tom*. Concretely, the web services that are

Table 2. User similarity between target user *Tom* and other users.

$user_i$	$user_1$	$user_2$	$user_3$	$user_4$	$user_5$
$Sim(Tom, user_i)$	**−0.894**	Null	−0.707	**−1**	Null

Table 3. User similarity between $user_1$ (or $user_4$) and other users.

$user_i$	$user_1$	$user_2$	$user_3$	$user_4$	$user_5$
$Sim(user_1, user_i)$	1	−0.967	0	Not necessary	0.318
$Sim(user_4, user_i)$	Not necessary	0	0.442	1	−0.952

recommended to *Tom* are listed in Table 4 (with remarked symbol "$\sqrt{}$"). Next, according to formula (4), the recommendation-credibility $Rec\_Credibility(ws_x)$ of each recommended web service ws_x is calculated, whose results are also shown in Table 4. Here, please note that web service ws_3 is recommended by *Tom*'s potential friends $user_2$ and $user_5$ simultaneously, therefore, ws_3's average $Rec\_Credibility(ws_3)$ is adopted here. Likewise, ws_4's average $Rec\_Credibility(ws_4)$ is calculated. Finally, according to average $Rec\_Credibility(ws_x)$, we rank services $\{ws_3, ws_4, ws_5, ws_6\}$ in descending order, i.e., $\{ws_6, ws_4, ws_3, ws_5\}$, and recommend them to *Tom*.

Table 4. Web services that are recommended to target user *Tom*.

		ws_1	ws_2	ws_3	ws_4	ws_5	ws_6
Tom's potential friends	$user_2$			$\sqrt{}$	$\sqrt{}$	$\sqrt{}$	
	$Rec\_Credibility(ws_x)$			0.864	0.864	0.648	
	$user_5$			$\sqrt{}$	$\sqrt{}$		$\sqrt{}$
	$Rec\_Credibility(ws_x)$			0.476	0.952		0.952
average $Rec\_Credibility(ws_x)$				0.67	0.908	0.648	0.952

5   Experiment Analyses

In this section, a set of experiments are deployed to validate the feasibility of our proposed *SBT-SR* approach. Afterwards, the time complexity of *SBT-SR* is analyzed.

5.1   The Dataset and Experiment Deployment

To the best of our knowledge, the real user ratings on web services are few. Therefore, in this paper, the well-known MovieLens-10M [11] dataset is recruited for simulation purpose. The dataset contains 10 million ratings applied to 10,000 movies by 72,000 users. Our proposed *SBT-SR* approach mainly focuses on the service recommendation situations when no similar neighbors and similar services are present for the target user; therefore, we select appropriate experiment data to simulate the above recommendation situations. For each user in the dataset, we split his/her ratings into two parts with the split ratio of r: $100 - r$ ($0 < r \leq 20$). Namely, we use r % ratings of the user to predict the remaining $(100 - r)\%$ ratings, and the prediction accuracy could be measured by the well-known *Mean Absolute Error* (*MAE*) in (5). Here, $R_{target-j}$ and $\overline{R_{target-j}}$ denote $user_{target}$'s real and predicted ratings on service ws_j, respectively; $Num\_Ser(user_{target})$ denotes the number of web services that are rated by $user_{target}$.

$$MAE_{target} = \sum_{j=1}^{Num\_Ser(user_{target})} \frac{|R_{target-j} - \overline{R_{target-j}}|}{Num\_Ser(user_{target})} \qquad (5)$$

The experiments were conducted on a Dell laptop with 2.80 GHz processors and 2.0 GB RAM. The machine is running under Windows XP and JAVA 1.5. Each experiment was carried out 10 times and the average results were adopted.

5.2   Experiment Results

Concretely, three evaluation profiles are tested and compared, which will be introduced in detail respectively. Here, N denotes the number of users and n denotes the number of services.

Profile 1: Prediction Accuracy Comparison with Other Approaches. In this profile, we test the prediction accuracy of *SBT-SR* and compare it with other three approaches, i.e., *WSRec* [8], *MCCP* [12] and *K-means* [13]. Concretely, parameters $w_u = w_i = 0.5$ holds in *WSRec*, $\alpha = 0.8$ and $N = 1000$ holds in *MCCP*, $K = 3$ holds in *K-means* and $P = -0.5$ holds in our *SBT-SR*. Besides, $N = 1000$ and $n = 1000$ hold for all the four approaches, and the density of user-service matrix is varied from 4 % to 20 %, i.e., $r = 4, 8, 12, 16, 20$. The experiment results are shown in Fig. 4. As Fig. 4 shows, the *MAE* values (the smaller, the better) of four approaches all decrease with the growth of r, and our *SBT-SR* outperforms the other three approaches when r is small (i.e., $r = 4$, 8, 12); this is because when the user-service matrix is very sparse, *WSRec*, *MCCP* and *K-means* can only make approximate and coarse prediction, while *SBT-SR* can still find the target user's potential friends as well as their preferred services, through the "enemy's enemy is friend" rule.

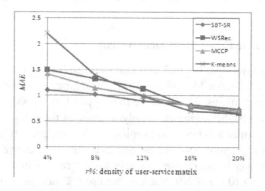

Fig. 4. *MAE* comparison between *SBT-SR* and other approaches.

Besides, it can also be observed from Fig. 4 that the prediction accuracy of *SBT-SR* is not as good as the other three approaches, when *r* is big (i.e., *r* = 16, 20); this is because when the user-service matrix is dense, *WSRec*, *MCCP* and *K-means* can find the target user's friends in a "direct" manner, while our *SBT-SR* can only find the target user's friends in an "indirect" manner, which loses the prediction accuracy.

Profile 2: Execution Efficiency of SBT-SR with N. In this profile, we test the execution efficiency of *SBT-SR* with the number of users *N*. Concretely, *N* is varied from 200 to 1000, the number of web services, i.e., *n* = 1000 holds, the density of user-service matrix is 10 % and *P* = −0.5 holds. The experiment result is shown in Fig. 5(a). As can be seen from Fig. 5(a), the time cost of *SBT-SR* increases faster and faster with the growth of *N*; this is because each user is considered twice (once for determining the target user's enemy, once for determining the target user's enemy's enemy), when determining the target user's potential friends. However, as Fig. 5(a) indicates, the efficiency of *SBT-SR* is still acceptable (at "millisecond" level).

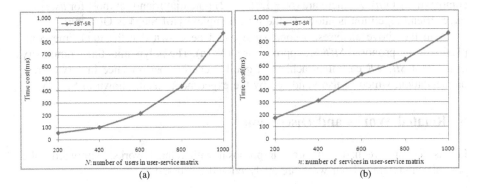

Fig. 5. Execution efficiency of *SBT-SR* approach

Profile 3: Execution Efficiency of SBT-SR with n. In this profile, we test the execution efficiency of *SBT-SR* with the number of web services, i.e., *n*. Concretely, *n* is varied from 200 to 1000, the number of users, i.e., *N* = 1000 holds, the density of user-service matrix is 10 % and *P* = −0.5 holds. The experiment result is shown in Fig. 5(b). As can be seen from Fig. 5(b), the time cost of our proposed *SBT-SR* increases approximately linearly with the growth of *n*; this is because for each user, all his/her invoked services should be considered for service recommendation in *SBT-SR*. Similar with Profile 2, the execution efficiency of *SBT-SR* is acceptable in most service recommendation situations (at "millisecond" level).

5.3   Complexity Analyses

Suppose there are *N* users and *n* web services in the user-service invocation network.

Step1: according to (1), we can calculate the similarity between target user and other users, whose time complexity is O(*N* *n). Then with the derived user similarity, the enemies of target user $user_{target}$ could be determined based on (2), whose time complexity is $O(N)$. Therefore, the time complexity of Step1 is $O(N *n)$.

Step2: For each enemy $user_i$ of $user_{target}$, we can determine his/her enemy $user_k$ by (1), whose time complexity is $O(N *n)$. As there are $N - 1$ enemies at most for $user_{target}$, the time complexity of Step2 is $O(N^2 *n)$.

Step3: According to the "enemy's enemy is friend" rule, we can determine target user $user_{target}$'s potential friends, whose time complexity is $O(1)$. Next, we can calculate the credibility that $user_k$ is a friend of $user_{target}$ by (3), whose time complexity is $O(1)$. As there are $(N - 1)*(N - 2)$ potential friends at most for $user_{target}$ (here, friend repetition is allowed), the time complexity of Step3 is $O(N^2)$.

Step4: For each potential friend $user_k$ of $user_{target}$, we can determine all his/her recommended services, as well as their recommendation-credibility by (4), whose time complexity is $O(n)$. As there are $(N - 1)*(N - 2)$ potential friends at most for $user_{target}$ (here, friend repetition is allowed), the time complexity of Step4 is $O(N^2*n)$.

With the above analyses, a conclusion could be drawn that the total time complexity of our proposed *SBT-SR* approach is $O(N^2*n)$. The polynomial time complexity means that *SBT-SR* can often achieve a good execution performance in service recommendation, which is also validated by the experiment results shown in Sect. 5.2.

6 Related Works and Discussion

In this section, we first compare our proposal with related works. Afterwards, the possible limitations of our work are analyzed and discussed.

6.1 Related Works and Comparison Analyses

Service recommendation is now considered as a promising way, to quickly find the target user's interested web services that are tailored to user preference, especially when the user-service invocation network is very large and dynamic. At present, many researchers have studied the service recommendation problem and give their valuable resolutions.

A two-phase K-means clustering approach is introduced in [13] to make service quality prediction and service recommendation. However, this approach calls for a denser user-service invocation matrix, and hence cannot be applied in service recommendation with sparse data. A Collaborative Filtering-based recommendation approach is proposed in [4], which recommends appropriate web services to the target user, based on the services preferred by target user's similar neighbors. However, when the target user has no similar neighbors, the prediction accuracy is not as good as expected. A bidirectional service recommendation approach *WSRec* is brought forth in [8], which combines user-based collaborative filtering and service-based collaborative filtering, for better service recommendation. However, when the target user has no similar neighbors and the target user's invoked services have no similar services, *WSRec* can only make approximate prediction for the missing service quality, by considering the target user's average rating on all his/her invoked services, as well as a recommended service's average rating from all users. In [12], a *MCCP* approach is put forward to model and capture various users' preferences towards different services; however, only similar neighbors are recruited in *MCCP* for service quality prediction, which drops some valuable information existing in the sparse user-service invocation network.

In summary, the above approaches are not suitable in service recommendation situations when both similar neighbors and similar services (corresponding to the target user) are absent from the user-service invocation network. In view of this shortcoming, a novel service recommendation approach named *SBT-SR* is proposed in this paper. By utilizing the "enemy's enemy is friend" rule in Social Balance Theory, we can determine the potential friends of the target user, and further recommend the services preferred by potential friends to the target user. Through a case study and a set of experiments, we validate the feasibility of our proposal in terms of recommendation accuracy and efficiency.

6.2   Further Discussion

There are still some shortcomings in our paper, which are discussed as below.

(1) In our proposed *SBT-SR* approach, the employed user similarity threshold, i.e., P is set manually, which does not satisfy the requirement of automated service recommendation in social network. In the future, we will study this problem and investigate more automated parameter setting manner for similarity threshold.

(2) Service quality is often dynamic and varied with time; correspondingly, user's ratings on services are also time-aware. However, for simplicity, the time factor is not discussed in this paper. In the future, we will include the time factor into our service recommendation model for better recommendation quality.

(3) We are forced to adopt the "enemy's enemy is friend" rule for service recommendation, when the user-service invocation network owns no similar friends and similar services for target user; while obviously, the recommendation effect of "enemy's enemy is friend" rule is weaker than "friend's friend is friend" rule that is widely adopted in traditional CF-based service recommendation approaches. Therefore, our proposed *SBT-SR* approach can only be considered as a beneficial supplement to the traditional CF-based approaches, instead of replacing the traditional ones. In the future, we will study how to combine them for better service recommendation.

7   Conclusions

In this paper, a novel recommendation approach *SBT-SR* is put forward in this paper, to make service recommendation for the target user with no similar neighbors and similar services. Instead of looking for similar neighbors in traditional service recommendation approaches, we look for dissimilar enemy of the target user, and further determine the target user's potential friends based on the "enemy's enemy is friend" rule in Social Balance Theory. Through a case study and a set of experiments, we validate the feasibility of our proposal.

In the future, we will investigate more automated setting manner for similarity threshold P, and introduce the time factor into our service recommendation model, so as to accommodate the automated and dynamic service recommendation requirements; besides, we will combine the "enemy's enemy is friend" rule and "friend's friend is friend" rule, for better service recommendation.

Acknowledgements. This paper is supported by National Natural Science Foundation of China (No. 61402258, 61402167), China Postdoctoral Science Foundation (No. 2015M571739), Open Project of State Key Laboratory of Software Engineering (No. SKLSE2014-10-03), Open Project of State Key Lab. for Novel Software (No. KFKT2015A03), DRF (No. BSQD20110123) of QFNU.

References

1. Wang, H., Chen, X., Wu, Q., Yu, Q., Zheng, Z.: Integrating on-policy reinforcement learning with multi-agent techniques for adaptive service composition. In: 12th International Conference on Service Oriented Computing, pp. 154–168. ACM Press, New York (2014)
2. Yang, J., Lin, W., Dou, W.: An adaptive service selection method for cross-cloud service composition. Concurrency Comput. Pract. Experience **25**(18), 2435–2454 (2014)
3. Chen, L., Wang, Y., Yu, Q., Zheng, Z., Wu, J.: WT-LDA: user tagging augmented LDA for web service clustering. In: 11th International Conference on Service Oriented Computing, pp. 162–176. ACM Press, New York (2013)
4. Lin, S.Y., Lai, C.H., Wu, C.H., Lo, C.C.: A trustworthy QoS-based collaborative filtering approach for web service discovery. J. Syst. Softw. **93**, 217–228 (2014)
5. Jiang, D., Guo, X., Gao, Y., Liu, J., Li, H., Cheng, J.: Locations recommendation based on check-in data from location-based social network. In: 22nd International Conference on Geoinformatics, pp. 1–4. IEEE Press, New York (2014)
6. Tang, M., Xu, Y., Liu, J., Zheng, Z., Liu, F.: Combining global and local trust for service recommendation. In: 21st IEEE International Conference on Web Services, pp. 305–312. IEEE Press, New York (2014)
7. Cao, B., Liu, J., Tang, M., Zheng, Z., Wang, G.: Mashup service recommendation based on user interest and social network. In: 20th IEEE International Conference on Web Services, pp. 99–106. IEEE Press, New York (2013)
8. Zheng, Z., Ma, H., Lyu, M.R., King, I.: QoS-aware web service recommendation by collaborative filtering. IEEE Trans. Serv. Comput. **4**(2), 140–152 (2011)
9. Cartwright, D., Harary, F.: Structural balance: a generalization of Heider's theory. Psychol. Rev. **63**(5), 277 (1956)
10. Rodgers, J.L., Nicewander, W.A.: Thirteen ways to look at the correlation coefficient. Am. Stat. **42**(1), 59–66 (1988)
11. http://www.grouplens.org/datasets/movielens/
12. Rong, Y., Wen, X., Cheng, H.: A Monte Carlo algorithm for cold start recommendation. In: 23rd International Conference on World Wide Web, pp. 327–336. ACM Press, New York (2014)
13. Wu, C., Qiu, W., Zheng, Z., Wang, X., Yang, X.: QoS prediction of web services based on two-phase k-means clustering. In: 22nd IEEE International Conference on Web Services. IEEE Press, New York (2015)

A Software-Defined Cloud Resource Management Framework

Aaqif Afzaal Abbasi, Hai Jin[(⊠)], and Song Wu

Services Computing Technology and System Lab, Cluster and Grid Computing Lab,
School of Computer Science and Technology,
Huazhong University of Science and Technology, Wuhan 430074, China
{aaqif,hjin,wusong}@hust.edu.cn

Abstract. Network systems employ policies that are inherently dynamic in nature and that depend on temporal conditions defined in terms of external events such as the measurement of bandwidth, use of hosts, intrusion detection or specific time events. *Software-defined networking* (SDN) offers the opportunity to make networks easier to configure by providing richer configuration methods. To reduce network monitoring costs and traffic overheads, herein, we propose a software-defined cloud resource management framework that uses a *Fuzzy Analytical Hierarchy Process* (Fuzzy-AHP) to customize the network resource allocation. The framework can be incorporated into SDN-enabled cloud infrastructures by using an *Application Program Interface* (API). Using real-time data, we demonstrate that our framework can improve network resource management and is capable of handling increasing traffic requests. We also validate our framework efficiency through simulations.

Keywords: Cloud computing · Software-defined networking · Fuzzy Analytical Hierarchy Process (Fuzzy-AHP) · Network management · Resource management · Scheduling

1 Introduction

Cloud computing [1] has emerged as a major computing paradigm built around the concept of high computing performance but with the additional benefits of reduced investment requirements and lower operational costs, whilst facilitating on-demand service provision and options for pay per usage. Despite attaining maturity in many major areas of service provision, cloud computing is still developing. However, even as it develops, gaps are emerging due to the evolution of technologies, and these are addressed by different work groups, alliances, industries and standard bodies.

Cloud network resource management continues to be challenging. These networks typically comprise a large number of switches, routers, firewalls and numerous types of middleboxes, and many types of events and processes occurring simultaneously. Network operators are responsible for configuring network resources to enforce various high-level policies and for responding to a wide range

© Springer International Publishing Switzerland 2015
L. Yao et al. (Eds.): APSCC 2015, LNCS 9464, pp. 61–75, 2015.
DOI: 10.1007/978-3-319-26979-5_5

of network events. As demand for resources increases, networks are experiencing increasing resource management issues including the allocation, provisioning, requirement mapping, adaptation, discovery, brokering, estimation and modeling of resource needs [1]. Solving these problems would provide benefits like increased scalability, quality of service, optimal utilization, reduced overheads, improved throughput, reduced latency, specialized environment, cost effectiveness and simplified interfaces.

However, cloud network resource management implementation remains incredibly difficult because high-level policies need to be specified in terms of distributed low-level configuration. Today's networks provide little or no mechanism for automatically responding to this wide range of issues. Nowadays, network operators must implement increasingly sophisticated policies and complex tasks with a limited and highly constrained toolset basically consisting of low-level device configuration commands in a *command line interface* (CLI) environment [9]. Not only are network policies low-level, they are also not well equipped to react to the continually changing network conditions and today's state-of-the-art network configuration methods can only implement a network policy that deals with a single snapshot of the network state.

In cloud environments, network states change continually and operators must manually adjust the network configuration in response to change network conditions. Due to this limitation, operators use external tools or build ad hoc scripts to dynamically reconfigure network devices when events occur. As a result, configuration changes are frequent, and lead to frequent misconfiguration. Unfortunately, state-of-the-art networks typically involve the integration and interconnection of many proprietary, vertically integrated devices [9]. This vertical integration makes it incredibly difficult for operators to specify high-level network-wide policies using current technologies. Innovation in network management has therefore been limited to stop-gap techniques and measures, such as tools that analyze low-level configurations to detect errors or to otherwise respond to network events. Proprietary software and closed developments in network devices by a handful of vendors make it extremely difficult to introduce and deploy new protocols.

SDN [13] is a paradigm where a central software program, called a controller, dictates the overall network control behaviour. In SDN, network elements become simple packet forwarding devices (the infrastructure layer), while the "brain" or control logic is implemented in the controller (the control layer). This paradigm shift offers great benefits when compared to legacy methods. SDN makes it much easier to introduce new ideas into the network through a software program, as it is easier to change and manipulate than using a fixed set of commands with proprietary network devices. It also introduces the benefits of a centralized approach to network configuration, where operators do not have to configure all the network devices individually to make changes in network behaviour, but instead can make network-wide traffic-forwarding decisions in a logically single location i.e., the controller with global knowledge of the network state. Cloud-based network management systems deliver a great deal of business value to enterprises, but they also add in additional complexity. Today, in order to support SDN-enabled network management operations, vendors offer proprietary solutions of specialized hardware, operating systems and control programs (network applications). Managing network infrastructure is time-intensive, costly and has traditionally required expensive, third-party applications to effectively manage larger networks.

1.1 Our Approach

The specific objective of this paper is to establish the necessary baseline for a tool-supported decision support method aimed at facilitating selection of cloud services in a multi-cloud environment. This paper presents the main results of the recent efforts towards development of a SDN-enabled cloud resource management framework for multi-cloud environments. We employ Fuzzy-AHP [16] to assign priorities to users on the basis of their requests. Fuzzy-AHP can alleviate the lack of network management functionalities by assigning priorities to received traffic demands. Our proposed framework is flexible and easy to implement. It uses attributes associated with resource management functions and evaluates the execution costs of individual applications. Based on the method proposed, we elaborate the suitability of both the method proposed and the state of the art for analyzing risks as well as for ensuring quality and cost in the multi-cloud context.

1.2 Evaluations and Contributions

The present paper proposes, presents and evaluates a novel paradigm in the field of SDN-enabled cloud computing. First, in this paper, we highlight the resource management issues in cloud environments. Next, we present a Fuzzy-AHP based cloud management framework that can administer cloud resource management functionalities. We also propose adaptive algorithms for admission control and network resource scheduling. The main contributions of this paper are four-fold:

– We analyze resource administration challenges in cloud environments.
– We present SDN-enabled cloud resource management framework.
– We present admission control and scheduling algorithms for our framework.
– We implement and evaluate the proposed framework's performance.

1.3 Outline

The rest of the paper is organized as follows. Section 2 briefly outlines the background details and motivation behind the performed research work. Section 3 underlines the research problem and gives a statement of the problem. Next, we extensively elaborate and discuss the proposed framework's design architecture in Sect. 4. Section 5 provides comprehensive details about implementation and also evaluates the test results compared to existing techniques. Section 6 discusses the related work in the field. Finally, Sect. 7 concludes the paper.

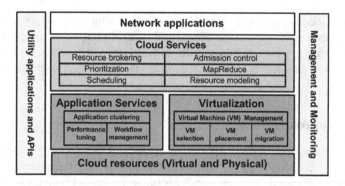

Fig. 1. Functional components of cloud resource management system

2   Background and Motivation

In this section, we first discuss the basics of cloud resource allocation schemes. Then, we take a quick look at the challenges in cloud resource administration.

2.1   Cloud Resource Management

Resource management is a core function of clouds, and affects the three very basic criteria for system evaluation, i.e., performance efficiency, functional programming and operational cost. The use of an inefficient resource management scheme impacts adversely upon system performance and reliability. Conventional cloud systems provide several cloud management solutions, some of which may sometimes be considered too expensive. Cloud infrastructures, being complex, are difficult to manage. They require complex policies and decision structures to achieve multi-objective optimization, and failure to implement these can result in inaccurate global state information. The design objective of cloud resource management is to provide a suite of tools that facilitate computational resource sharing and enhance application-to-infrastructure mapping. Some of the (many) functions performed by a cloud resource manager are shown in Fig. 1.

In general, cloud resources can be divided into 2 broad categories:

- **Physical Resources:** Consisting of tangible devices such as the CPU, memory, storage, workstation, network elements.
- **Logical Resources:** Comprising a pool of system abstractions lying over the physical resources, e.g., APIs, network operating system, storage pool, network capacity, processing capacity.

Resource management systems ensure that cloud services and their workloads agree to a common platform, where the physical and logical resource behaviours match each other and hence improve the resource utilization [5].

2.2   Challenges in Cloud Resource Management

The major resource management issues in cloud systems are resource modelling, resource mapping, resource provisioning and resource scheduling. In order to effectively cope with these issues, we need to manage a number of challenges, such as scheduling, dynamicity and monitoring.

- **Scheduling:** Administering millions of cores in a cloud environment can result in the need to deal with multiple service requests, where resource demand may be higher than available system capacity. This requires a scalable and intelligent method to analyze and schedule requests by applying smart optimization rules.
- **Dynamicity:** In order to adapt to dynamic changes in the usage, load and configuration of data centre machines, a real-time decision-making approach is necessary to achieve optimal performance metrics.
- **Monitoring:** Managing data centres would result in a huge volume of monitoring data, produced across multiple management platforms, which could limit conventional approaches to the process scheduling of miscellaneous applications.

2.3 Motivation

With the advent of SDN in general and *software-defined clouds* (SDCs) specifically, network operators can configure all static network elements to enforce high-level policies, thereby addressing the longstanding wide range of network limitations.

In order to fully realize the potential of SDCs, all infrastructure disciplines must be virtualized, and put under automated control. The opportunity for IT industries is to now fully appreciate the potential of emerging storage models, and to take the practical steps today that are needed to ease the transition into the future. To achieve this goal, we must overcome the challenges of working in virtual environments. Moreover, merely extending SDN features to only one aspect of the cloud (i.e., datacentres) may not ease all traffic management issues. We therefore believe that applying SDN control concepts to SDCs in resource management can vigorously improve the current network resource management capabilities and overcome the present challenges.

3 Problem Statement

Software-defined cloud infrastructures enable users to take advantage of IT services in the most efficient way by optimizing resource utilization for cost reduction. From a purely theoretical perspective, cloud decisions are based on a comparison of internal and external IT services. More precisely, the transaction overheads caused by these services can be reduced by improving decision structuring. Fuzzy-AHP can be employed to optimize cloud service in a distributed environment by repetitively searching for the best combination of available resource from a resource pool.

However, in order to make the best use of SDCs, a comprehensive portfolio of management tools is required that can dynamically manage workloads.

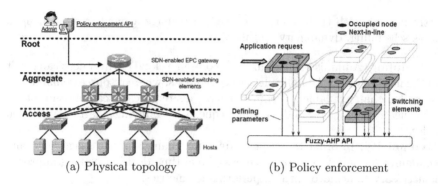

(a) Physical topology   (b) Policy enforcement

Fig. 2. Proposed system model

By transforming IT infrastructure into a workload-intensive environment, we aim to develop the current rigid architecture of the cloud into an SDN-enabled environment, where cloud resource management functionalities can be broken into solvable problems and can be addressed independently.

We want to enhance the resource allocation capabilities of individual services based on user experiences. We also intend to reduce the admission control and scheduling constraints faced by these services.

4 System Design and Architecture

In this section, we describe the functioning of our framework. We begin by explaining our adopted methodology, followed by outlining its design considerations and architecture details.

4.1 Methodology

In our proposed methodology, we enable system users to define weights and criteria according to their own needs. This is required to cope with the varying network requirements. Restricting users purely according to predefined criteria will violate the basic concept of programmability features in SDN. We present the physical topology and policy enforcement scheme of our proposed resource management framework in Fig. 2. In our framework, policies are enforced on the host through a mesh of SDN-enabled switching elements. An open modular network operating system with the ability to respond in real-time to both internal and external control operations is required to support SDN features. Below, we explain the key points of our proposed framework.

– **Managing Bottlenecks:** In SDCs, the controller may become the bottleneck due to increasing network size. Therefore, the question arises as how to allocate memory across a set of switches to support a given accuracy. As the resources connected can now reside anywhere: within or between datacentres, our framework uses admission control policy to underline available resource capacity, whereas Fuzzy-AHP based decision elements define individual application criteria. This helps in reducing bottlenecks to a greater extent.

- **Defining Threshold Capacity:** Although cloud computing systems provide a better way to carry out the submitted tasks in terms of responsiveness, scalability and flexibility, most job and task scheduling problems on cloud computing systems require customized risk management thresholds based on a predictive algorithm. Our proposed scheduling algorithm (Algorithm 2) address these concerns by allocating network service requests to those switching elements with the highest available packet forwarding capacity. This improves the scheduling of incoming requests and reduces network load management constraints at the hardware level.
- **Overview:** To sum up our framework's methodology, first, we begin by assigning weights and criteria to the network applications. We then use our admission control technique (Algorithm 1) to ensure that the admittance of new service requests in the network will not degrade the performance of existing service requests. Our scheduling algorithm (Algorithm 2) then selects the switching device with the maximum capacity to entertain the received service requests. Finally, if there are no inter-domain routing conflicts, the service request is sent to its respective host for processing.

4.2 Design Considerations in Cloud Resource Administration

In this section, we identify the principles that facilitate computational resource management and enhance the application-to-infrastructure mapping of service requests.

- **Resource Instrumentation:** An efficient resource management system requires the detailed and timely tracking of the resources used by each tenant and request. This creates an ideal environment to administer system resource constraints that can efficiently handle imminent resource demands.
- **Controlling Overheads:** *Service Level Agreement* (SLA)-based policy implementations are of fundamental concern in cloud resource management systems. They are fine tuned to ensure that their execution will not yield in overheads, which could in turn affect the overall system performance.
- **Resource Contention:** As cloud systems comprise millions of nodes hosting a large number of processes across distributed datacentres, the tenants contribute variable load to the system. This results in resource contention among processes where a tenant might use more resources than its required share fair. Cloud resource management systems are designed to reduce these constraints by allocating a fixed pool of resources to individual service requests.

4.3 Architecture

Our proposed framework uses a Fuzzy-AHP based API over an SDN-enabled switch. In SDCs, open APIs refer to the software interfaces that lie between the software modules of the controller platform and the SDN applications running

Table 1. Priority criteria in AHP

Scale	Definition	Description
1	*Equally important*	*Two elements contribute equally*
3	*Moderately important*	*Slightly favours one over another*
5	*Strongly important*	*Strongly favours one over another*
7	*Very important*	*Very strongly favours one over another*
9	*Extremely important*	*Extremely favours one over another*
2,4,6,8,10	*Intermediate values*	*Intermediate values*

atop the network platform. They, together with SDN controller, form an important part of the open SDN ecosystem of customers and partners. The SDN controller exposes API (northbound), which allows the deployment of a wide range of off-the-shelf and custom-built network applications many of which are fundamentally not feasible prior to the advent of SDN. We assign criteria, weights and parameters to different applications based on our own preference. In this paper, we translate Saaty's [14] developed 1 to 9 scale (Table 1) to describe the preferences between alternatives as being either equally, moderately, strongly, very strongly or extremely preferred. Our proposed framework consists of four major functional elements: decision element, workload evaluator, admission controller and scheduler. We discuss them below in detail.

Decision Element. In decision element, we perform decision-making through a fuzzy comprehensive evaluation. This is an application of fuzzy mathematics that uses principles of fuzzy transformation and maximum membership degree. In order to make a comprehensive decision, it evaluates all the functions that may influence resource management concerns. The process takes place in five steps.

(1) Create an evaluation index (U) to determine the factor and sub-factor weights that have to be calculated for the cloud services.

(2) Create a set of comments (V) to describe the evaluation of cloud services by using phrases like "Acceptable", "Constrained", etc.

(3) Create an evaluation matrix (R) from U to V where each factor $u_i (i \leq n)$ can be written as fuzzy vector $R_i \in \mu(V)$. Mathematically, this fuzzy relationship can be expressed as

$$\mathbf{R} = (r_{ij})_{nm} = \begin{pmatrix} r_{11} & r_{12} & \cdots & r_{1m} \\ r_{21} & r_{22} & \cdots & r_{2m} \\ \vdots & \vdots & \ddots & \vdots \\ r_{n1} & r_{n2} & \cdots & r_{nm} \end{pmatrix} \tag{1}$$

The evaluated result of Eq. 1 should match the normalized conditions, because the sum of the weight of the vector is 1 (i.e., for every $i, r_{i1} + r_{i2} + r_{i3} + ... + r_{im} = 1$).

(4) Determine factor weight (FW). FW denotes the proportion of each factor in the evaluation index (U) and is based on its relative importance.

(5) Obtain evaluation results (E) through the product of the factor weight (FW) and the evaluation matrix (R). It can be denoted as $E = FW(R) = (E_1, E_2, E_3, ..., E_m)$. Finally, the evaluated weight can be assigned to their respective application. The decision element provides weights of individual applications to workload evaluator which helps the software-defined cloud infrastructure in decision structuring.

Workload Evaluation. Performing a thorough workload analysis can significantly improve cloud performance; whereas selecting policies that are incompatible with the workload can lead to wasted resource time and excessive charges.

To evaluate an individual cloud node workload, we express our cloud model as an arrangement of 2 interdependent row vectors. We consider a cloud scenario where the number of tasks performed by the cloud is represented as a row vector $\mathbf{a} = [a_1, a_2, a_3, ..., a_n]$ of a resource class c. We then represent the resource occupation of individual tasks by another row vector $\mathbf{v} = [v_1, v_2, v_3, ..., v_n]$. Each member of the row vector $\mathbf{v}$ represents resource usage for its respective element in compute nodes row vector $\mathbf{a}$. Here, we consider a case, where the number of service requests accepted by the cloud environment does not exceed its aggregate compute nodes. If N represents the number of aggregate compute nodes, the condition can be expressed mathematically as

$$\mathbf{a} \cdot \mathbf{v} = \sum_{i=1}^{n} a_c v_c \leq N \tag{2}$$

Due to the varying allocation models and schemes, workload evaluation for cloud services is increasingly complex. However, applying probability to schemes with varying behaviours can address these concerns. This will help in analyzing the relationship between the application deployment topology, its workload oscillation over time, and the expected performance. We therefore employ the recursive methodology approach presented in [4,8] to calculate the probability p of an individual cloud node capacity $\mathbf{q}$ as

$$p(q) = \sum_{a:av=n} \frac{\alpha_1{}^{a_1}}{\alpha_1!}, \cdots, \frac{\alpha_1{}^{a_n}}{\alpha_n!} \quad where \quad n = 0, 1, 2, \cdots, N \tag{3}$$

The measured node occupancy probability can be input to Fuzzy-AHP for fine tuning its decision infrastructure using Eq. 3. Our reason for adopting a recursive approach is because the performance of multirate systems can be expressed in terms of flow throughput. Apart from topology, a workload evaluation also depends on traffic characteristics (like the traffic intensity) as well as the way link capacity shared between ongoing flows in case of congestion. Now, we can evaluate the submitted workload occupancy probability in our cloud system. In the next section, we explain our admission control algorithm, which helps in upholding the workload occupancy of the entire cloud system under the cloud's permissible resource threshold limits.

Algorithm 1. Admission Control

1: Let *ava* denote the available resources
2: Let *req* denote the requested resources
3: Let *queue* denote the request queue
4: /* Traverse the queue to process request */
5: **while** !queue.isEmpty() **do**
6:     /* Evaluate resources required by first request in the queue */
7:     req=queue.firstRequestResource();
8:     **if** *req* <= *ava* **then**
9:         ava=ava - req;
10:         queue.pop();
11:     **else**
12:         /* If the request is not satisfied, prompt user */
13:         **sendMessage**("resources deficient");
14:         queue.pop();
15:     **end if**
16: **end while**

Admission Control. Cloud admission control is a validation process, whereby a check is performed before a connection is established to see if the current resources are sufficient for the resource allocation request.

Our admission control strategy can cope with horizontal elasticity issues to handle network overloading issues concerning SDCs. This reduces the risk of resource congestion, which ultimately leads to SLA violations. Admission control schemes in SDCs are of paramount importance. Usually admission control methods consider multiple criteria for effective service deliverance. In our case, the objective of admission control is to determine the number of application instances that can fit the system resource capacity (i.e., the number of active applications) to maximize performance. Our admission control algorithm (Algorithm 1) ensures that sufficient resources are available to entertain a request. If resources are scarce, a message about unavailability of resources is prompted. The conditional approach used in our algorithm makes it simple yet effective to implement. In our framework, we consider resources as a set of the cloud's physical resources, computational resources, memory and bandwidth.

Scheduling. In conventional networks, a centralized scheduler knows all the aggregate traffic demands. This assists the network administrators in assigning scheduling priorities. On arrival, traffic flows have to be routed to their destined routes. The ingress, egress and middlebox traversal rules need to follow a scheduler, which can help in monitoring the overall traffic flow concerns. Due to the emergence of SDN [9], the dynamic management of routing paths for flows has become possible in practice. In OpenFlow networks [12] (the first standard communications interface defined between the control and forwarding layers of an SDN architecture), the controller schedules the routing paths and then manipulates the flow tables in the switches via the OpenFlow protocol.

Algorithm 2. Scheduling

1: /*Traverse the switch list to satisfy application requests*/
2: Let $switch[n]=$ The list of SDN enabled switches
3: Let $UsedCapacity[i]=$ The used capacity of switch i
4: Let $MaxCapSwitch[i]=$ The maximum capacity of switch i
5: Let $AvaCapacity[i]=$ The available capacity of switch i
6: Let $ApplicationReq=$ The request capacity of an application
7: /* sort switch according to $AvaCapacity$ in ascending order*/
8: **Sort**($Switch[n]$);
9: /* traverse switch list, choose switch with sufficient $Avacapacity$ */
10: **for** $i \leftarrow 0, n$ **do**
11:     **if** $UsedCapacity[i] + ApplicationReq < MaxCapSwitch[i]$ **then**
12:         /* choose switch i to allocate resources */
13:         **AllocateCap**($Switch[i], ApplicationReq$);
14:         **break**;
15:     **end if**
16: **end for**
17: /* if allocation failed, insert the request to the switch buffer */
18: **if** $i == n$ **then**
19:     /* $Switch[max]$ is the switch with largest $AvaCapacity$ */
20:     **BufferReq**($Switch[max], ApplicationReq$);
21: **end if**

We extend the heuristic approach provided in [12] to develop a case where a traffic flow is aware of its required resources and only executes after being guaranteed that sufficient resources are available for its execution. By taking into consideration the QoS requirements, each scheduling request is assigned to the device with sufficient switching capacity. Therefore, the prioritized flow-scheduling enables us to allocate the best switching devices to the scheduled service requests. Our proposed scheduling algorithm (Algorithm 2) traverses all the available switches and sorts them according to their available capacity. Since the incoming request's resource requirements are already known (at the admission controller stage, see Algorithm 1), the scheduling algorithm selects a switch with sufficient capacity to handle the process flow. This improves scheduling and also largely eliminates network load management constraints at the hardware level.

4.4 Summary of Functions

In a nutshell, the proposed framework's functioning initiates apriority assignment to individual applications and services. In the next step, the workload evaluator examines an individual service's influence on the system resources. This is done by measuring the current node occupancy of the system. On the assurance that sufficient resources are available (by the admission controller), the service request is forwarded for scheduling, where it is assigned to the switching element with the maximum available capacity to handle the process flow.

5   Evaluation and Results

5.1   Implementation

We conduct the performance evaluation of the proposed framework initially by simulating the physical resources of a data centre and then by emulating their execution on a fixed number of application and services. We generate a mix of services and applications that, to the best of our knowledge, are representative of those within a cloud environment. We emulate their behaviour by carrying out discrete event simulations on the resources that execute them.

The cloud infrastructure used in our testing scenario consists of eight hosts connected over seven SDN-enabled switches. We deploy our framework using an API that uses the SDN-enabled EPC Gateway to communicate with the SDN-enabled switching mesh. Figure 2(a) illustrates our proposed framework's network topology in detail. The simulations on the network devices and hosts are conducted by using the OpenDayLight controller and Mininet v.2.2.0 on an AMD Opteron 6300 Series Platform with 16-core x86 processor.

5.2   Discussion and Results

We compare the framework's performance to existing static and aggregated resource reservation methods. In the static resource reservation method, resources can be assembled on the basis of simple heuristics or historical demand patterns, whereas aggregated resource reservation methods are commonly used to address scalability concerns in resource allocation. However, the biggest challenge with aggregated resource reservation is that it may underutilize resources in bulk, while other similar requests may be turned down due to lack of resources.

In our experimental settings, there are seven switching elements, which are connected by means of an SDN-enabled EPC Gateway. After allocating Fuzzy-AHP-based priorities and criteria to the applications and services, we calculate the communication delay between the switch and controller. In the next step, we compare the bandwidth consumed by these services for similar data loads. We extended the scope of experimentation by running multiple instances of the service requests to measure the inter-switch delay. We then compare the performance of our proposed solution with the static and aggregated methods. The acquired results demonstrate (Fig. 3) that our framework's performance is better than the two conventional approaches in terms of delay and bandwidth usage.

We believe that the improvement in our results is mainly possible due to the Fuzzy-AHP. By enabling users to assign criteria and priorities to applications, the Fuzzy-AHP assists in the decision structuring process. It improves planning and administration over cloud network resources. Similarly, our admission control and scheduling algorithms also ensure the provision of sufficient resources to individual service requests. This makes our proposed framework a simple, feasible, yet practical cloud resource management solution.

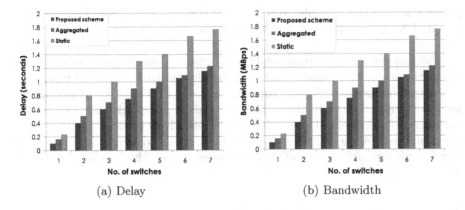

(a) Delay   (b) Bandwidth

Fig. 3. Framework performance evaluation

6   Related Work

Our research is related to cloud resource management in SDCs. In the following section, we highlight some important work performed in this area.

6.1   SDN Support for Datacentre Management

Software-defined network solutions focus on the need for the use of programmability in the network. This helps networks to reap benefits like cost reduction and ease of management [7]. The *Open Network Foundation* (ONF) workgroup on Configuration and Management is working to enhance the operation, administration and management capabilities of the OpenFlow protocol. Proposals such as from Procera [9], Frentic [6] and Meridian [2] encourage the use SDN to enhance network management capabilities.

6.2   Developments in Software-Defined Clouds

Software-defined networking is a key factor driving the SDC research community. The concept for outsourcing enterprise middlebox processing to the cloud has been proposed in [15]. The proposed system is called *Appliance for Outsourcing Middleboxes* (APLOMB). It delegates middlebox tasks to the cloud to ensure the ease of management and *capital expenditure* (CAPEX) functions.

In multi-tenant datacentres, tenants need a flexible solution for migration to ensure that during the migration of unmodified workloads from enterprise networks to the service provider network, the network configurations do not change. These concerns are addressed by the *Network Virtualization Platform* (NVP) in [10]. NVP uses SDN to decouple network management from the physical infrastructure in two profound ways: first, by decoupling the tenant control planes from the physical infrastructure, and then by decoupling the implementation of NVP from the physical infrastructure by using software switching at the edge.

The easy creation and manipulation of software-defined cloud environments enable the model-based deployment of upcoming standards like the *Topology and Orchestration Specification for Cloud Applications* (TOSCA) [3]. It also allows model description to be built for different kinds of workloads such as topology patterns and management plans. In order to effectively monitor service-centric cloud environments, a unified solution has been presented [11], which can rapidly migrate and optimize the use of cloud resources in ways that are consistent with user service profiles.

7   Conclusions and Future Work

We propose a software-defined cloud resource management framework to address the admission control and scheduling concerns of modern cloud datacentres. The framework can be implemented using an API. It employs Fuzzy-AHP as its decision structuring technique. The proposed framework's admission control and scheduling features can reduce performance degradation and improve application load processing. We evaluate the performance efficiency of our framework by using real-life workloads. The results demonstrate that our proposed framework improves resource utilization and allocation capacities of an SDC infrastructure, with an acceptable level of performance degradation.

In future, we plan to study the effects of topology change for the given parameters. We also plan to develop an improved overbooking model for our framework.

Acknowledgement. We thank Haibao Chen and Youchuang Jia for their feedback on data loads. This research is supported by EU FP7 MONICA project under grant no. 295222.

References

1. Armbrust, M., Fox, A., Griffith, R., Joseph, A.D., Katz, R., Konwinski, A., Lee, G., Patterson, D., Rabkin, A., Stoica, I., Zaharia, M.: A view of cloud computing. Commun. ACM **53**(4), 50–58 (2010)
2. Banikazemi, M., Olshefski, D., Shaikh, A., Tracey, J., Wang, G.: Meridian: an SDN platform for cloud network services. IEEE Commun. Mag. **51**(2), 120–127 (2013)
3. Binz, T., Breiter, G., Leyman, F., Spatzier, T.: Portable cloud services using TOSCA. IEEE Internet Comput. **3**, 80–85 (2012)
4. Bonald, T., Virtamo, J.: A recursive formula for multirate systems with elastic traffic. IEEE Commun. Lett. **9**(8), 753–755 (2005)
5. Buyya, R., Yeo, C.S., Venugopal, S., Broberg, J., Brandic, I.: Cloud computing and emerging IT platforms: vision, hype, and reality for delivering computing as the 5th utility. Future Gener. Comput. Syst. **25**(6), 599–616 (2009)

6. Foster, N., Guha, A., Reitblatt, M., Story, A., Freedman, M.J., Katta, N.P., Monsanto, C., Reich, J., Rexford, J., Schlesinger, C., Story, A., Walker, D.: Languages for software-defined networks. IEEE Commun. Mag. **51**(2), 128–134 (2013)
7. Katta, N.P., Rexford, J., Walker, D.: Incremental consistent updates. In: Proceedings of the Second ACM SIGCOMM Workshop on Hot Topics in Software Defined Networking, pp. 49–54. ACM (2013)
8. Kaufman, J.S.: Blocking in a shared resource environment. IEEE Trans. Commun. **29**(10), 1474–1481 (1981)
9. Kim, H., Feamster, N.: Improving network management with software defined networking. IEEE Commun. Mag. **51**(2), 114–119 (2013)
10. Koponen, T., Amidon, K., Balland, P., Casado, M., Chanda, A., Fulton, B., Ganichev, I., Gross, J., Gude, N., Ingram, P., Jackson, E., Lambeth, A., Lenglet, R., Li, S.H., Padmanabhan, A., Pettit, J., Pfaff, B., Ramanathan, R., Shenker, S., Shieh, A., Stribling, J., Thakkar, P., Wendlandt, D., Yip, A., Zhang, R.: Network virtualization in multi-tenant datacenters. In: Proceedings of the 11th USENIX Symposium on Networked Systems Design and Implementation, pp. 203–216 (2014)
11. Mahindru, R., Sarkar, R., Viswanathan, M.: Software defined unified monitoring and management of clouds. IBM J. Res. Develop. **58**(2/3), 12:1–12:11 (2014)
12. McKeown, N., Anderson, T., Balakrishnan, H., Parulkar, G., Peterson, L., Rexford, J., Shenker, S., Turner, J.: OpenFlow: enabling innovation in campus networks. ACM SIGCOMM Comput. Commun. Rev. **38**(2), 69–74 (2008)
13. Rubio-Loyola, J., Galis, A., Astorga, A., Serrat, J., Lefevre, L., Fischer, A., Paler, A., Meer, H.: Scalable service deployment on software-defined networks. IEEE Commun. Mag. **49**(12), 84–93 (2011)
14. Saaty, T.L.: What is the analytic hierarchy process? In: Mitra, G., Greenberg, H.J., Lootsma, F.A., Rijkaert, M.J., Zimmermann, H.J. (eds.) Mathematical Models for Decision Support. NATO ASI Series, vol. 48, pp. 109–121. Springer, Heidelberg (1988)
15. Sherry, J., Hasan, S., Scott, C., Krishnamurthy, A., Ratnasamy, S., Sekar, V.: Making middleboxes someone else's problem: network processing as a cloud service. ACM SIGCOMM Comput. Commun. Rev. **42**(4), 13–24 (2012)
16. Zadeh, L.A.: Fuzzy sets. Inf. Control **8**(3), 338–353 (1965)

Automated Clarification of Constraints in Web Services for Accurate Service Reuse

Xiaocao Hu, Zhiyong Feng, Shizhan Chen, and Keman Huang[✉]

Tianjin Key Laboratory of Cognitive Computing and Application,
School of Computer Science and Technology, Tianjin University, Tianjin, China
{huxiaocao,zyfeng,shizhan,keman.huang}@tju.edu.cn

Abstract. Service reuse must follow certain constraints in order to correctly interact with Web Services. Violations of constraints can cause fatal errors or incorrect results in the service reuse. However, constraints are often not formally specified and are thus not available in the service reuse. To address this issue, this paper focuses on two common types of such constraints, including location constraints on Web Services and object constraints on simple parameters. An approach is proposed to clarify the two constraints automatically, via a hybrid analysis of heterogeneous information, including the social tags and the service documentations. Then an improved method is presented to identify collaborative relations among Web Services, integrating constraints compatibility into semantic matching. One experiment is carried out on 509 Web Services crawled from the Internet to evaluate the effectiveness of our approach. The other experiment is conducted on the same dataset to assess impacts of the two constraints on service relations. Experimental results show that our approach can clarify the two types of constraints effectively and achieve adequate recall and precision. Moreover, it is indicated that the two types of constraints, especially object constraints, have significant impacts on improving the quality of identified service relations, thus provide a strong guarantee for accurate service reuse.

Keywords: Web Services · Location constraints · Object constraints · Opencyc · Constraints compatibility · Service relations

1 Introduction

With the advent of Web Services [1], the software industry is evolving from developing specific functionality from scratch to reusing functionalities off the shelf, and more and more new useful solutions are generated by composing existing Web Services into more complex services. The service reuse should follow certain constraints in order to correctly interact with Web Services. For example, the CDYNE Weather API[1] allows users to retrieve weather information only for US cities. When requesting the operation "*SendEmail*" provided by the Tiscali Email Services API[2], a *from* parameter and a *to*

---

[1] CDYNE Weather API: available at http://wsf.cdyne.com/WeatherWS/Weather.asmx.
[2] Tiscali Email Services API: available at http://webservices.tiscali.com/EmailServices.asmx.

© Springer International Publishing Switzerland 2015
L. Yao et al. (Eds.): APSCC 2015, LNCS 9464, pp. 76–91, 2015.
DOI: 10.1007/978-3-319-26979-5_6

parameter are required, and valid objects of the *from* parameter and the *to* parameter can only be email addresses. Violations of such constraints can cause fatal errors or incorrect results in the service reuse. For example, it is in vain to acquire weather information for China cities via the CDYNE Weather API.

In fact, according to an empirical investigation conducted on Programmable.com, a quarter of Web Services have the same type of constraints as the CDYNE Weather API does, and three quarters of Web Services have the same type of constraints as the Tiscali Email Services API does. These constraints are mainly expressed or implied in the service descriptions. To correctly reuse functionalities off the shelf, developers must first read through the service descriptions, and memorize the constraints. However, as the number of services available over the Web increases dramatically, it is unrealistic for developers to browse so many Web Services.

To make computers understand Web Services and achieve automatic service reuse, Semantic Web Services [2] are advanced and a variety of conceptual models have been proposed, such as SAWSDL, OWL-S and WSMO. And many researches are carried out to transform Web Services into Semantic Web Services based on domain ontologies. However, these researches are not comprehensive enough to formally specify the constraints. For example, the *zip* parameter in the CDYNE Weather API is usually associated with concept ZipCode instead of USZipCode (if concept USZipCode exists in the domain ontology), and the *from* and *to* parameters in the Tiscali Email Services API are likely associated with other concepts instead of concept Email.

To address this issue, this paper proposes an approach to automatically clarify constraints in Web Services. In particular, we focus on two types of constraints that commonly exist in Web Services, including location constraints on Web Services and object constraints on simple parameters. Given a Web Service, the location constraint describes the service coverage area. Given a simple parameter, the object constraint describes the context for accurate interpretation of the parameter. Then we present an improved method to identify the collaborative relations among Web Services. To evaluate our approach, it is applied to 509 Web Services crawled from the Internet. In addition, to assess impacts of the two types of constraints on service reuse, a controlled experiment is conducted to identify collaborative relations among these Web Services. In summary, this paper makes the following main contributions:

- An empirical investigation of Web Services to show the non-trivial presence of location constraints and object constraints.
- An approach to clarify the two types of constraints, via a hybrid analysis of heterogeneous information, including social tags and service documentations.
- An improved method to identify the collaborative relations among Web Services, integrating constraints compatibility into semantic matching.

The rest of this paper is organized as follows. Section 2 presents the empirical investigation of Web Services. Section 3 describes the approach for constraints clarification. Section 4 conducts experiments. Section 5 discusses the related work and Sect. 6 concludes.

2 Empirical Investigation

2.1 Dataset

ProgrammableWeb.com[3], an online social platform, provides services that allow users to publish different kinds of OpenAPIs. The site lists and profiles OpenAPIs, and also categorizes OpenAPIs through a provided taxonomy and tags.

We crawl 509 Web Services, included in 49 categories, from the Programmable Web.com. The selection of Web Services is guided by three principles:

1. A Web Service should be available.
2. The profile of a Web Service should provide the link to its WSDL documentation.
3. The WSDL documentation of a Web Service should be accessible.

As ProgrammableWeb.com retains profiles for Web Services that are no longer available, the first principle guarantees that those no longer available Web Services are excluded in the empirical investigation. The second principle is designed for reducing the workload of locating WSDL documentations. And the third principle ensures that we can successfully access and obtain WSDL documentations.

The profiles of the 509 Web Services are crawled from the ProgrammableWeb.com. In addition, the corresponding WSDL documentations are obtained. With a WSDL parser [3], 7996 operations are acquired along with 33959 complex parameters and 107510 simple parameters.

2.2 Location Constraint

A profile advertises a Web Service, including functional description and so on. Therefore, location constraints are contained in the profiles. Then we manually investigate profiles of these 509 Web Services provided by the ProgrammableWeb.com. From the investigation, we observe that 23.58 % of Web Services have location constraints.

We next discuss the reason for location constraints. A large number of Web Services are designed for a certain place. Typically, if a Web Service is designed and applied for a certain place, the Web Service can be considered as a service with location constraint. Meanwhile, parameters in Web Services are usually not able to reflect location constraints. For example, the CDYNE Weather API utilizes the *zip* parameter instead of the *USZip* parameter. Therefore, additional location constraints are required to restrict the coverage area of Web Services.

2.3 Object Constraint

Unlike location constraints, object constraints are not reflected in the profiles of Web Services. Since simple parameters are responsible for carrying values when requesting Web Services, we perform statistics on simple parameters to find out how often a certain parameter appears in these Web Services. For simplicity, all parameters refer to simple parameters unless otherwise specified.

---

[3] ProgrammableWeb.com: available at http://www.programmableweb.com/.

Through the statistics, we find that a total of 12815 distinct names are utilized for the 107510 parameters. We also observe that it is common to utilize the same parameter names in different Web Services. For the top 2000 most commonly used parameter names, each name appears in one or more Web Services, while for the remaining parameter names, each name merely appears in one Web Service. Figure 1 shows the statistical results of the top 50 most commonly used parameters. The horizontal axis represents the top 50 most commonly used parameter names, and the vertical axis displays the number of Web Services that utilize a certain parameter name. In these 50 parameters, the *name* parameter appears in 152 Web Services, and records the highest frequency with the percentage of 29.86 %. The *todate* parameter appears in 25 Web Services, accounting for 4.91 %.

In addition, we observe that developers tend to utilize simple words as parameter names, and most of their meanings have no sufficient discriminative power, such as *name, type, string*. As a consequence, the interpretation of such parameters should be combined with additional object constraints. The observations also explain the reason for the existence of object constraints.

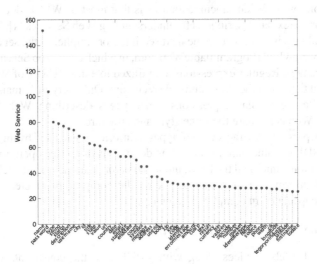

Fig. 1. Statistical results of the top 50 most commonly used parameter names.

3   Constraints Clarification Approach

As Fig. 2 illustrates, our approach starts with a preparatory analysis, which collects social tags and WSDL documentations for Web Services, extracts necessary information from WSDL documentations. The approach then proceeds in two modules. In the location clarification module, social tags are analyzed and location constraints are generated. In the object clarification module, object constraints are produced.

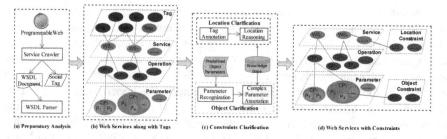

Fig. 2. Clarifying location constraints and object constraints in Web Services.

3.1 Preparing Web Services Along with Tags

The preparatory analysis stage generates Web Services along with social tags, including two parts, service crawler and WSDL parser. The service crawler fetches WSDL documentations and social tags for Web Services from the Internet. The reason that we obtain social tags along with WSDL documentations is that most of WSDL documentations contain little or no textual descriptions for understanding Web Services [4]. The specific usage for social tags is explained in the next section. For simplicity, the service crawler is simplified by crawling ProgrammableWeb.com, in which each Web Service is labeled by one or more tags. Regular expressions are utilized to extract URLs of WSDL documentations and tags from the ProgrammableWeb.com. Only very little manual effort is needed to write the regular expressions, since pages describing Web Services in ProgrammableWeb.com share the same style and structure.

The WSDL parser [3] extracts various types of information needed by the subsequent steps from WSDL documentations, including definitions of service operations, e.g., the available operation names and the list of input and output parameters for each operation, and structural information of parameters, e.g., the complex parameters and simple parameters encapsulated in complex parameters.

3.2 Clarifying Web Services with Constraints

On the basis of Web Services along with social tags, the constraints clarification produces Web Services with constraints. In particular, the clarification concentrates on the acquisition of location constraints and object constraints, as illustrated in Fig. 2(c). With the semantic support of general knowledge base, such as DBpedia [5], OpenCyc [6], YAGO [7], intelligible semantics is introduced into the location clarification and object clarification. Reasons that we utilize general knowledge base as the semantic support are that building knowledge base by ourselves is omitted, and general knowledge base contains much more facts than domain ontology.

Location Clarification Generally, collaborative tagging provides a convenient way for users to annotate Web Services with any keyword. Social tags not only are useful when browsing a large number of Web Services, but also help users gain a quick understanding of each Web Service. For example, SSW Australia Postcode API provides the postcode check service in Australia. Four tags are labeled on the SSW Australia Postcode API, including address, tool, Australia, and mail. From users' perspectives, tags address and mail describe the functionality of the Web Service, tag tool states the particular class of the Web Service, and tag Australia tells the coverage area of the Web Service. In addition, Pan et al. [8] finds six cohesive semantic communities by clustering social tags, in which location is one important topic.

Inspired by this, we utilize tags to acquire location constraints on Web Services. The basic idea of location clarification is to separate tags that describe locations from other tags. To achieve this, intelligible semantics is brought into tags with the support of general knowledge base. We next present the approach for location clarification.

Firstly, tags are annotated with concepts from general knowledge base. Algorithm 1 presents the annotation process for a given word. The tag is firstly reduced to its root form (Line 1). And then concept candidates are obtained from the general knowledge base by acquiring concepts whose denotation matches the root form (Line 3). Finally, the tag is compared against all concepts in the candidate list (Line 5) and the concept with the biggest match score is regarded as the best mapping (Lines 6–8).

```
Algorithm 1: Annotation
Input: Word w
Output: Concept c
1 r = stemmer(w)
2 max = 0
3 CC = getDenotsOfString(r)
4 for i = 1 to CC.size
5   s = getMatchScore(r, cc_i)
6   if(s > max)
7     max = s
8     c = cc_i
```

Secondly, logical reasoning is applied for identifying tags describing locations. Algorithm 2 gives the detailed process. According to our manual investigation on tags, tags that describe locations can be classified into three types, namely a country or a continent or a populated place. Table 1 lists the three types of locations. For each tag,

Table 1. Three types of locations

Type	Description
Country	A nation-state having its own territory, population, and government
Continent	One of the seven land masses is traditionally considered to be a continent
Populated place	A place or area with clustered or scattered buildings and a permanent human population, including cities, settlements, towns, and villages

three sentences are constructed and queried to see if true (Line 3–8). If any of the three sentences returns a true value, which means that the tag describes a location, the tag is regarded as a location constraint for the corresponding Web Service (Line 9–10).

```
Algorithm 2: Location identification
Input: Tag-Concept TC
Output: Location Constraint on a Web Service LC
1 LC = {}
2 for i = 1 to TC.size
    // the predicate isa denotes "is an instance of"
3    cs₁ = makeSentence("(isa"+tcᵢ+"Country)")
4    isTrue₁ = isQueryTrue(cs₁)
5    cs₂ = makeSentence("(isa"+tcᵢ+"Continent)")
6    isTrue₂ = isQueryTrue(cs₂)
7    cs₃ = makeSentence("(isa"+tcᵢ+"PopulatedPlace)")
8    isTrue₃ = isQueryTrue(cs₃)
9    if(isTrue₁ ∨ isTrue₂ ∨ isTrue₃)
10       LC = LC ∪ {tcᵢ}
```

To illustrate how the location clarification works, the location constraint on SSW Australia Postcode API is acquired step by step with the semantic support of OpenCyc. The four tags are associated with #$Address-ContactInfoString[4], #$Tool, #$Australia, and #$MailingLocation respectively. For tags address, tool and mail, the three sentences all return false. For tag Australia, the sentence (#$isa #$Australia #$Country) returns true. Therefore, tag Australia is separated from the other three tags and #$Australia is acquired as the location constraint on SSW Australia Postcode API.

Object Clarification   Intuitively, the name of a parameter describes what the parameter means. Nevertheless, as shown in the empirical investigation, a large number of parameters are named with simple words whose meaning does not have sufficient discriminative power, thus demanding additional object constraints to restrict the interpretation of such parameters.

Typically, an operation should provide textual descriptions to explain each parameter. However, a recent study [4] showed that most of WSDL documentations contain little or no textual descriptions, let alone provide descriptions for parameters. Fortunately, almost all parameters are encapsulated in complex parameters, which provide the context for the interpretation. The primary difference between simple parameters and complex parameters is that simple parameters are responsible for carrying values, while complex parameters are designed to encapsulate a set of simple parameters in an object. Therefore, complex parameters are utilized to obtain the object constraints on simple parameters. We next present the approach for object clarification.

Firstly, simple parameters that demand object constraints are recognized in a Web Service. To achieve this, a parameter collection is predefined to enumerate commonly

---

[4] For each term prefixed by the string '#$', it denotes a specific CycL constant of OpenCyc. For example, #$Country denotes the collection of all countries.

used simple parameters whose meaning does not have sufficient discriminative power. As mentioned above, distinct names for parameters outside the top 2000 all appear in only one Web Service. Therefore, we carefully analyze the top 2000 distinct parameter names and select 206 distinct names from them to predefine the parameter collection. For each parameter in the Web Service, it is matched against parameters in the predefined object parameter collection.

Secondly, corresponding complex parameters are annotated with concepts from general knowledge base. Due to the rule defined in the WSDL standard, compound words are usually utilized for complex parameters. Therefore, complex parameters are preprocessed firstly, including separating the complex parameter into a set of meaning words, assigning parts of speech to each word, extracting nouns, and removing stop words. And then intelligible semantics is brought into complex parameters with the semantic support of general knowledge base by applying Algorithm 1. The concept that describes a complex parameter is obtained as the object constraint for the corresponding simple parameters.

Algorithm 3 illustrates the detailed process of object identification. Line 2 checks whether a simple parameter is contained in the predefined parameter collection. Line 3 acquires the corresponding complex parameter. Lines 4–7 process the complex parameter, and line 8 obtains the object constraint by annotating the complex parameter.

```
Algorithm 3: Object identification
Input: Parameter p, Predefined parameter collection PPC
Output: Object Constraint on parameter p OC
1 OC = Φ
2 if(PPC.contains(p))
3   cp = p.father()
4   ws = segment(cp)
5   pos = POSTagger(ws)
6   ns = filterNoun(pos)
7   ns = removeStopWord(ns)
8   OC = annotate(ns)
```

To illustrate how the object clarification works, the complex parameter *SendEmail* in the Tiscali Email Services API is used as an example. The complex parameter *SendEmail* contains seven simple parameters, i.e., *from*, *to*, *cc*, *bcc*, *subject*, *body*, and *smtpServer*. Firstly, four simple parameters *from*, *to*, *subject* and *body* are recognized as parameters that demand additional object constraint. Secondly, the complex parameter *SendEmail* is utilized to obtain the object constraint on the four simple parameters. The word segmentation decomposes the compound word SendEmail into two words Send and Email. Then the POS tagger identifies Send as a verb and Email as a noun. The noun Email is passed to the stop-word elimination. Finally, Email is annotated with the concept #$Email. Therefore, #$Email is obtained as the object constraint on the four simple parameters.

4 Evaluation

To evaluate the effectiveness of our approach, it is applied to the 509 Web Services to clarify the location constraints and object constraints. OpenCyc is utilized as the semantic support. In addition, to assess the impacts of the two types of constraints on service reuse, a controlled experiment is conducted to identify collaborative relations among the 509 Web Services.

Our evaluations address the following research questions:

RQ1: How effectively can the approach clarify the location constraints and the object constraints?

RQ2: How much can the improved method for relation identification benefit from the constraints compatibility?

The first research question concerns the effectiveness of our approach, while the second concerns the benefits of integrating constraints compatibility into the relation identification. The answer for the first question is given in Sect. 4.1, and the answer for the second question is presented by the controlled experiment in Sect. 4.2.

4.1 Effectiveness of Constraints Clarification Approach

As the constraints clarification approach concentrates on two different types of constraints, we measure the effectiveness of the approach by evaluating the two modules separately. As discussed above, location constraints are attached to whole Web Services, while object constraints are attached to a certain parameters in Web Services. Therefore, quantitative analysis is utilized to evaluate the location clarification module, while qualitative analysis is utilized to evaluate the object clarification module.

For the location clarification module, the quality measure is defined by comparing the location constraints acquired by the module to the true location constraints obtained in the manual investigation. Precision and recall are utilized to measure the quality. Precision is the fraction of real location constraints among proposed location constraints, while recall is the fraction of real location constraints that is proposed by the module. Table 2 presents the evaluation results. Column "Real" lists the number of Web Services with location constraints obtained in the manual investigation. Column "Detected" lists the total number of Web Services with location constraints acquired by the module. Column "Tag" lists the total number of tags acquired by the module. Column "Location" lists the number of tags that are identified as locations. Columns "Precision" and "Recall" list the precision and recall respectively.

Table 2. Manual evaluation of the location clarification

Real	Detected	Tag	Location	Precision	Recall
120	122	1561	136	80.33 %	81.67 %

From results in Table 2, we have following observations. First, the location clarification can achieve adequate precision (80.33 %) and recall (81.67 %) on these Web

Services. We will later show examples of false positives and negatives, and analyze reasons for producing them. Second, a Web Service may have more than one location in its location constraints, since the number of tags that represent locations is larger than the one of Web Services with location constraints acquired by the location clarification.

We next analyze reasons for producing false positives and negatives. All the false positives and negatives are caused by the inconsistency between tags and service descriptions. In addition, false positives are due to that tags labeled on a Web Service contain locations, while service descriptions of the Web Service do not mention the service coverage area. For example, for the Northtext API, four tags are labeled on it, including messaging, campaigns, marketing, USA and other. As a result, USA is identified as a location, and the API is regarded as a service with location constraint. However, the service description does not mention that the API is merely available for communication in USA. False negatives are contrary to false positives. For example, for the CDYNE Weather API, only a weather tag is labeled on it. Since no tag is identified as a location, the API is regarded as a service without location constraint. However, the service description describes that the API is to retrieve weather information for US cities.

Since the number of parameters is pretty large, it is time-consuming to manually checking all the object constraints. Therefore, for the object clarification module, we aim to assess the quality of object constraints by qualitative analysis, namely statistics and manual check of a random sample. Table 3 lists the statistical results of object constraints, in which all the three values are acquired by the module. Column "Parameter" lists the number of parameters with object constraints acquired by object clarification, columns "Operation" and "Web Service" list the number of operations and Web Services that utilize parameters with clarified object constraints respectively.

Table 3. Statistical results of the object clarification

	Parameter	Operation	Web Service
Object constraint	24296	6068	398

From results in Table 3, we have following observations. First, with the help of the 206 predefined parameters, 24296 parameters are recognized as parameters that demand additional object constraints, indicating the efficiency of the predefined parameters. Second, parameters with object constraints account for 22.59 % of all simple parameters, while operations and Web Services that utilize such parameters account for 75.89 % of all operations and 78.19 % of all Web Services respectively. The large percentages of operations and Web Services echo the inference in the empirical investigation, i.e., a non-trivial percentage of Web Services have object constraints.

We then manually check object constraints for a random sample of 3000 parameters from the 24296 parameters. A criterion is that an object constraint is appropriate for a parameter, if the object constraint describes the precise interpretation of the parameter, or the object constraint represents the general interpretation of the parameter. For example, both of concepts EmailAddress and Email are regarded as appropriate object constraints for the *from* and *to* parameters in the Tiscali Email Services API, since concepts EmailAddress and Email respectively describe the precise meaning and the general meaning of the two parameters. Accuracy is utilized to measure the fraction of appropriate object constraints. The number of parameters with appropriate object constraints is 2683, accounting for 89.43 percent of the sample. Therefore, the precision and recall of the sample are 89.43 % and 100 % respectively. As a consequence, it is inferred that the object clarification module can achieve adequate precision and recall.

4.2 Impacts of Constraints on Relation Identification

Collaborative relations are widely used in the service composition. And the widely-adopted method for identifying collaborative relations is the semantic matching between two Web Services, as described by Algorithm 4. Specifically, suppose that OP_1 and OP_2 denote two operations from Web Services WS_1 and WS_2 respectively, outputs of OP_1 are semantically matched with inputs of OP_2. If each input in OP_2 can be satisfied by an output in OP_1, the relation between OP_1 and OP_2 is defined as completely collaborative (Line 3–4). Otherwise, if several inputs in OP_2 can be satisfied by outputs in OP_1, the relation is defined as partially collaborative (Line 5–6).

```
Algorithm 4: Widely-adopted relation identification
Input: Web Service WS₁, Web Service WS₂
Output: Collaborative Relation CR
1 for each operation OP₁∈ WS₁.operation
2    for each operation OP₂∈ WS₂.operation
      // E(o₁,i₂): o₁ is equivalent to i₂
      // K(o₁,i₂): o₁ is a subclass of i₂
      // I(o₁,i₂): o₁ is an instance of i₂
3      if (∀i₂∈ OP₂.input, ∃o₁∈ OP₁.output  s.t.
           E(o₁,i₂) ∨ K(o₁,i₂) ∨ I(o₁,i₂)
        // →c : completely collaborative
4        CR = CR ⋃ {OP₁ →c OP₂}
5      if (∃i₂∈ OP₂.input, ∃o₁∈ OP₁.output  s.t.
           E(o₁,i₂) ∨ K(o₁,i₂) ∨ I(o₁,i₂)
        // →p : partially collaborative
6        CR = CR ⋃ {OP₁ →p OP₂}
```

The main issue in Algorithm 4 is that the semantics of parameters is not enough. For example, it is meaningless to identify the collaborative relation between a BookSelling service located in Shanghai and a Shipping service located in North America. To address this issue, we present an improved method for identifying collaborative relations, integrating constraints compatibility into Algorithm 4.

Algorithm 5 describes the improved method. Location constraints of two Web Services are compared firstly to check whether the intersection of service coverage areas of the two Web Services is not null. If the intersection of two location constraints is null, it is meaningless to identify the collaborative relation between the two Web Services, thus terminating the relation identification (Line 2–4). If the intersection of two location constraints is not null, outputs of OP_1 are compared with inputs of OP_2 subsequently. More specifically, for the case that both of two parameters have object constraints, not only concepts of the two parameters are compared, but also object constraints are matched. In other words, their object constraints should also be compatible. For example, suppose that both of parameters o_1 and i_2 are associated with concept From (if concept From exists), however, the object constraint of o_1 is Email and the object constraint of i_2 is Place. Since object constraints of o_1 and i_2 are not compatible, Algorithm 5 will not regard o_1 as a potential match for i_2. And then if all inputs in OP_2 are matched by a specific or equivalent output in OP_1, a completely collaborative relation is established between OP_1 and OP_2 (Line 7–8). If several inputs in OP_2 are matched by a specific or equivalent output in OP_1, a partially collaborative relation is established between OP_1 and OP_2 (Line 9–10).

```
Algorithm 5: Improved relation identification
Input: Web Service WS₁, Web Service WS₂
Output: Collaborative Relation CR
1 LC₁ = WS₁.location; LC₂ = WS₂.location
2 if(LC₁ ≠ {} and LC₂ ≠ {})
3     if(LC₁ ∩ LC₂ == Φ)
4         exit(0)
5 for each operation OP₁∈ WS₁.operation
6     for each operation OP₂∈ WS₂.operation
        // p: i₂.object != null; q: o₁.object != null
        // r: E(o₁.object,i₂.object); s: K(o₁.object,i₂.object)
        // t: I(o₁.object,i₂.object)
7         if (∀i₂∈ OP₂.input, ∃o₁∈ OP₁.output  s.t.
                (E(o₁,i₂) ∨ K(o₁,i₂) ∨ I(o₁,i₂)) ∧ (p∧q→r∨s∨t))
8             CR = CR ∪ {OP₁→꜀OP₂}
9         if (∃i₂∈ OP₂.input, ∃o₁∈ OP₁.output  s.t.
                (E(o₁,i₂) ∨ K(o₁,i₂) ∨ I(o₁,i₂)) ∧ (p∧q→r∨s∨t))
10            CR = CR ∪ {OP₁ →ₚ OP₂}
```

And then a controlled experiment is conducted to assess the impacts of the two types of constraints on relation identification. In particular, Algorithms 4 and 5 are respectively applied to the 509 Web Services to identify collaborative relations. Results identified by Algorithms 4 and 5 are respectively marked as the control group and the experimental group. For relations in the control group but not in the experimental group, we further check whether they are removed by the location constraint compatibility or by the object constraint compatibility. Since semantics is the basis of service relations, parameters in the 509 Web Services are annotated based on OpenCyc.

Table 4 shows the overall results of collaborative relations in the controlled experiment. Columns "Total", "Completely" and "Partially" represent all collaborative relations, completely collaborative relations and partially collaborative relations respectively. For statistics concerning the control group, the three columns list the number of collaborative relations in the control group. For statistics concerning the experimental group, row "Removed by location" lists the number of inappro-priate collaborative relations removed by location constraints compatibility, row "Intermediate results" lists the number of collaborative relations after removing relations that fail the location constraints compatibility, row "Removed by object" lists the number of inappropriate collaborative relations removed by object constraints compatibility, and row "Final results" lists the number of collaborative relations after removing relations that fail either the location constraints or the object constraints compatibility.

Table 4. Overall results of the controlled experiment

		Total	Completely	Partially
Control group		2091651	431144	1660507
Experimental group	Removed by location	23125	3857	19268
	Intermediate results	2068526	427287	1641239
	Removed by object	1276773	295950	980823
	Final results	791753	108421	683332

From the overall results in Table 4, we have the following observations. First, more than half (62.15 %) of collaborative relations in the control group are removed in the experimental group through constraints compatibility, indicating that constraints compatibility plays an important role in the relation identification. Second, compared with location constraints compatibility, object constraints compatibility removes much more inappropriate collaborative relations. In general, 1.11 % of relations in the control group are removed by location constraints compatibility, while 61.04 % of relations in the control group are removed by object constraints compatibility, indicating that object constraints make much more significant contributions to relation identification. Third, 22916 completely collaborative relations in the control group are identified as partially collaborative relations in the experimental group after object constraints compatibility, indicating that object constraints can indeed identify more accurate service relations.

5   Related Work

To the best of our knowledge, only a few researches are carried out to automatically clarify formal constraints in Web Services. Wu et al. [9] concentrates on the dependency constraints on parameters, such as "a user_id parameter must be provided if a screen_name parameter is not provided". Bertolino et al. [10] focuses on the temporal constraints on operations, such as "a CartCreate operation should be invoked before a CartAdd opera-tion". Fisher et al. [11] concerns the occurrence constraint and type constraint on simple parameters, such as finding an optional parameter to be required, or an integer parameter to be positive Integer, and the dependency constraints on operations. Compared with these researches, we concentrate on two new and common types of constraints, location constraints on Web Services and object constraints on simple parameters.

Our work is also related to semantic annotation techniques. Most semantic annotation techniques [12–15] utilize manually built ontologies as their semantic support. On one hand, building ontologies manually is difficulty. On the other hand, these ontologies are not comprehensive enough to contain as many concepts as possible. The two main issues hamper the applicability of these semantic annotation techniques. Compared with these techniques, we utilize public open ontology from the linked open data as semantic support, which addresses the two main issues faced by [12–15].

We finally present some technically related researches concerning the service relations. Most researches [16, 17] consider two Web Services as composable, if the preceding service can satisfy the following service. The main issue is the absence of constraints. As the empirical investigation shows, there is a non-trivial presence of constraints. And our work takes the constraints into account in the relation identification. Tang et al. [18] notices the constraints and applies them in the service composition. For example, a BookSelling service located in Shanghai cannot cooperate correctly with a Shipping service located in North America. However, Tang assumes that the constraints are known in advance. In addition, Tang just mentions that it is necessary to check whether two services are constraint compatible, while does not gives the details. Compared with Tang, we concentrate on the clarification of constraints, and discuss the details in the relation identification based on constraints.

6 Conclusion

In this paper, we have proposed an approach for clarifying the location constraints and object constraints in Web Services. The benefits of constraints clarification are twofold. On one hand, location constraints and object constraints provide more accurate descriptions for Web Services. On the other hand, constraints clarification is an effective complement to the semantic annotation for Web Services. We concern two research questions in the evaluation. One concerns the effectiveness of our approach, the other concerns the benefits of constraints compatibility in relation identification.

On the dataset of 509 Web Services crawled from the ProgrammableWeb.com, we test the effectiveness of our approach by clarifying the location constraints and object constraints, and then assess the impacts of the two types of constraints on service reuse by identifying the collaborative relation among Web Services. The experimental results show that our approach can clarify the location constraints and object constraints effectively and achieve adequate recall and precision. Moreover, it is indicated that the two types of constraints, especially the object constraints, have significant impacts on improving the quality of service relations, thus provide a strong guarantee for accurate service reuse.

In the future, we will try to clarify more types of constraints, such as the time constraints. Furthermore, we plan to generate terminologies at different granularity level based on constraints to enrich the conceptualizations of domain ontologies.

Acknowledge. This work is supported by the National Natural Science Foundation of China grant 61373035, 61173155, the Tianjin Research Program of Application Foundation and Advanced Technology grant 14JCYBJC15600.

References

1. Vaughan-Nichols, S.J.: Web services: beyond the hype. Computer **35**, 18–21 (2002)
2. McIlraith, S.A., Son, T.C., Zeng, H.L.: Semantic Web Services. IEEE Intell. Syst. **16**, 46–53 (2001)
3. Hu, X., Chen, S., Feng, Z.: Semi-automatic acquisition and formal representation of OpenAPI. In: Khachidze, V., Wang, T., Siddiqui, S., Liu, V., Cappuccio, S., Lim, A. (eds.) iCETS 2012. CCIS, vol. 332, pp. 85–96. Springer, Heidelberg (2012)
4. Wang, L.J., Liu, F., Zhang, L.J., Li, G., Xie, B.: Enriching descriptions for public web services using information captured from related web pages on the internet. In: 5th IEEE International Symposium on Service-Oriented System Engineering, pp. 141–150. IEEE Computer Society Press, Washington (2010)
5. Bizer, C., Lehmann, J., Kobilarov, G., Auer, S., Becker, C., Cyganiak, R., Hellmann, S.: DBpedia - a crystallization point for the web of data. J. Web. Semant. **7**(3), 154–165 (2009)
6. Lenat, D.B.: CYC: a large-scale investment in knowledge infrastructure. Commun. ACM **38**(11), 33–38 (1995)
7. Suchanek, F., Kasneci, G., Weikum, G.: Yago: a core of semantic knowledge. In: 16th International Conference on World Wide Web, pp. 697–706. ACM (2007)
8. Pan, W.S., Chen, S.Z., Feng, Z.Y.: Automatic clustering of social tag using community detection. Appl. Math. Inf. Sci. **7**, 675–681 (2013)
9. Wu, Q., Wu, L., Liang, G.T., Xie, T., Mei, H.: Inferring Dependency constraints on parameters for web services. In: 22nd International Conference on World Wide Web, pp. 1421–1432. International World Wide Web Conferences Steering Committee, Republic and Canton of Geneva (2013)
10. Bertolino, A., Inverardi, P., Pelliccione, P., Tivoli, M.: Automatic synthesis of behavior protocols for composable web-services. In: 7th Joint Meeting of the European Software Engineering Conference and the ACM SIGSOFT Symposium on the Foundations of Software Engineering, pp. 141–150. ACM, New York (2009)
11. Fisher, M., Elbaum, S., Rothermel, G.: Automated Refinement and augmentation of web service description files. Technical report, University of Nebraska-Lincoln (2007)
12. Bouchiha, D., Malki, M.: Semantic annotation of web services. In: 4th International Conference on Web and Information Technologies, pp. 60–69 (2012)
13. Kungas, P., Dumas, M.: Cost-effective semantic annotation of XML schemas and web service interfaces. In: IEEE International Conference on Services Computing, pp. 372–379. IEEE Computer Society Press, Washington (2009)
14. Lerman, K., Plangprasopchok, A., Knoblock, C.A.: Automatically labeling the inputs and outputs of web services. In: 21st National Conference on Artificial Intelligence, pp. 1363–1368. AAAI Press (2006)
15. Patil, A.A., Oundhakar, S.A., Sheth, A.P., Verma, K.: METEOR-S web service annotation framework. In: 13th International Conference on World Wide Web, pp. 553–562. ACM, New York (2004)
16. Lee, D., Kwon, J., Lee, S., Park, S., Hong, B.: Scalable and efficient web services composition based on a relational database. J. Syst. Softw. **84**(12), 2139–2155 (2011)

17. Shin, D.H., Lee, K.H., Suda, T.: Automated generation of composite web services based on functional semantics. J. Web. Semant. **7**(4), 332–343 (2009)
18. Tang, X.F., Jiang, C.J., Zhou, M.C.: Automatic web service composition based on horn clauses and petri nets. Expert Syst. Appl. **38**(10), 13024–13031 (2011)

Common Topic Group Mining for Web Service Discovery

Jian Wang, Panpan Gao, Yutao Ma[✉], and Keqing He

State Key Laboratory of Software Engineering, Wuhan University, Wuhan 430072, China
ytma@whu.edu.cn

Abstract. Recent years have witnessed an increasing number of services published on the Internet. How to find suitable services according to user queries remains a challenging issue in the services computing field. Many prior studies have been reported towards this direction. In this paper, we propose a novel service discovery approach by mining and matching common topic groups. In our approach, we mine the common topic groups based on the service-topic distribution matrix generated by topic modeling, and the extracted common topic groups can then be leveraged to match user queries to relevant services, so as to make a better trade-off between the number of available services and the accuracy of service discovery. The results of experiments conducted on a publicly available data set show that compared with other widely used methods, our approach can improve the performance of service discovery by decreasing the number of candidate services.

Keywords: Web services discovery · Common topic group · Topic model

1 Introduction

With the rapid development of the service-oriented computing, more and more distributed software applications are composed by reusing the operations offered by existing services, which can be discovered and appropriately orchestrated to deliver the desired functionality [1]. The number of published Web services is therefore increasing continuously on the Internet [2]. For example, Programmableweb[1] has published over 13,400 Web services by May 10, 2015, almost increased to three times as compared with three years ago. Currently, since most public UDDI (Universal Description Discovery and Integration) registries are no longer available, Web service search engines become increasingly popular. However, Web service search engines based on keyword matching always suffer from lacking keywords in Web service descriptions or using synonyms or variations of predefined keywords [2]. Although many semantic Web service discovery approaches that can retrieve more appropriate services have been proposed [3, 4], semantically annotating attributes of Web services using ontologies is a tedious task, which brings difficulties in utilizing the approaches in practice. How to find suitable Web services for users remains a challenging issue in the services computing field.

Service discovery involves finding services that can match and meet users' functional requirements as well as nonfunctional requirements. In this paper, we focus our attention

---

[1] http://www.programmableweb.com/

© Springer International Publishing Switzerland 2015
L. Yao et al. (Eds.): APSCC 2015, LNCS 9464, pp. 92–107, 2015.
DOI: 10.1007/978-3-319-26979-5_7

on the matching of users' functional requirements and available services. Generally speaking, the functionality of a Web service is expressed by its description document. Thus, the problem of Web service discovery can be formally defined as follows. $S = \{s_1, s_2, \ldots, s_n\}$ is a set of Web services in a repository, where s_i ($i \in [1, n]$) is a Web service represented by its attributes such as description, input, and output. Given a user query q, how to retrieve the most relevant services of q from S?

The basic solution towards this issue is to calculate the similarity between q and each service s_i. Since the number of services in the repository is usually very large and may continually increases over time, it will be time-consuming and computationally expensive to compare a user query to each service in the repository. Clustering technologies that enable similar services to group together have been widely used in Web service discovery [2, 13, 26]. By organizing service descriptions into clusters in advance, service clustering can improve the efficiency of service discovery. As a widely used clustering technology, topic models such as LDA (Latent Dirichlet Allocation) [7] are able to extract unobserved groups that explain why some parts of the documents are similar and capture the underlying domain semantics. Although topic models can effectively support service discovery, considering that the numbers of topics and services remain very large, the performance of service discovery still has more room to improve.

In this paper, we introduce a new concept named as *Common Topic Group* (*CTG*) to further organize the clustered services and thereby improve the performance of service discovery. The basic idea of CTG is that if two services share more similar distribution probabilities over multiple topics, the similarity between them will be higher. We use CTG to denote the services that have the same probability grade under each topic within a specified scope. Based on CTG, those similar services can be organized according to their distribution probabilities over topics. Suppose that the distribution probabilities (mapped into four grades $g_1 \sim g_4$) of four services over five topics ($t_1 \sim t_5$) are as follows: $s_1 = (g_1, 0, g_3, 0, g_2)$, $s_2 = (0, 0, 0, g_2, g_1)$, $s_3 = (g_1, g_1, g_3, 0, g_2)$, $s_4 = (g_2, 0, g_1, 0, g_1)$. Services s_1 and s_3 can be grouped together to form a ctg since they have the same probability grade over topics t_1, t_3 and t_5. If the topic distribution probabilities are mapped into two grades, i.e., 0 and 1, s_4 can be merged into the group that consists of s_1 and s_3.

In this way, the available services in a given repository can be organized by the common topic groups that they share. Given a service query with the topic distribution probability $q = (g_1, 0, g_3, 0, g_2)$, s_1 and s_3 will be retrieved as relevant services based on CTG matching, and s_4 will also be viewed as a relevant service with a relative lower ranking position using the ctg generated in terms of two grades. In this case, we do not need to further compute the similarity between each service with the query. Only if the number of services retrieved based on CTG matching is less than the expected number, shall we compute the similarities between the remainder services with the query based on the service-topic distribution and the topic-word distribution generated by topic models. Therefore, our approach aims to reduce the search space of services even more using CTG matching, which can improve the response time of service discovery.

In brief, the contributions of this work are outlined as follows:

- We introduce a new concept *Common Topic Group* (*CTG*) to organize Web services based on topic models. The services that share similar distribution probabilities over multiple topics can be grouped together, and the size of a ctg is easy to adjust. An effective algorithm to mine ctgs is presented in detail.

- We propose a novel service discovery approach based on CTG matching. Experimental results show that the approach can decrease the number of candidate services, and therefore improves the response time of service discovery while keeping the accuracy of service discovery.

The rest of this paper is structured as follows. In Sect. 2, we discuss the related work. Section 3 presents the problem definition and gives an overview of the proposed approach. Section 4 presents how to mine common topic groups in detail. Section 5 introduces the CTG-based service matching and ranking approach. The experiment results and analysis are given in Sect. 6. Finally, Sect. 7 concludes this paper and puts forward future work.

2   Related Work

The work presented in this paper is closely related to Web service discovery, which is a hot topic in the services computing field in the past decade. Many service discovery approaches have been reported to consider both functional and non-functional (also known as quality of service, QoS) properties of services. Since our work focuses mainly on service discovery based on functional requirements, in this paper we only discuss the related work in this direction. Earlier studies in this area usually exploit the structure of WSDL (Web Service Description Language) documents to perform profile-based service matching. Due to the limitations of keyword-based service matching such as low accuracy, many progresses have been made to improve the results of this type of service discovery methods. Among these efforts, semantics-based service discovery and machine learning based service discovery are the two major directions in this area.

The semantics-based approaches incorporate with Semantic Web technologies to improve the performance of service matchmaking. For example, Klusch et al. [4] present a hybrid semantic matchmaker for SAWSDL (a semantic extension of WSDL), which considers both semantic similarity and structural similarity between services. Bipartite graphs have also been used to calculate the similarity between semantic Web services [20] to achieve the above goal. Besides, OWLS-MX [10] and WSMO-MX [11] are proposed to integrate logic-based reasoning and syntactic concept similarity into OWL-S (Web Ontology Language for Services) and WSMO (Web Service Modeling Ontology), respectively. The limitation of existing semantics-based approaches is that it is intractable over large service datasets [12] due to the complexity of the discovery process and heavy dependencies on domain ontologies.

In recent years, various machine learning techniques such as classification and clustering algorithms have been widely applied in service discovery. For example, Elgazzar et al. [22] propose an approach to extract features including content, types, messages, ports, and service name from WSDL documents, and then cluster the services with similar features into a group; Liu and Wong [13] apply text mining techniques to extract service features from WSDL documents for clustering; Chen et al. [14] and Wu et al. [21] propose a service clustering approach by integrating service tags and WSDL documents through augmented LDA; Aznag et al. [15] propose an approach based on correlated topic models to extract topics from service descriptions and organize hierarchical clusters to search services.

Besides these two types of service discovery approaches, many other efforts have also been reported to improve the performance of service discovery. For example, Pantazoglou and Tsalgatidou [1] present a fuzzy-based query evaluation mechanism that supports the service matchmaking process by employing and combining existing similarity metrics; Wang et al. [23] propose a service discovery approach by extracting domain-specific service goals to match with users' intentional requests.

In this paper, inspired by our previous work on context-aware role mining [5], we introduce the concept of a common topic group to organize services, and propose a new strategy for service matchmaking based on the extracted common topic groups. Our work can be categorized into the machine learning based approaches, which is an extension of existing topic models by grouping together all services that share similar distribution probabilities over multiple topics. Although several context-aware role mining approaches, e.g., [5, 18, 19], have been proposed, this paper presents an improved mining approach based on our previous work [5], which can decrease the mining time of common topic groups.

3 Problem Definition and Solution Overview

In this section, we describe the definition of common topic group mining, and give an overview of the proposed solution.

3.1 Problem Definition

As mentioned before, common topic groups are mined based on a service-topic distribution matrix, which is the output of topic modeling for service descriptions. The probabilities in the matrix should be mapped into different grades in advance.

Definition 1 (Tile). A *tile* is a sub-matrix of the service-topic distribution matrix, where all cells are non-zero and the cells in the same column have the same probability grade. The area of a tile is the number of cells within the tile.

Definition 2 (Common Topic Group). A *common topic group* (*ctg*) is a tile that has at least MINS(≥ 2) rows and MINT(≥ 2) columns, where the values of MINS and MINT can be manually set. Clearly the area of a ctg is no less than MINS $\times$ MINT.

Definition 3 (Common Topic Group Mining). Given a service-topic distribution matrix $STM = <S, T, G, STG>$, where S is a set of services, T is a set of topics, G is a set of probability grades, and $STG \subseteq S \times T \times (G \bigcup \{0\})$ is a set of the service-topic-grade relations, the common topic group mining problem is to find a state $<CTG, SA, TGA>$ that is consistent with STM, such that for any $<CTG', SA', TGA'>$ which is consistent with STM, #$CTG \leq$ #CTG', where CTG and CTG' are the set of common topic groups, $SA, SA' \subseteq S \times CTG$ are the service-ctg assignment matrix, and TGA,

96   J. Wang et al.

$TGA' \subseteq CTG \times T \times (G \bigcup \{0\})$ are the ctg-topic assignment matrix. Please note that a state is consistent with *STM*, if every service in *S* has the same topic-grade assignment as in *STM*.

For example, suppose that $S = \{s1, s2, s3, s4, s5\}$, $T = \{t1, t2, t3, t4, t5\}$, $G = \{g1, g2, g3\}$, MINS = MINT = 2. The original service-topic distribution matrix *M* is shown in Fig. 1(a), where the value g in cell $\{i, j\}$ represents that the probability grade of service *si* under topic *tj* is g. Figure 1(b) shows three tiles extracted from *M*, as shown in the shaded regions with dashed lines. Since it is impossible to find a tiling that covers *M* with less than 3 tiles, these 3 tiles are viewed as three ctgs. Once the process of CTG mining is finished, we can obtain a state $<CTG, SA, TGA>$, where $CTG = \{ctg1, ctg2, ctg3\}$, *SA* and *TGA* are shown in Fig. 1(c) and (d), respectively.

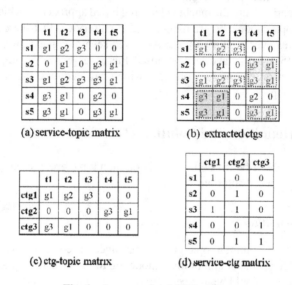

	t1	t2	t3	t4	t5
s1	g1	g2	g3	0	0
s2	0	g1	0	g3	g1
s3	g1	g2	g3	g3	g1
s4	g3	g1	0	g2	0
s5	g3	g1	0	g3	g1

(a) service-topic matrix

	t1	t2	t3	t4	t5
s1	g1	g2	g3	0	0
s2	0	g1	0	g3	g1
s3	g1	g2	g3	g3	g1
s4	g3	g1	0	g2	0
s5	g3	g1	0	g3	g1

(b) extracted ctgs

	t1	t2	t3	t4	t5
ctg1	g1	g2	g3	0	0
ctg2	0	0	0	g3	g1
ctg3	g3	g1	0	0	0

(c) ctg-topic matrix

	ctg1	ctg2	ctg3
s1	1	0	0
s2	0	1	0
s3	1	1	0
s4	0	0	1
s5	0	1	1

(d) service-ctg matrix

Fig. 1. An example of CTG mining

3.2 Overall Framework

As shown in Fig. 2, the framework of CTG based Web service discovery consists of three parts: data preprocessing, service clustering and organization, and service matching and ranking.

In the first part, data preprocessing is used to extract meaningful words from description documents as feature words. Currently, there are two mainstream types of service descriptions: WSDL-based and short text based service descriptions. For WSDLs, all features that describe a web service, such as service name, content, types, messages, and ports, will be extracted. For short texts, all the content in the description will be considered. Afterwards, we take the following steps to preprocess the service descriptions:

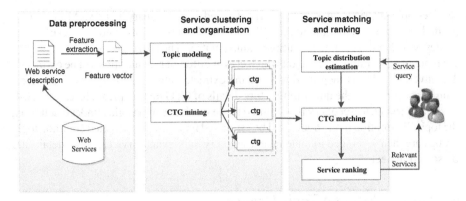

Fig. 2. The overall framework for web service discovery

- **Tokenization.** We perform tokenization over service descriptions to produce the original content vector by using NLTK[2], a Python package for natural language processing.
- **Function Words and General Words Removal.** Function words, e.g., *a*, *the*, as well as WSDL specific words, e.g., *request*, *response*, will be removed. In addition, those words with high occurrence frequencies in each document such as *api*, *format*, are too general to distinguish among Web services. We leverage the inverse document frequency to remove these types of words.
- **Stemming.** We use the Porter stemming algorithm in NLTK for prefix and affix removal.

The extracted feature words are used to construct a vector space, which will be the input of topic modeling.

In the second part, two steps will be taken to cluster and organize the services available. One is leveraging existing topic modeling technologies to discover the latent topics from the vector space, and the other is extracting common topic groups, which is the main contribution of this paper. LDA [7] is one of the most widely used topic models. Many extensions of LDA that are suitable to deal with short texts have also been proposed. For example, Correlated Topic Model (CTM) [16] extends LDA in modeling relations between topics by using the logistic normal distribution instead of the Dirichlet. Biterm Topic Model (BTM) [17] is another latest progress in extending LDA, which deals with the short message problem by modeling the generation of co-occurrence patterns in the whole corpus. In this paper, since the preprocessed service descriptions are usually very short, we need to test these three approaches to find the best one for the experimental dataset we used.

In order to perform CTG mining, the probabilities in the service-topic distribution matrix generated by topic modeling should be mapped into different grades first. The basic and simple mapping is to map the probabilities into two grades. That is, the probabilities that exceed a certain threshold are viewed as "1", otherwise "0". Thus, the

---

[2] http://www.nltk.org/

service-topic distribution matrix is transformed into a binary matrix. Moreover, if needed, the values belonging to the "1" grade can be further classified into several sub-grades, which will generate a multiple-valued matrix. The two types of matrices will be the input of CTG mining, based on which ctgs at different granularities can be extracted. The inverted index is then used to store the extracted ctgs and the topics of services, aiming at decreasing the query time. More details of CTG mining please refer to Sect. 4.

Finally, when a user query is submitted to the service search engine under discussion, the topic distribution of the query will be estimated first, and the extracted common topic groups are then leveraged to match the user query to relevant services. More details are presented in Sect. 5.

4   Mining Common Topic Groups

We employ the context-aware role mining method proposed in [5] and improve its core algorithm to enhance efficiency in mining ctgs. In [5], the idea of tiling database [9] is adopted in context-aware role mining. The CTG mining problem can be reduced to the minimum tiling problem, which seeks for a tiling of which the area equals to the total number of non-zero cells in the matrix and consists of the minimum number of tiles. Next, we introduce the algorithm in detail.

Algorithm 1   Minimal Common Topic Groups (CTGs) mining

Input: service-topic matrix M_{st}

Output: ctg-topic matrix M_{ct} and service-ctg matrix M_{sc}

1 $Tiles$ = ExtractTileSet(M_{st});

2 **for** each tile $tile \in Tiles$ **do**

3 | create a new common topic group ctg;

4 | extract topic set T and grade set D from $tile$;

5 | **for** each topic $t \in T$ and each grade $d \in D$ **do**

6 | | add assignment $\{ctg, t, d\}$ to M_{ct};

7 | extract service set S from $tile$;

8 | **for** each service $s \in S$ **do**

9 | | add assignment $\{s, ctg\}$ to M_{sc};

10 **return** M_{ct} and M_{sc}

Algorithm 1 shows the overall procedure of CTGs mining. Line 1 calls Algorithm 2 to find a minimum tile set for a given service-topic matrix M_{st}. Based on the tiles mined from M_{st}, Lines 3–10 generate new ctgs, as well as the ctg-topic matrix M_{ct} and the service-ctg matrix M_{sc}.

To extract tiles from M_{st}, we firstly group together the services that have the same probability grade over each topic by transforming the service-topic matrix M_{st} into a grade-topic matrix M_{gt} (see Line 2 in Algorithm 2). Table 1 shows the transformation result of the matrix in Fig. 1(a). For each cell in the grade-topic matrix, we record its coordinate (<grade, topic>) and the services associated with the cell, e.g., for cell c_{11}

in Table 1, c_{11}.grade = {g1}, c_{11}.topic = {t1}, c_{11}.services = {s1, s3}. Note that since the number of services is much larger than that of probability grades, the size of the matrix to be analyzed can be largely decreased.

Table 1. The grade-topic matrix tranformed from Fig. 1(a)

	t1	t2	t3	t4	t5
g1	{s1, s3}	{s2, s4, s5}			{s2, s3, s5}
g2		{s1, s3}		{s4}	
g3	{s4, s5}		{s1, s3}	{s2, s3, s5}	

As mentioned in Subsect. 3.1, MINS and MINT denote the minimal numbers of services and topics, respectively, which are used to form a ctg. Therefore, a cell in the grade-topic matrix which contains less than MINS services is impossible to constitute a ctg, and thus we only extract the cells that contain more than MINS services (see Line 3 in Algorithm 2), which will be stored in array $Cell[nc]$.

Algorithm 2 ExtractTileSet

Input: a service-topic matrix M_{st}

Output: extracted tiles $TileSet$

1 $TileSet$ = {};
2 Transform M_{st} to a grade-topic matrix M_{gt} ;
3 $Cell[nc] \leftarrow \{c_{ij} \in M_{gt} \mid \#c_{ij}.services \geq MINS\}$;
4 **for** $i = 1$ to nc **do**
5    $commonServiceSet$ = {}; $cTileSet$ = {};
6   **for** $j = i+1$ to nc **do**
7   **if** $Cell[i].topic \neq Cell[j].topic$ **then**
8          $commonServiceSet = Cell[i].services \cap Cell[j].services$;
9          $tile$ = {}; $tile.services$ = {};
10   **if** $commonServiceSet.size() > MINS$ **then**
11            $tile = tile \cup <Cell[j].grade, Cell[j].topic>$;
12            $tile.services = tile.services \cup commonServiceSet$;
13   CompareAndUpdate($tile$, $cTileSet$);
14   **for** each tile t in $cTileSet$ **do**
15       $t = t \cup <Cell[i].grade, Cell[i].topic>$;
16    $TileSet = TileSet \cup cTileSet$;
17 RemoveDuplication($TileSet$);
18 **return** $TileSet$

Next, we merge the elements in $Cell[nc]$ to generate tiles. For each element in $Cell[nc]$, we compare it with its subsequent cells to find qualified tiles (see Lines 4–16 in Algorithm 2). Arrays $commonServiceSet$ and $cTileSet$ are used to store the intersection of services included in any two cells, and the candidate tile set generated from each cell, respectively. Because each service has one and only one probability under each topic, it is not necessary to compare the cells with the same topic (see Line 7 in Algorithm 2). For any two cells, only if the size of the intersection between their related services exceeds MINS, they can be considered to constitute a tile. Any new tile constituted by two cells should be compared with tile set *cTileSet* to check whether *cTileSet* will be updated (see Algorithm 3). Then, the tile sets created from each cell will be merged together. Since there may be some duplicated tiles in the merged tile set, the final step is to remove the duplications. The tiles that are the subset of other existing tiles will be removed (see Line 17 in Algorithm 2).

Algorithm 3 CompareAndUpdateTile

Input: tile *tile*, candidate tile set *cTileSet*
Output: candidate tile set *cTileSet*
1 **if** *cTileSet*.isEmpty() **then**
2 $\quad$ *cTileSet* = *cTileSet* $\cup$ {*tile*};
3 **else**
4 $\quad$ *equalToTile* = 0;
5 $\quad$ **for** each tile *t* in *cTileSet* **do**
6 $\quad\quad$ **if** *t*.services = *tile*.services **then**
7 $\quad\quad\quad$ *t* = *t* $\cup$ *tile*;
8 $\quad\quad\quad$ *equalToTile* = 1;
9 $\quad\quad$ **else if** *t*.services $\subset$ *tile*.services **then**
10 $\quad\quad\quad$ *t* = *t* $\cup$ *tile*;
11 $\quad\quad$ **else if** *tile*.services $\subset$ *t*.services **then**
12 $\quad\quad\quad$ *tile* = *t* $\cup$ *tile*;
13 $\quad\quad$ **else**
14 $\quad\quad\quad$ *commonServiceSet* = *tile*.services $\cap$ *t*.services;
15 $\quad\quad\quad$ **if** *commonServiceSet*.size() > MINS **then**
16 $\quad\quad\quad\quad$ *newTile* = *tile* $\cup$ *t*;
17 $\quad\quad\quad\quad$ *newTile*.services = *commonServiceSet*;
18 $\quad\quad\quad\quad$ *cTileSet* = *cTileSet* $\cup$ {*newTile*};
19 $\quad$ **if** *equalToTile* = 0 **then**
20 $\quad\quad$ *cTileSet* = *cTileSet* $\cup$ {*tile*};
21 **return** *cTileSet*;

Algorithm 3 updates a candidate tile set *cTileSet* by comparing it with a tile *tile*. We compare *tile* with each tile *t* in *cTileSet* when considering the following cases. Firstly, if the service sets included in *t* and *tile* are the same, *t* will be updated by adding the cells in *tile* (see Lines 6–7 in Algorithm 3). Note that the variable *equalToTile* is defined to record whether the service set included in *tile* is identical to that of any tile in *cTileset*. If so, *equalToTile* is assigned to 1. Secondly, if the service set included in *t* is a proper subset of that of *tile*, *t* will be updated by adding the cells in *tile* (see Lines 9–10 in Algorithm 3). Thirdly, on the contrary to the above, *tile* will be updated by adding the cells in *t* (see Lines 11–12 in Algorithm 3). Finally, we judge whether the size of the set of common services between t and *tile* exceeds MINS. If it satisfies the minimum requirements of common services, a new tile can be created and added to *cTileSet* (see Lines 13–18 in Algorithm 3). In addition, if *equalToTile* equals to 0, *tile* will be added to *cTileSet*.

5   Web Service Matching and Ranking Based on CTGs

When a user query is submitted to the service search engine, the query is first preprocessed using the preprocessing steps mentioned in Section 3. In this way, a service query will be represented as a set of words $SQ = \{w_1, w_2, ..., w_{Nsq}\}$. Next, we estimate the topic distribution of the query. According to the Gibbs sampling approach [6], the distribution probability of SQ under topic t_i can be computed as follows:

$$P\left(SQ|t_i\right) = \frac{n_i^{SQ} + \alpha}{\sum_{j=1}^{T}\left(n_j^{SQ} + \alpha\right)}, \tag{1}$$

where n_i^{SQ} is the number of words in SQ assigned to topic t_i, which can be calculated based on the topic-word matrix, $\sum_{j=1}^{T} n_j^{SQ}$ is the total number of words in SQ, T is the number of topics, and α is the document-topic Dirichlet parameter.

The topic distribution of a query will be mapped into different grades similar to those of services, and afterwards we match the query with the extracted ctgs to find candidate services by taking the following steps.

- Step 1: if there exists a ctg whose topic distribution exactly matches with that of the query, i.e., if $\exists ctg$ (*ctg*.topic = *SQ*.topic), the services that contain the ctg will be returned;
- Step 2: if there exists a ctg whose topic distribution covers that of the query, i.e., if $\exists ctg$ (*SQ*.topic $\subset$ *ctg*.topic), the services that contain the ctg will be returned;
- Step 3: if there exists a ctg whose topic distribution is a subset of that of the query, i.e., if $\exists ctg$ (*ctg*.topic $\subset$ *SQ*.topic), the services related to the ctg will be returned;
- Step 4: the inverted index is leveraged to retrieve the services that share at least one topic with SQ.

Note that not all these steps should be taken for a Top-k service query. Once the number of candidate services reaches k, the matching process will stop. The services selected in the first two steps can match with the query very well and thus are ranked as the preferred candidate services. It is not necessary to further compute their similarities with the query. For those services selected from the other two steps, we adopt a classical LDA-based information retrieval approach [8] to rank them. In Eq. (2), each service s_i is scored by the likelihood of its model generating query SQ.

$$P\left(SQs_i\right) = \prod_{w_j \in SQ} P\left(w_j s_i\right), \tag{2}$$

$$P\left(w_j, s_i\right) = \sum_{k=1}^{T} P\left(w_j | t_k\right) P\left(t_k | s_i\right), \qquad (3)$$

where $P(w_j | t_k)$ and $P(t_k | s_i)$ are retrieved from the topic-word distribution matrix and the service-topic distribution matrix, respectively. In this way, the candidate services are ranked for each query, and the preferred services will be returned to users.

6   Experiments and Result Analysis

In this section, we evaluate the performance of the proposed approach using an open dataset. All the algorithms are developed in Java and conducted on a PC with 2 GHz Intel Core T7300 CPU and 4 GB RAM, running Windows 7 OS.

The dataset we used is a WSDL service retrieval test collection: SAWSDL-TC[3], which consists of 1080 WSDL Web services, as well as 42 queries represented in WSDL documents. A graded relevance set for each query is also provided in the dataset, which serves as the benchmark of relevant services. There are three relevance levels: "1" denotes that a service is potentially relevant to the query, "2" denotes a relevant relation, and "3" denotes high relevance.

6.1   Evaluation of Service Discovery Based on Topic Models

Since our approach depends on topic models, we will first evaluate the performance of service discovery based on existing topic models including LDA, CTM, and BTM to test which one achieves the best result towards the dataset. We use TMSD to denote the service discovery approach based on topic models such as LDA, CTM, and BTM. Note that the procedure of TMSD is similar to our proposed approach in Sect. 5 except that the common topic group matching process is not included in TMSD. That is, it uses Eqs. (2) and (3) to rank all the candidate services according to a given service query.

Precision at k is one of the most intuitive metrics for search engines. For each user query in the test set, this metric means the fraction of the first k retrieved services that also appear in the query's relevant service set. As mentioned above, each query has three types of relevant services, but we only consider two cases: relevance = 3, suggesting that only highly relevant services are involved in the evaluation, and relevance = 2, which means that both highly relevant services and relevant services are involved.

The number of topics is an important factor that influences the quality of service clustering. The work [25] has reported that LDA can get the best clustering performance for the SAWSDL-TC dataset when the number of topics is set to 90. Our experimental results also show that each topic model can achieve the best result when the number of topics is 90. Since only the top ranked retrieved services will be selected by users, up to the first 20 services among all retrieved results are taken into account. Figure 3(a) and (b) show the results of measuring Precision at k using LDA, CTM, and BTM when relevance = 3 and relevance = 2, respectively. The results show the average value of Precision at k over all 42 queries. As shown in Fig. 3, these three approaches can get

---

[3] http://www.semwebcentral.org/projects/sawsdl-tc

similar precision, and BTM performs relatively better than CTM and LDA. Since the relevant services of some queries are less than 20, the precision is slightly decreased with the increase of k.

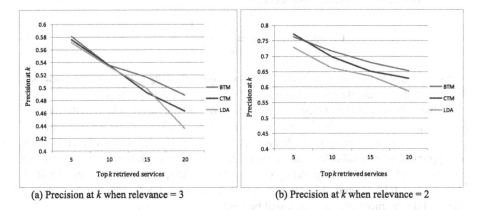

(a) Precision at k when relevance = 3   (b) Precision at k when relevance = 2

Fig. 3. Comparison of topic model based service discovery

6.2   Evaluation of CTG-Based Service Discovery

Since BTM is the best approach towards the experimental dataset, the service-topic distribution matrix generated by BTM is used as the input of our CTG mining approach. The probabilities are mapped into different grades first. We have tested two different scales. Since the number of topics is 90, the average probability of each topic is 1/90 (about 0.111). According to our observation, the boundary between the "0" grade and the "1" grade in the binary scale is chosen to be 0.014. In another multiple-valued scale, we further classify the probabilities belonging to the "1" grade into three sub-grades using the boundaries 0.02 and 0.1.

Our approach (CTG) differs from the TMSD approach in that ours can reduce the search space by CTG matching. TMSD will compute the similarity between a query and each service in the service set, while ours only ranks the candidate services filtered by CTG matching. Therefore, the precision of ours will not be better than that of TMSD. Our approach aims to minimize the number of candidate services while keeping the precision of service discovery. We evaluate the difference in the precision of service discovery between ours and TMSD, which is calculated by the coverage rate at k and the decreased percentage of candidate services using CTG, also known as the compression rate.

Coverage rate at k denotes the fraction of the top k services retrieved using CTG that also appear in the top k list using TMSD.

$$\text{Coverage rate} @ k = \frac{|S_{CTG} \cap S_{TMSD}|}{|S_{TMSD}|}, \tag{4}$$

where S_{CTG} and S_{TMSD} denote the top k services obtained by CTG and TMSD, respectively.

Compression rate indicates the proportion of the candidate services obtained based on CTG matching to all candidate services.

$$\text{Compression rate} = \frac{|CS_{CTG}|}{|CS|}, \tag{5}$$

where CS_{CTG} denotes the candidate services extracted using CTG matching and CS denotes all the candidate services (i.e., 1080 services in this experiment).

Table 2 shows the average coverage rate at k over all queries provided in the dataset. More than 90 percent of Web services retrieved by TMSD will be returned by our method. That is, most of services that are relevant to the queries will not be omitted by our approach, which indicates that the precision of our approach is very close to that of TMSD. In addition, the average compression rate is 0.1354, which suggests that 86.46 % of services will be filtered out from the candidate service set before the service ranking stage, while the most relevant ones will be kept.

Table 2. Average coverage rate at k over 42 queries

	Top 5	Top 10	Top 15	Top 20
Coverage rate	95.68 %	93.24 %	91.58 %	90.28 %

Table 3 shows the average query response time of TMSD and CTG. Compared with TMSD, the query response time of our approach is, on average, decreased by about 32.8 %, which is caused by decreasing the number of candidate services based on CTG matching.

Table 3. Comparasion of average query response time

	Time (*ms*)
TMSD	662.9
CTG	445.5

Agglomerative hierarchical clustering (AHC) and Formal Concept Analysis (FCA) are two widely used clustering approaches that can also be used to organize services based on their distributions over topics. We compare the computing time of our approach with those of AHC and FCA. In the experiment, we use 2 service-topic distribution matrices, with the sizes of 1080*90*2 and 1080*90*4, respectively. According to Fig. 4, the time to mine ctgs using CTG is less than the computing time of FCA and AHC. Note that the computing time of our approach will decrease when the scale of the probability grades increases from 2 to 4, since the number of tiles will decrease with the increase of the number of grades. Moreover, compared with FCA and AHC, our approach is more flexible in setting the area of a common topic group.

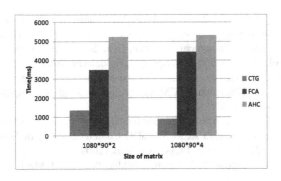

Fig. 4. Comparison of computing time

7 Conclusions and Future Work

In this paper, we present an approach to service discovery based on common topic groups, as well as the algorithms to mine common topic groups based on the service-topic distribution matrix generated by topic modeling. The extracted common topic groups can contribute to minimizing the number of candidate services in service discovery. Experiments on an open dataset show that compared with the service discovery approach that ranks all services based on topic models, the proposed approach can significantly decrease the size of candidate service set while maintaining high accuracy of service discovery.

We will extend our approach in two directions in the future. On the one hand, we plan to leverage more domain knowledge during CTG mining. For example, the knowledge mentioned in [24] such as must-link, which indicates that two words should be in the same topic, and cannot-link, which indicates that two words should not be in the same topic, will be considered. On the other hand, we will consider how to further improve the performance of CTG mining.

Acknowledgements. The work is supported by the National Basic Research Program of China under grant No. 2014CB340404, the National Natural Science Foundation of China under grant Nos. 61202031, 61272111, and 61373037, and the central grant funded Cloud Computing demonstration project of China undertaken by Kingdee Software (China).

References

1. Pantazoglou, M., Tsalgatidou, A.: A generic query model for the unified discovery of heterogeneous services. IEEE Trans. Serv. Comput. **6**, 201–213 (2013)
2. Yu, Q., Liu, X., Bouguettaya, A., Medjahed, B.: Deploying and managing web services: issues, solutions, and directions. VLDB J. **17**, 537–572 (2008)
3. Paliwal, A.V., Shafiq, B., Vaidya, J., Xiong, H., Adam, N.: Semantics-based automated service discovery. IEEE Trans. Serv. Comput. **5**, 260–275 (2012)

4. Klusch, M., Kapahnke, P., Zinnikus, I.: Hybrid adaptive web service selection with SAWSDL-MX and WSDL-analyzer. In: Aroyo, L., Traverso, P., Ciravegna, F., Cimiano, P., Heath, T., Hyvönen, E., Mizoguchi, R., Oren, Eyal, Sabou, M., Simperl, E. (eds.) ESWC 2009. LNCS, vol. 5554, pp. 550–564. Springer, Heidelberg (2009)

5. Wang, J., Zeng, C., He, C., et al.: Context-aware role mining for mobile service recommendation. In: 27th Annual ACM Symposium on Applied Computing, pp. 173–178. ACM Press, New York (2012)

6. Yao, L., Mimno, D., McCallum, A.: Efficient methods for topic model inference on streaming document collections. In: 15th ACM SIGKDD Conference on Knowledge Discovery and Data Mining, pp. 937–946. ACM Press, New York (2009)

7. Blei, D.M., Ng, A.Y., Jordan, M.I.: Latent dirichlet allocation. J. Mach. Learn. Res. **3**, 993–1022 (2003)

8. Wei, X., Croft, W.B.: LDA-based document models for ad-hoc retrieval. In: 29th Annual International ACM SIGIR Conference on Research and Development in Information Retrieval, pp. 178–185. ACM Press, New York (2006)

9. Vaidya, J., Atluri, V., Guo, Q.: The role mining problem: finding a minimal descriptive set of roles. In: 12th ACM Symposium on Access Control Models and Technologies, pp. 175–184. ACM Press, New York (2007)

10. Klusch, M., Fries, B., Sycara, K.: OWLS-MX: a hybrid semantic web service matchmaker for OWL-S services. J. Web Semant. **7**, 121–133 (2009)

11. Klusch, M., Kaufer, F.: WSMO-MX: a hybrid semantic web service matchmaker. Web Intell. Agent Syst. **7**, 23–42 (2009)

12. Mohebbi, K., Ibrahim, S., Khezrian, M., et al.: A comparative evaluation of semantic web service discovery approaches. In: 12th International Conference on Information Integration and Web-based Applications and Services, pp. 33–39. ACM Press, New York (2010)

13. Liu, W., Wong, W.: Web service clustering using text mining techniques. Int. J. Agent Oriented Softw. Eng. **3**, 6–26 (2009)

14. Chen, L., Wang, Y., Yu, Q., Zheng, Z., Wu, J.: WT-LDA: user tagging augmented LDA for web service clustering. In: Basu, S., Pautasso, C., Zhang, L., Fu, X. (eds.) ICSOC 2013. LNCS, vol. 8274, pp. 162–176. Springer, Heidelberg (2013)

15. Aznag, M., Quafafou, M., Jarir, Z.: Leveraging formal concept analysis with topic correlation for service clustering and discovery. In: 2014 IEEE International Conference on Web Services, pp. 153–160. IEEE Press, New York (2014)

16. Blei, D.M., Lafferty, J.D.: Correlated topic models. Adv. Neural Inf. Process. Syst. **18**, 147–154 (2006)

17. Cheng, X., Yan, X., Lan, Y., Guo, J.: BTM: topic modeling over short texts. IEEE Trans. Knowl. Data Eng. **26**, 2928–2941 (2014)

18. Chu, V.W., Wong, R.K., Chi, C.H.: Online role mining without over-fitting for service recommendation. In: 20th IEEE International Conference on Web Services, pp. 58–65. IEEE Press, New York (2013)

19. Wong, R.K., Chu, V.W., Hao, T.: Online role mining for context-aware mobile service recommendation. Pers. Ubiquit. Comput. **18**, 1029–1046 (2014)

20. Bellur, U., Kulkarni, R.: Improved matchmaking algorithm for semantic web services based on bipartite graph matching. In: 2007 IEEE International Conference on Web Service, pp. 86–93. IEEE Press, New York (2007)

21. Wu, J., Chen, L., Zheng, Z., Lyu, M.R., Wu, Z.: Clustering web services to facilitate service discovery. Int. J. Knowl. Inf. Syst. **38**, 207–229 (2014)

22. Elgazzar, K., Hassan, A.E., Martin, P.: Clustering WSDL documents to bootstrap the discovery of web services. In: 2009 IEEE International Conference on Web Services, pp. 147–154. IEEE Press, New York (2009)

23. Wang, J., Zhang, N., Zeng, C., Li, Z., He, K.Q.: Towards services discovery based on service goal extraction and recommendation. In: 2013 IEEE International Conference on Services Computing, pp. 65–72. IEEE Press, New York (2013)

24. Chen, Z., Liu, B.: Mining topics in documents: standing on the shoulders of big data. In: 20th ACM SIGKDD International Conference on Knowledge Discovery and Data Mining, pp. 1116–1125. ACM Press, New York (2014)

25. Aznag, M., Quafafou, M., Jarir, Z.: Correlated topic model for web services ranking. Int. J. Adv. Comput. Sci. Appl. **4**, 283–291 (2013)

26. Yu, Q.: Place semantics into context: service community discovery from the WSDL corpus. In: Kappel, G., Maamar, Z., Motahari-Nezhad, H.R. (eds.) ICSOC 2011. LNCS, vol. 7084, pp. 188–203. Springer, Heidelberg (2011)

Context-Aware Web Services Recommendation Based on User Preference Expansion

Yakun Hu[1], Xiaoliang Fan[1,2,3(✉)], Ruisheng Zhang[1], and Wenbo Chen[1]

[1] School of Information Science and Engineering,
Lanzhou University, Lanzhou 730000, China
{huyk14, fanxiaoliang, zhangrs, chenwb}@lzu.edu.cn
[2] Fujian Key Laboratory of Sensing and Computing for Smart City,
School of Information Science and Engineering, Xiamen University,
Xiamen 361000, China
[3] State Key Laboratory for Novel Software Technology,
Nanjing University, Nanjing 210023, China

Abstract. Context-Aware Recommender System is known to not only recommend items or services similar to those already rated with the highest score, but also consider the current contexts for personalized Web services recommendation. Specifically, a key step for CARS methods refers to previous service invocation experiences under the similar context of the user to make Quality of Services prediction. Existing works either considered the influence of regional correlations on user preference, or combined the location-aware context with the matrix factorization method. However, the user preference expansion triggered by instant update of user location is not fully observed. For instance, when making Web service recommendation for a user, it is expected to be aware of rapid change of the user location immediately and the expansion of user preference as well. In this paper, we propose a Web services recommendation approach dubbed as CASR-UPE (Context-aware Web Services Recommendation based on User Preference Expansion). First, we model the influence of user location update on user preference. Second, we perform the context-aware similarity mining for updated location. Third, we predict the Quality of Services by Bayesian inference, and thus recommend the best Web service for the user subsequently. Finally, we evaluate the CASR-UPE method on WS-Dream dataset by evaluation matrices such as RMSE and MAE. Experimental results show that our approach outperforms several benchmark methods with a significant margin.

Keywords: Context awareness · Web service · Recommender system · QoS · Preference expansion

1 Introduction

Context-Aware Recommender System (CARS) for Web services aims to recommend Web services not only similar to those already rated with the highest score, but also that could combine the contextual information with the recommendation process [1, 4, 10].

© Springer International Publishing Switzerland 2015
L. Yao et al. (Eds.): APSCC 2015, LNCS 9464, pp. 108–120, 2015.
DOI: 10.1007/978-3-319-26979-5_8

In recent years, preliminary benefits have been seen in Web services recommendation considering various contextual factors [2, 3]. For example, temporal [7, 8], spatial [9, 12, 13] and social [5, 6, 11, 14, 16] contexts are widely extracted separately for personalized Web services recommendation.

Specifically, one of the key steps for CARS method is referring to previous service invocation experiences under the similar location of the current user to make Quality of Services (QoS) prediction [17]. Existing works mainly discussed the influence of regional correlations on user preference [14]. There are also several novel methods combining the location-aware contexts with matrix factorization methods [13]. However, the user preference expansion triggered by instant update of user location is not fully observed for personalized recommendation. For instance, when making recommendation for a user, we are expected to be aware of the rapid change of the user location immediately, and thus of the expansion of user preference as well.

In this paper, we propose a Web services recommendation approach dubbed as CASR-UPE. Our approach consists of three steps: (1) model the influence of user location update on user preference; (2) perform the context-aware similarity mining for updated location; (3) predict the QoS of Web services by Bayesian inference and recommend the best Web service for the user subsequently. Finally, we evaluate the CASR-UPE algorithm on WS-Dream dataset [17] by evaluation matrices of both RMSE and MAE. Experimental results show that our approach outperforms the six state-of-the-art benchmark methods with a significant margin.

Hereafter, the paper is organized as follows. Section 2 introduces the related works. Section 3 shows a motivating example. Section 4 gives the details of CASR-UPE method. Section 5 shows the experimental results and discussion. Finally, the general conclusion as well as perspectives in Sect. 6 closes this paper.

2 Related Works

Context-aware recommender system has gained significant momentum in recent years. The development of mobile devices and their crowd sensing capabilities have enabled the collection of rich contextual information on time and location [9, 11–13].

In the CARS methods, the temporal contexts [7, 8] have been widely used in conventional CARS methods. Another widely discussed context information is the location context [9, 12, 13], especially in LBSN [5, 6]. A location-aware services recommendation method is presented in [9, 10] by referring to previous service invocation experiences under the similar location with the current user. However, these methods merely consider the location as a filter to make recommendation to the current user. The influence of regional correlation on user preference is also considered in [13]. In addition, a location-based hierarchical matrix factorization (HMF) method [15] is proposed to perform personalized QoS prediction. In short, the above location-based service recommendation methods overlook the user preference expansion triggered by the instant update of user location.

Furthermore, temporal location correlations [20, 21] have been studied for location recommendations in location-based social network (LBSN). However, the location recommendation in LBSN is different from Web services recommendation and the temporal location effects seem to be not suitable for Web services recommendation.

Among the current works of considering location-aware context in CARS, we propose a Web services recommendation approach, considering the influence of user location update on user preference, and then have an updated location similarity mining.

3   A Motivating Example

Figure 1 is a scenario of use to recommend weather forecast services considering the user preference expansion. The upper part represents a Web services repository $(S_1, S_2, \ldots, S_n)$ and many service users $(u_1, u_2, \ldots, u_m)$, where services and users are distributed all over the world. Suppose S_1 = "Weather China[1]"; S_2 = "Moji Weather China[2]"); S_3 = "US National Weather Service[3]"; S_4 = "Le Figaro météo in France[4]". The underlying part illustrates the distributed networks. The curves link users and services to their corresponding geographic positions.

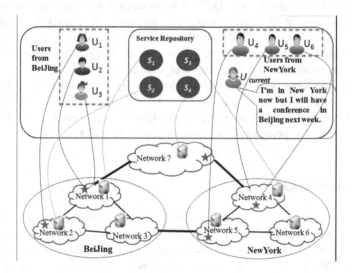

Fig. 1. A scenario of weather forecast services recommendation considering user preference expansion

Firstly, in this scenario of Fig. 1, $U_{current}$ is in NYC at present time. As we all know, the accuracy of weather forecast services is highly relevant to the location and it is natural to believe that a user prefers the service either located in his/her city. So when making recommendations to $U_{current}$ at present time, we should consider her current

---

[1] Weather China, http://en.weather.com.cn/.

[2] Moji Weather China, http://www.moweather.com/.

[3] US national weather service, http://www.weather.gov/.

[4] Le Figaro météo weather forecasting service, http://www.lefigaro.fr/meteo/france/index.php.

location (New York) and recommend S_3 to her. However, if we know that the user will have a conference in Beijing next week (e.g. from his/her mobile phone calendar), we should consider the influence of location update (Beijing) on user preference and recommend S_1 to $U_{current}$ when recommending weather services for next week. In a word, when making recommendations to users, we should consider the influence of user location update on user preference.

Secondly, when making recommendations to $U_{current}$, we should also consider the set of users in the same location with $U_{current}$ because of "the more similar between the current user and another user's context (e.g. location), the more probability of the two users will have a same preference". The user location update will lead to different sets of similar users. For instance, $U_{current}$ is in New York at present time, her similar set of users is made of U_4, U_5, U_6 (all from New York). But $U_{current}$ will be in Beijing next week, her similar set of users will become U_1, U_2, U_3 (all from Beijing). So we should also consider the influence of user location update on the set of similar users.

Above all, we consider the expansion of user preference triggered by instant update of user location when making recommendation to the current user.

4 CASR-UPE Algorithm: Context-Aware Web Services Recommendation Based on User Preference Expansion

4.1 Problem Definition

In order to help readers understanding our algorithm better, the following definitions are given.

We assume that there are a set of users $U = \{u_1, u_2, \ldots, u_n\}$ and a set of Web services $S = \{s_1, s_2, \ldots, s_m\}$. $u_i (1 \leq i \leq n)$ in the context-aware web services recommender system. A service user from U must have invoked a service from S at least once.

$L_{U,t} = \{l_{u_i,t}\} (1 \leq i \leq n)$ is the set of $l_{u_i,t}$ that is the temporal location of user u_i.

$L_S = \{l_{s_k}\}$ is the set of l_{s_k}, which is the network location of the service s_k.

$R = \{r_{u_i,s_k}\}$ is the set of rating records on the Web service s_k by the user u_i, where $(1 \leq i \leq n)$ and $(1 \leq k \leq m)$.

$\bar{R} = \{\bar{r}_1, \bar{r}_2, \ldots, \bar{r}_i, \ldots, \bar{r}_n\}$ is the set of mean rating of all Web services invoked by $U = \{u_1, u_2, \ldots, u_n\}$.

When a service s_k is invoked by the user u_j, it will present a set of QoS properties. We will have $Q_{i,j} = <q_1^{i,j}, q_2^{i,j}, \ldots, q_k^{i,j}>$, which is a l-tuple denoting service invocation records of s_k invoked by the user u_j, where $q_l^{k,j} (1 \leq k \leq m, 1 \leq j \leq n)$ denotes the value of l-th property recorded during the invocation of s_k called by u_j.

4.2 Modeling the Influence of User Location Update on User Preference

We know that the user's locations are different as time goes on. As described in Sect. 3, for region-related service (e.g. weather forecast services), the accuracy of recommendation is highly relevant to the specific region in real-time. The impact of regional correlation on user preference is defined as:

$$P_{RCL(t)} = \begin{cases} 1 & \text{if web service is related to regio} \\ 0 & \text{if web service is not related to region} \end{cases}, \tag{1}$$

For region-irrelated services, it's also reasonable for users to have a preference to services near to his/her region because the network distance between users and services (mainly because of transfer delay) have an obvious effect on Internet application performance (e.g. response time and throughput). The network distance's influence on user preference can be defined as:

$$P_{NDL(t)} = P_0 Dis\left(l_{u_i,t}, l_{s_k}\right)_{nor}, \tag{2}$$

Here we assign 1 to it according to our need. $Dis\left(l_{u_i,t}, l_{s_k}\right)$ is the network distance between $l_{u_i,t}$ (the user's network location) and l_{s_k} (the service's network location). Also, network distance measurement technology can help us to get the $Dis\left(l_{u_i,t}, l_{s_k}\right)$, which should be normalized as $Dis\left(l_{u_i,t}, l_{s_k}\right)_{nor}$ to have a same evaluation criterion

In addition, different weights are assigned to the impact of both regional correlation and network distance (w_1 to $P_{RCL(t)}$ and w_2 to $P_{NDL(t)}$). Thus, w_1 and w_2 will be used to represent the influence of user location update on user preference as follows:

$$P_{L(t)} = w_1 P_{RCL(t)} + w_2 P_0 Dis\left(l_{u_i,t}, l_{s_k}\right)_{nor}, \tag{3}$$

Finally, we can have a data filtering results based on $P_{L(t)}$ and get the services which correspond to the current preference of a user.

4.3   Context-Aware Similarity Mining for Updated Location

In this step, it is assumed that for location-based services recommendation, the more similar between the current user and another user's context (e.g. location), the more probability of the two users will have similar QoS on the same Web service. The set of similar users with the current user in terms of location will be got after this step.

We use the Euclidean distance later to describe the similarity between the two users' locations. The nearer the distance is, the more similar they are. The following formula is the presentation of Euclidean distance between $l_{u_i,t}$ and $l_{u_j,t}$:

$$Sim\left(l_{u_i,t}, l_{u_j,t}\right) = \sqrt{\sum_{k=1}^{N} \left(l_{i,t,k} - l_{j,t,k}\right)^2}, \tag{4}$$

Furthermore, we could calculate the distances between current user's location and other users' locations, thus and get the set of users who are the closest with the current user.

4.4 QoS Predication and Services Recommendation

In the final step, we use the Bayesian inference to make QoS prediction and services recommendation based on the past invocation records filtered from the above steps. The formula of Bayesian inference is defined as:

$$P(OS = 1|s_i) = \frac{P(s_i|OS = 1) * P(OS = 1)}{P(s_i)}, \tag{5}$$

In order to explain our formula, we give an example in Table 1. In the example, we suppose that a threshold (e.g. $q = 0.7$) just for the explanation of Bayesian inference (When we have experiment, we will set different values of q to get different results so that we can find the q which lead to the best result). That is to say, if $QoS > 0.7$, we will say the service satisfies the user who invoked it, while if $QoS < 0.7$, we will say the service is not satisfied. We use 1 to donate "satisfied" while 0 to donate "not satisfied".

Table 1. Example of Bayes inference

Record	QoS	OS
$<s_1, u_1, 1>$	0.85	1
$<s_1, u_1, 2>$	0.75	1
$<s_1, u_1, 3>$	0.45	0
$<s_2, u_1, 1>$	0.80	1
$<s_2, u_1, 2>$	0.50	0
$<s_2, u_1, 3>$	0.60	0
$<s_3, u_1 1>$	0.75	1
$<s_3, u_1, 2>$	0.55	0

Table 1 shows an example of service invocation records, where each triple s_i, u_j, n represents a n-th service invocation of s_i by the user u_j. According to formula (6), the $P(OS = 1|s_i)$ donates the prediction QoS of the current user to the Web service s_i, $P(OS = 1)$ donates the probability of the satisfactory ones in all the web service, $P(s_i|OS = 1)$ donates the probability of Web service s_i in the satisfactory ones. The maximum result represents the best service. Thus, we can recommend the *top n* Web services to the current user. The approaches to calculate $P(OS = 1|s_i)$ are:

$$P(OS = 1|s_1) = \frac{P((s_1|OS = 1) * P(OS = 1))}{P(s_1)} = \frac{\frac{1}{2} * \frac{1}{2}}{\frac{3}{8}} = \frac{2}{3}$$

$$P(OS = 1|s_2) = \frac{P((s_2|OS = 1)) * P(OS = 1)}{P(s_2)} = \frac{\frac{1}{4} * \frac{1}{2}}{\frac{3}{8}} = \frac{1}{3}$$

$$P(OS = 1|s_3) = \frac{P((s_3|OS = 1)) * P(OS = 1)}{P(s_3)} = \frac{\frac{1}{4} * \frac{1}{2}}{\frac{2}{8}} = \frac{1}{2}$$

Finally, we can make the QoS prediction for the user s_1 by 2/3. Later, according to the results of each Web service, we could rank the value from the higher to the lower. Hence, we conclude that s_1 would be recommended to the current user compared with s_2 and s_3. The entire procedure of CASR-UPE algorithm is shown as follows.

Algorithm: Context-aware Web Services Recommendation based on User Preference Expansion (CASR-UPE)

```
Input: q, u, t, dataset
    // q is the threshold of the QoS; u is the test user;
    t is the time of recommendation; dataset is the training
```

```
    dataset. //
Output: A group of values of MAE and RMSE with different
    QoS threshold q.
    1.    Start
    2.       When (t ==t₁)
```
3. $P_{L(t_1)} = w_1 P_{RC\ L(t_1)} + w_2 P_0 \mathrm{Dis}(l_{u_i,t_1}, l_{s_k})_{\mathrm{nor}};$
```
    4.          // get the set P_{L(t₁)}of Web services corresponding
    to user current preference determined by L(t₁), the lo-
    cation of the user at time t₁//
    5.             dataset2= filtered(P_{L(t₁)}, dataset);
    6.       //get the filtered dataset2 according to the pref-
    erence set P_{L(t₁)}//
```
7. $U_{L(t_1)} = Sim(u);$
```
    8. //get similar user set U_{L(t₁)} in location //
    9. dataset3=filtered(U_{L(t₁)}, dataset2);
    10.          //get the filtered dataset3 according to the
    similar set U_{L(t₁)} //
    11.             for (different q)
    12.                 preQoS =Beyesian (q, dataset3);
    13.             end for
    14.             mae = MAE (preQoS); rmse=RMSE(preQoS);
    15.       When (t ==t₂)
```
16. $P_{L(t_2)} = w_1 P_{RC\ L(t_2)} + w_2 P_0 \mathrm{Dis}(l_{u_i,t_2}, l_{s_k})_{\mathrm{nor}};$
```
    17.             // get the set P_{L(t₂)} of Web services cor-
    responding to user current preference //
    18.             dataset2= filtered(P_{L(t₂)}, dataset);
    19.             // get the filtered dataset2 according
    to the preference set P_{L(t₂)}//
```
20. $U_{L(t_2)} = Sim(u);$
```
    21.             // get similar user set U_{L(t₂)} in location //
    22.         dataset3=filtered(U_{L(t₂)}, dataset2);
    23.          //get the filtered dataset3 according to the
    similar set U_{L(t₂)} //
    24.             for (different q)
    25.                 preQoS =Beyesian (q, dataset3);
    26.             end for
    27.             mae = MAE (preQoS); rmse=RMSE(preQoS);
    28.       // For different preQoS, we can calculate MAE
    and RMSE values of q. //
    29.             End When
```
---

5 Experiments

In this section, we choose six algorithms to compare with CASR-UPE algorithm on WS-Dream dataset by evaluation metrics of both MAE and RMSE.

5.1 Datasets and Data Processing

WS-Dream [17] dataset 1[5] is adopted in our experiments, which contains 1,542,884 Web services invocation records executed by 150 distributed service users on 100 Web services. Approximately, every user invokes a Web service 100 times. Each invocation record contains 6 parameters: IP address, WSID (ID of web service), RTT (round-trip time), Data Size, Response HTTP Code, and Response HTTP Message. Since Response HTTP Code and Message are highly related, here we omit the property Response HTTP Code.

The raw data must be normalized before use. Gaussian approach is used to normalize QoS data, due to its well-balanced distribution. The normalization rule for Response HTTP Message is as follows: if the message is "OK", the normalized value is 1, otherwise it is 0. The normalization rule for RTT and Data Size is defined as:

$$r_l^{k,j} = 0.5 + \left(r_l^{k,j} - \overline{r_l^j}\right)/(2*3\sigma_j), \tag{6}$$

Where σ_j is the standard deviation of user u_j's QoS data on the l-th property and $\overline{q_k^l}$ denotes the arithmetic mean of QoS data collected from user u_j on the l-th QoS property. Now we can simulate the feedback of a user after invoking a Web service by evaluating the overall QoS of a service. The weight QoS formula can be described as:

$$QoS = w_1 * v_{RTT} + w_2 * v_{DataSize} + w_3 * v_{RHTTPMessage}, \tag{7}$$

Where w_1, w_2 and w_3 are set to 0.35, 0.05 and 0.6 respectively according to their different significance. For example, Response HTTP Message shows that whether the invocation succeeded so it is a fundamental property and can be set 0.6. Thus the properties of RTT and Data Size are not that important and they can be set 0.35 and 0.05 respectively.

All experiments were developed with Matlab. They were performed on a Lenove desktop computer with the following configuration: Intel Core i5 2.50 GHz CPU, 2 GB RAM with the Windows 7 operating system.

5.2 Evaluation Metrics

The evaluation metrics [17] we use in our experiments are Mean Absolute Error (MAE) and Root Mean Square Error (RMSE):

---

[5] WS-Dream dataset, http://www.wsdream.net/dataset.html.

$$MAE = \frac{\sum_{u,s} |Q_{u,s} - \hat{Q}_{u,s}|}{N}, \tag{8}$$

$$RMSE = \frac{\sqrt{\sum_{u,s} (Q_{u,s} - \hat{Q}_{u,s})^2}}{N}, \tag{9}$$

In the formulas (8) and (9), $Q_{u,s}$ denotes actual QoS values of a Web service s observed by the user u, $\hat{Q}_{u,s}$ represents the predicted QoS values of service s for the user u, and N denotes the number of predicted value.

5.3 Evaluation

Comparative Algorithms. Six algorithms are compared with our CASR-UPE in this paper:

- **RBA** (Recommendation by all): recommend web services to a user collected by all users without a filtering.
- **UPCC** [22]: recommend web services collected from the users sharing the similar preference with the current user (PCC based on user profiles).
- **IPCC** [23]: recommend web services similar to the ones the current user preferred in the past (PCC based on services).
- **CASR** [9]: make recommendation based on the service invocation experiences under similar location context with the current user.
- **ITRP-WS** [24]: ITRP-WS considers the time decay effects in UPCC.
- **CASR-UP** [25]: make recommendation considering the user preference determined by user's location.

Performance Comparison. Figure 2 shows the results of MAE and RMSE for different algorithms. The results are generated in different threshold q (from 0.65 to 0.95 and the interval is 0.025) in the ratio 14:1 of training dataset and test dataset. From Fig. 2, we could make the conclusion that: (1) the MAEs and RMSEs of CASR-UPE are much better than other five algorithms when the threshold $0.725 \leq q \leq 0.925$; (2) it is abnormal when the threshold $q = 0.95$; and (3) When $q \leq 0.725$, the MAEs and RMSEs of the algorithms remain almost invariable. We can also see that the best q is 0.775. In Sect. 5.4, we will further explain the reason of both (2) and (3). In general, the results demonstrate that the significant of CASR-UPE algorithm in recommending web services considering the user preference expansion.

Figure 3 shows the average MAE/RMSE results of the six algorithms in different ratios (8:7, 9:6, 10:5, 11:4, 12:3, 13:2, and 14:1). What we could learn from the results are: (1) as the ratio of training and test data increases the MAE and RMSE results of six algorithms decrease; (2) in different ratios, the results of the CASR-UPE algorithm also performs better than the other six algorithms; and (3) CASR-UPE performs worse than CASR-UP in the ratio of 8:7, 9:6 and 10:5 adopting RMSE as the evaluation metric. In Sect. 5.4, we will further explain reasons for those three results above. In general,

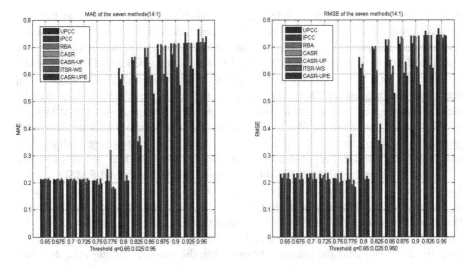

Fig. 2. MAE and RMSE results of compared methods (14:1)

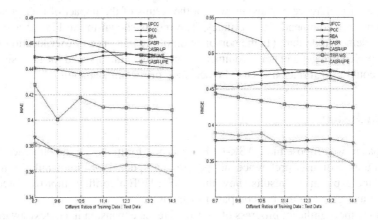

Fig. 3. MAE and RMSE results of compared methods (in various ratios)

the results in different ratios demonstrate that the significant of CASR-UPE algorithm in recommending web services considering the user preference expansion.

5.4 Discussion

In this subsection, we will discuss two aspects in our experiments to further explain the results in 5.3.

Trade-off Parameters: From the results of Fig. 2, we can infer that: (1) The MAEs and RMSEs of CASR-UPE are smaller than other algorithms when the threshold $0.725 \leq q \leq 0.925$, but why $q = 0.95$ is an exception? By analyzing CASR-UPE method, we can find that after selecting some Web services according to the user dynamic preference, the invocation records of these selected services are more useful. As the threshold q rises up to 0.95, most of positive services will be excluded and the results will be in a high and abnormal value. (3) Why is that when $q \leq 0.725$, the MAEs and RMSEs of the algorithms remain almost invariable? When the threshold q decreases, the request of QoS will decrease and many negative services will be included. When q is low enough, all Web services will be included, thus the MAEs and RMSEs remain invariable. (4) We could conclude that the threshold q for the calculated probability is highly relevant to the result. If q is too low, many negative Web services will be included, while if q is too high, many positive Web services will be excluded.

Figure 3 shows the influence of different ratios on the MAEs and RMSEs results. When the ratio of training dataset and test dataset arises, more data is used to train the algorithm and few data is used to test the results. Thus, the accuracy will be better. However, why does CASR-UPE perform worse than CASR-UP in the ratio of 8:7, 9:6 and 10:5 adopting RMSE as the evaluation metrics? The possible reason may come from the randomly changed training dataset when the ratio of training data: test data decreases.

Impact of User Preference Expansion. Comparing with other algorithms not considering user preference expansion, we got the results of the impact of user preference expansion on recommendation accuracy. The results shown in the Figs. 2 and 3 collectively demonstrate that: (1) the combination of the influence of user location update on user preference get a better recommendation; and (2) the updated location similarity mining could also further improve the accuracy of recommendation.

6   Conclusion

In this paper, we propose the CASR-UPE algorithm, for modeling the influence of user location update on user preference and performing updated location similarity mining. Finally, the experiments results show that CASR-UPE algorithm improves predictive accuracy and outperforms the compared methods.

Despite the significant progress of user preference expansion in context-aware Web services recommendation, there still remain numerous avenues to explore. Our future works include: (1) incorporate novel context properties, such as social context (interpersonal interest similarity, interpersonal influence among social network, etc.) to improve more personalized recommendation; and (2) focus on the correlations between context properties, such as temporal-spatial correlations to improve the accuracy of QoS prediction.

Acknowledgment. This work is supported by the grants from Natural Science Foundation of China (No. 61300232); Ministry of Education of China "Chunhui Plan" Cooperation and Research Project (No. Z2012114, Z2014141); Funds of State Key Laboratory for Novel Software Technology, Nanjing University (KFKT2014B09); Fundamental Research Funds for the Central Universities (lzujbky-2015-100); and China Telecom Corp. Gansu Branch Cuiying Funds (lzudxcy-2014-6).

References

1. Truong, H., Dustdar, S.: A survey on context-aware web service systems. Int. J. Web Inf. Syst. **5**(1), 5–31 (2009)
2. Dourish, P.: What we talk about when we talk about context. Pers. Ubiquit. Comput. **8**(1), 19–30 (2003)
3. Jannach, D., Dortmund, T.U., Friedrich, G.: Tutorial: recommender systems. In: International Joint Conference on Artificial Intelligence, Beijing (2013)
4. Staiano, J., Oliver, N., Lepri, B., de Oliveira, R., Caraviello, M.: Money walks: a human-centric study on the economics of personal mobile data. In: 2014 ACM Conference on Ubiquitous Computing, pp. 583–594. ACM, Seattle (2014)
5. Rossi, L., Musolesi, M.: It's the way you check-in: identifying users in location-based social network. In: 2nd ACM Conference on Online Social Networks, pp. 215–226. ACM, Dublin (2014)
6. Lima, A., Musolesi, M.: Spatial dissemination metrics for location-based social networks. In: 2012 ACM Conference on Ubiquitous Computing, pp. 972–979. ACM, Pittsburgh (2012)
7. Papapetrou, P., Roussos, G.: Social context discovery from temporal app use patterns. In: 2014 ACM International Joint Conference on Pervasive and Ubiquitous Computing: Adjunct Publication, pp. 397–402. ACM, Seattle (2014)
8. Lathia, N., Hailes, S., Capra, L., Amatriain, X.: Temporal diversity in recommender systems. In: 33rd international ACM SIGIR Conference on Research and development in information retrieval, pp. 210–217. ACM, Geneva (2010)
9. Xiong, H., Zhang, D., Gauthier, V.: Predicting mobile phone user locations by exploiting collective behavioral patterns. In: IEEE 9th International Conference on Ubiquitous Intelligence and Computing (UIC 2012), pp. 164–171. IEEE Press (2012)
10. Adomavicius, G., Tuzhilin, A.: Context-aware recommender systems. In: Ricci, F., Rokach, L., Shapira, B., Kantor, P.B. (eds.) Recommender Systems Handbook, pp. 217–253. Springer, New York (2011)
11. Guo, B., Chen, C., Zhang, D., Yu, Z., Chin, A.: Mobile crowd sensing and computing: when participatory sensing meets participatory social media. IEEE Commun. Mag. (2015)
12. Yu, Z., Feng, Y., Xu, H., Zhou, X.: Recommending travel packages based on mobile crowdsourced data. IEEE Commun. Mag. **52**(8), 56–62 (2014)
13. Yin, H., Cui, B., Chen, L.: Modeling location-based user rating profiles for personalized recommendation. ACM Trans. Knowl. Discov. Data **38** (2014)
14. Yang, D., Zhang, D., Yu, Z.: Fine-grained preference-aware location search leveraging crowdsourced digital footprints from LBSNs. In: 2013 ACM International Joint Conference on Pervasive and Ubiquitous Computing, pp. 479–488. ACM, Zurich (2013)
15. Yu, Q., Zheng, Z., Wang, H.: Trace norm regularized matrix factorization for service recommendation. In: 2013 IEEE International Conference on Web Services, pp. 34–41. IEEE press, Santa Clara Marriott (2013)
16. Cao, B., Liu, J., Tang, M., Zheng, Z., Wang, G.: Mashup service recommendation based on user interest and social network. In: IEEE International Conference on Web Services, pp. 99–106. IEEE Press, Santa Clara Marriott (2013)
17. Zheng, Z., Ma, H., Lyu, M.R., King, I.: QoS-aware web service recommendation by collaborative filtering. IEEE Trans. Serv. Comput. **4**(2), 140–152 (2010)
18. McCarthy, J.: Notes on formalizing context. In: International Joint Conference on Artificial Intelligence, pp. 555–560 (1993)

19. Brézillon, P.: Task-realization models in contextual graphs. In: Dey, A.K., Kokinov, B., Leake, D.B., Turner, R. (eds.) CONTEXT 2005. LNCS (LNAI), vol. 3554, pp. 55–68. Springer, Heidelberg (2005)

20. Gao, H., Tang, J., Hu, X., Liu, H.: Modeling temporal effects of human mobile behavior on location-based social networks. In: ACM International Conference on Information and Knowledge Management, pp. 1673–1678. ACM, Maui (2013)

21. Hu, B., Jamali, M., Ester, M.: Spatio-temporal topic modeling in mobile social media for location recommendation. In: IEEE International Conference on Data Mining, pp. 1073–1078. IEEE press (2013)

22. Breese, J., Heckerman, D., Kadie, C.: Empirical Analysis of Predictive Algorithms For Collaborative Filtering. Morgan Kaufmann Publishers Inc., San Francisco (1998)

23. Resnick, P., Iacovou, N., Suchak, M., Bergstrom, P., Riedl, J.: An open architecture for collaborative filtering of netnews. In: ACM Conference on Computer Supported Cooperative Work, pp. 175–186 (1994)

24. Cui, X., Yin, G., Han, Q., Dong, Y.: An improved time-effectiveness reliability prediction approach of web service. J. Comput. Inf. Syst. **10**(4), 1365–1374 (2014)

25. Fan, X., Hu, Y., Zhang, R.: Context-aware web services recommendation based on user preference. In: IEEE Asia-Pacific Services Computing Conference, pp. 55–61. IEEE press, Fuzhou (2014)

CPFirewall: A Novel Parallel Firewall Scheme for FWaaS in the Cloud Environment

Zhenfang Wang[✉], ZhiHui Lu, Jie Wu, and Kang Fan

School of Computer Science, Fudan University, Shanghai, China
{13210240114,1zh,jwu,111210240048}@fudan.edu.cn

Abstract. In cloud, resources are virtualized and the software delivery way is becoming something like a "service" to provide end user and operator benefits including on-demand self-service, resource pooling, rapid elasticity and service metering capability. As a part of network function virtualization, firewall virtualization can greatly increase the firewall configuration flexibility for the cloud environment. In this paper, we focus on FWaaS (Firewall as a Service) and we design a parallel firewall system called CPFirewall (Cloud Parallel Firewall System). In CPFirewall, the firewall resources are virtualized and multiple tenants can build up their own parallel firewall by renting virtual firewalls. This needs solve some challenges. We adopt a rule-splitting algorithm to build a rule anomaly set (We call it Wrapset.) for detecting rule anomaly. We design the rule-allocation algorithm to achieve the cloud-native features, including load balance and dynamic scale. And we also improve the system performance using Exponential Smoothing (ES) forecasting method. Experiment results have verified that CPFirewall has a higher efficiency than other firewall schemes and is much more suitable for the Cloud network environment.

Keywords: Cloud computing · FWaaS · Parallel firewall · NFV

1 Introduction

With the development and maturation of virtualization technology, the cloud computing trend is unstoppable. In the Cloud, all kind of physical resources are virtualized. A tenant can rent the resources in the virtual resource pool according to their requirements. As one of the most important parts of Network Function Virtualization in the Cloud, firewalls should be offered to the tenant flexibly, dynamically and efficiently to satisfy the users' requirements. This is also what the Cloud Computing concept stresses. There are four kinds of traditional firewalls, including packet filtering firewall, application firewall, proxy firewall and stateful firewall. However, traditional firewalls have no cloud computing features, as they do not conform to the flexibility requirement of the cloud environment. Therefore, we present a novel parallel firewall scheme to realize FWaaS (Firewall as a Service) in the cloud environment named CPFirewall to make the virtual firewall better apply to the cloud computing environment. In CPFirewall, firewall resources are virtualized and presented it to the tenant as a service (FWaaS, Firewall as a Service). Many tenants can build up their own parallel firewalls through renting virtual

© Springer International Publishing Switzerland 2015
L. Yao et al. (Eds.): APSCC 2015, LNCS 9464, pp. 121–136, 2015.
DOI: 10.1007/978-3-319-26979-5_9

firewalls in the virtual resource pool, which bring some new problems, such as firewall rules conflict, load imbalance, static scale. In CPFirewall, we solve these problems through anomaly detection and rule allocation methods. Therefore, this novel parallel system can be better applied to the Cloud environment.

The key contribution of this paper is to design the CPFirewall system which includes three key mechanisms:

1.1 Anomaly Detection

We detected and analyzed the firewall policies of many institutions, and found that almost all institutions' firewall strategies have different level conflict rules [1]. A loophole in the firewall policy may lead some unwanted packets to pass through the firewall, resulting in system security problems. In this part, we give an anomaly detection solution for the CPFirewall. This solution includes two steps. Firstly, we need to split the firewall rules into segments. Then, we build the rule-segment table and the Wrapset according to the result of last step. This part can figure out the anomalies among the rules to build a strong firewall system. At the same time, the second module will use the Wrapset built in this module to realize load balance.

1.2 Load Balance

To improve the efficiency of the system and apply it to the cloud environment, we realize load balance through redistributing the rules in the parallel system. In this module, we mainly use the Exponential Smoothing (ES) forecasting method to distribute rules into different single firewalls in the system to realize load balancing and rearrange the rules in the single firewall to improve its efficiency.

1.3 Dynamic Scale

The distinction of the cloud is to virtualize the resources to form a virtual resource pool, thus, tenants can apply for the resources on demand [2]. In this part, we present how to scale our designed firewall dynamically according to the workload changes over time in CPFirewall.

The remainder of this paper is organized as follows: In Sect. 2, we discuss related work on firewall techniques. In Sect. 3, we describe our cloud firewall system architecture-CPFirewall. Then we present CPFirewall implementation scheme in detail in Sect. 4. In Sect. 5, we carry out experiments and related analysis. In Sect. 6, we make a conclusion of this paper and give a prospect of future work.

2   Related Work

There have been some research works in the firewall filed. In these works, some are about traditional firewalls, some refer to cloud firewall techniques.

In [3], the authors found that a large number of data packets are filtered by a small part of firewall rules and they presented an algorithm to figure out the most important N rules to reduce the number of rules in the firewall. But these N rules can only ensure to filter packets as many as possible and there are still a lot of packets which cannot be filtered. Al–Shaer and other authors did a lot of research work in the filed of firewall policy conflict detection. In [4], the author proposed five kinds of relationships of the firewall rules: Exactly Matched, Inclusively Matched, Completely Disjoint, Partially Disjoint, and Correlated. Based on the five kinds of relationships, the author draws out the relationship state diagram between two firewall rules and then puts forward the rule anomaly detection method based on state diagram. Al-Shaer expanded the results to the distributed firewall again in [5] and studied the strategy conflict detection among the firewalls. However, the results are based on the traditional firewalls and cannot be directly used in the cloud environment. In [6] the author put forward a parallel firewall model. This model is based on packet classification. It can allocate the firewall rules to different firewalls according to a certain classification method. For example, if we classify packets according to the protocol type, the data packets will get through corresponding firewalls according to their own protocol type (e.g. TCP or UDP). This design can greatly improves the efficiency of the firewall. But the author just presents simple classifying methods which do not consider the characteristics of the data traffic. At the same time this kind of system cannot extend dynamically. In [7], the author proposed that unwanted packets should be filtered out as early as possible to improve the firewall efficiency. In this case the author extracted the "Deny" firewall rules and called the rule set Rejection. Rejection rules can filter the packets as early as possible, and thus can reduce the duration time of the packet within the firewalls. Rejection can be adjusted dynamically according to the data traffic to achieve better effect. In this model, if there are a lot of unwanted data packets, there will be a big improvement in the firewall performance. But if there are few such data packets, there will be no big improvement in performance. Moreover, the paper also didn't consider the packets which cannot match all the rules. Although the author did not use first matching method, the thought that we should filter unwanted packets as early as possible is worth using for reference. In [8], the authors investigated several kinds of firewall rule conflict detection methods. Then they presented that "irrelevant anomaly" was still a research difficulty. This is because a firewall would add or reduce rules according to the dynamic changes of the data packet. But for this kind of conflict, we can achieve a better effect through rearranging firewall rules dynamically.

For firewalls in the cloud environment, the author put forward the problems which were caused by moving the traditional service from the local server to the cloud in [9]. The paper first discussed the various restrictions which should be obeyed when administrators migrate services to the cloud, and then discussed the firewall rule migration during the service migration process. In this paper, the original services are in the same local area network. After the migration, they are in different LANs. If administrators want to allow the services in different LANs to communicate with each other, the local and the cloud firewall strategies will need to be updated to get the data packets through the firewalls. However, the firewall mentioned in the paper rarely has characteristics of the cloud. In [10], the author proposed that in order to reduce the cost of the firewall, the firewalls could be deployed through the Internet service provider to use their filtering service. This is similar to the thought of "service" in cloud computing, but this kind of firewall cannot be applied to the cloud environment. The author also thought the firewall can be deployed on the data center in [11]. In order to isolate the virtual machine communication, the cloud manager should ensure that data packets have to pass through the firewall between the virtual machines. As a result, a lot of packets need to be routed. Although this scheme deployed the firewall as a device, it did not reflect the characteristic of virtualization in the cloud environment. In [12], the author presented a kind of firewall applied to the cloud environment. The author expressed that administrators should package the firewall into multiple modules and each module has its specific function. So, tenants can build their own firewalls through combining different modules. This kind of firewall is linear in essence and it can't scale dynamically.

In our scheme, we consider the factors of anomaly detection, load balance, and dynamic scale to design an efficient parallel firewall system for the cloud environment.

3 System Architecture

As a firewall system for the cloud environment, the system should be scalable, load-balanced, flexible and efficient. Here, we present CPFirewall: a novel parallel firewall scheme to realize Firewall as a Service (FWaaS) in the cloud environment. In CPFirewall, we achieve the above features using the four components as shown in Fig. 1. In this figure, R denotes the single firewall of the parallel system, and r denotes the filtering rule in the single firewall.

- Anomaly detection. As described in [4], a firewall controls the traversal of packets across the boundaries of a secured network based on a specific security policy which is a list of ordered filtering rules that define the actions performed on matching packets. A rule is composed of filtering fileds such as protocol type, source IP address, destination IP address, source port and destination port, and a filter action filed. Filtering actions are either to accept, which passes the packet into or from the secure network, or to deny, which causes the packet to be discarded. The packet is accepted or denied by a specific rule if the packet header information matches all the network fileds of this rule. Otherwise, the next following rule is used to test the matching with this packet again. Similarly, this process is repeated until a matching rule is found or the default policy action is performed. A firewall policy anomaly is defined as the existence of two or more different filtering rules that match the same packet. In the cloud, the network gets larger and larger and the firewall rules increase in number. Tenants want to build their own parallel firewalls conveniently and should not find rule anomalies in these firewalls. It is the complex interdependence among the rules that makes firewall policy management a very challenging job. For this reason, anomaly detection is crucial for CPFirewall. This component will analyze the rules through splitting rules to detect anomalies. We will build the Wrapset according this step to help rule allocation component to realize load balance.
- Rule allocation. In this component, we count the number of packets passing through the firewall system according to their certain filtering filed value. Then we averagely

divide the packets into t (parallel level) groups according to the statistic and distribute them into t single firewalls. After finishing rule allocation, we will adjust the rule order in single firewall to improve the efficiency of our system. We also use this module to realize virtual firewall resource scaling dynamically through reallocating the rules according to the new system workload.

- Classifier. The classifier will send the packets to the corresponding single firewall according to certain filtering filed. This is different from the classifier of the model in [6], because it considers the characteristics of the data traffic according to the result of the rule allocation mechanism. Therefore, it can process t (parallel level) packets at the same time. This can greatly improve the efficiency.
- Monitor. This component mainly monitors the workload status of the system. Therefore, we can make adjustments according to the statistic to realize virtual firewall resource scaling dynamically.

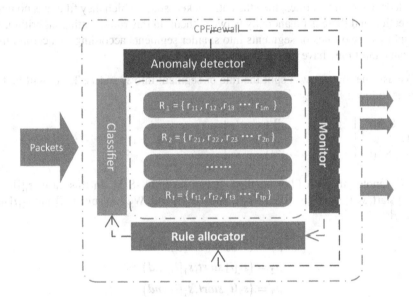

Fig. 1. CPFirewall architecture

According to this design, we implement CPFirewall with Python.

4 CPFirewall Implementation

As we can see in Fig. 1, CPFirewall has the four key components. In the following, we introduce these modules in detail.

4.1 Anomaly Detector

Generally, there are four kinds of rule anomalies [4]. (1) Shadowing anomaly: A rule is shadowed when a previous rule matches all the packets that match this rule, such that the shadowed rule will never be evaluated. If the shadowed rule is removed, the security policy will not be affected. (2) Correlation anomaly: Two rules are correlated if the first rule in order matches some packets that match the second rule and the second rule matches some packets that match the first rule. Rule Rx and rule Ry have a correlation anomaly if Rx and Ry are correlated and the actions of Rx and Ry are different. (3) Redundancy anomaly: A redundant rule performs the same action on the same packets as another rule such that if the redundant rule is removed, the security policy will not be affected. (4) Generalization anomaly: A rule is a generalization of another rule if the first rule matches all the packets that the second one could match but not the opposite. In CPFirewall, we adopt a rule-splitting algorithm to detect these anomalies and use its result to build a Wrapset to realize rule allocation.

(1) *Rule-splitting.*

We can easily prove that two firewall rules have overlapping data packet space if and only if all of their filtering fileds have intersection. If there are disjoint filtering fileds between two rules, then the data packets space which they filter has no intersection and there is no anomaly between them. In this rule-splitting algorithm, we split any two rules or segments into smaller segments according to a certain filed value until they have no intersection on this filed.

We assume that we will split the rule or segment into s_1 and s_2. There will be five cases as follows:

(1) $S_1[i]$ P   P

        $S_2[i]$ P   P

S[i] denotes the i-th filtering filed. It can be S 1] -S[5]. In this case, $s_1[i]$.start $< s_2[i]$.start, $s_2[i]$.start $\leq s_1[i]$.end, $s_1[i]$.end $< s_2[i]$.end. We can spilt $s_1[i]$ and $s_2[i]$ into four parts:

$$p_1 = \left(s_1[i].start, s_2[i].start\right)$$
$$p_2 = \left(s_2[i].start, s_1[i].end\right)$$
$$p_3 = \left(s_2[i].start, s_1[i].end\right)$$
$$p_4 = \left(s_1[i].start, s_2[i].end\right)$$

This produces four segments:

$$c_1 = \left(s_1[1], \ldots s_1[i-1], p_1, s_1[i+1], \ldots\right)$$
$$c_2 = \left(s_1[1], \ldots s_1[i-1], p_2, s_1[i+1], \ldots\right)$$
$$c_3 = \left(s_2[1], \ldots s_2[i-1], p_3, s_2[i+1], \ldots\right)$$
$$c_4 = \left(s_2[1], \ldots s_2[i-1], p_4, s_2[i+1], \ldots\right)$$

Obviously, s_1 is split into two sub-segments c_1 and c_2, and s_2 is split into two sub-segments c_3 and c_4. We can split following four cases as follows:

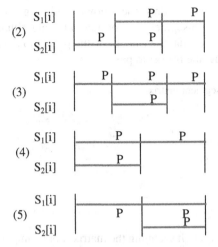

The effect of the algorithm is to split the rules into disjoint sub segments. The detailed mechanism of this algorithm is shown in Algorithm 1:

Algorithm 1: Rule splitting
Input Rule set-R
Output Segment set-S
1: for each rule in R do
2:    for each segment in S do
3:        if rule and segment are intersected but not equal then
4:        (seg_{rule}, $seg_{segment}$) = Devide(rule, segment)
5:            R.append(seg_{rule})
6:            S.append($seg_{segment}$)
7:            S.remove(segment)
8:            break
9:        S.append(rule)
10: return S

According to the rule splitting algorithm, any firewall rule can be represented by a union set of some sub segment sets. We will use a two-dimensional matrix to express the relationship between the rules and sub segments. In the matrix, the ordinate axis stands for firewall rule and horizontal axis stands for sub-segment. For example, we have the following rules which are listed in Table 1.

Table 1. Rules example

id	protocol	src_ip	src_port	dst_ip	dst_port	action
0	TCP	10.10.82.*	any	192.168.75.7	80	accept
1	TCP	10.10.82.110	80	192.168.75.*	any	deny
2	TCP	10.10.84.*	any	192.168.75.7	80	accept
3	TCP	10.10.82.110	80	192.168.75.*	any	deny
4	TCP	10.10.82.110	any	192.168.75.*	any	deny

After we split the rule into sub segments, we can draw the following rule-segment matrix table. The five rules are split into ten sub-segments according to the algorithm as shown in Table 2. In the matrix, "A" denotes that the rule allow the packet to pass while "N" denotes that the rule does not allow the packet to pass.

Table 2. Rule-segment matrix

	s_0	s_1	s_2	s_3	s_4	s_5	s_6	s_7	s_8	s_9	s_{10}
r_0		A		A	A		A				A
r_1	N		N						N	N	N
r_2						A					
r_3	N		N						N	N	N
r_4								N			

We can find the four kinds of anomalies through scanning the matrix. For example, r_1 and r_3 have the same segment set and all the actions of the corresponding segments are the same. So it is a redundancy anomaly. We can also see that r_0 and r_1 can match some same packets. As they include segment s_{10} at the same time. But the action is different for these packets which s_{10} can matches. Because r_0 is in the front of r_1, so this is a type of shadowing anomaly. Similarly, we can figure out the other two anomalies through scanning this matrix. When we find the anomalies, we will remind user to deal with these anomalies.

(2) *Building Wrapset*

For a redundancy anomaly, we can remove one of the rules. This does not influence the correctness of the system. However, for other anomalies, the ordering of filtering rules in a security policy is very crucial in determining the firewall policy because the firewall packet filtering process is performed by sequentially matching the packet against filtering rules until a match is found. If filtering rules are independent, the ordering of the rules is insignificant. But when the rules have the same sub-segment, if the relative rule ordering is not carefully assigned, some rules may always be screened by other rules producing an incorrect security policy and action. To solve this problem, we build the Wrapset. We put the rules which have at least one identical segment into this set according to the rule-segment matrix. The user can then only consider the rule ordering in the Wrapset. Also, when we allocate the rules in the Rule allocation module, we must take the Wrapset as one object to keep the logic order of these rules in them.

Now, we will analyze the algorithm which builds the Wrapsets. For each sub segment, if there are multiple rules corresponding with it, then, the algorithm will put these rules into a Wrapset. If some rule has existed in other Wrapsets, the algorithm will put these Wrapset into the new Wrapsets. The algorithm is shown in Algorithm 2:

Algorithm 2: Building Wrapset

Input Rule-segment matrix – RS
Output Wrapset
1: for each segment in segments do
2:   for each rule in RS[*][segment] do
3:   if rule is not visited then
4:   tempest.add(rule)
5:   else
6:   for each wrapset in wrapsets do
7:   if rule is in wrapset
8:   tempest = tempest U wrapset
9:   wrapsets.remove(wrapset)
10:   wrapsets.append(tempest)
11: return wrapsets

4.2 Rule Allocator

In this module, we introduce how to realize load balance and dynamic scale.

(1) Load Balance
    In this section, we first introduce the Exponential smoothing (ES). And then we
    will use ES to allocate and rearrange rules.

- Exponential smoothing
  Exponential smoothing (ES) is presented by Robert G.Brown in [13]. Exponential
  smoothing is a technique that can be applied to time series data, either to produce
  smoothed data for presentation, or to make forecasts. The time series data themselves
  are a sequence of observations. The observed phenomenon may be an essentially random
  process, or it may be an orderly, but noisy, process. Whereas in the same forecasting
  method the past observations are weighted equally, exponential smoothing assigns
  exponentially decreasing weights over time. Exponential smoothing is commonly
  applied to financial market and economic data, but it can be used with any discrete set
  of repeated measurements. The raw data sequence is often represented by $\{x_t\}$ beginning at time t = 0, and the output of the exponential smoothing algorithm is commonly
  written as $\{s_t\}$, which may be regarded as a best estimate of what the next value of x
  will be. When the sequence of observations begins at time t = 0, the simplest form of
  exponential smoothing is given by the formulae:

$$s_0 = x_0$$
$$s_t = \alpha x_{t-1} + (1 - \alpha)s_{t-1}, t > 0 \tag{1}$$

  Where α is the smoothing factor, and $0 < \alpha < 1$.

- Rule allocation
  In this section, we sort the rules according to a certain filed value. The rules in the
  same Wrapset should be taken as one rule object to keep the logic order when the

rules are reallocated in the following steps. Then we divide these rule objects into t (the number of single firewalls in parallel system) groups according to their average number of matching packets as shown in Fig. 2:

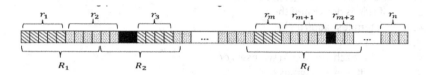

Fig. 2. Rule allocation

In this figure, every unit represents a value of this filed, r means rule and R means the rule group. In every group, the filed values of these rules are in the same range and every group almost matches the same number of packets. At last we allocate the t rule groups to t single firewalls. Then the classifier will send the packets to different single firewalls by judging which group it belongs to according to the filed value.

If the filed value of the packet is any value (expressed by '*'), the classifier will make t copies and send it to all the single firewalls. The monitor will decide the action according to the t results.

In order to reduce noise and improve correctness, we calculate the ES values for every packet number matching the filed and take it as the basis to divide group.

- Rearranging rules
  In a single firewall, the firewall packet filtering process is performed by sequentially matching the packet against filtering rules until a match is found. Therefore, if we put the rule which matches the most packets in the front of the rule queue, the efficiency of this firewall will be improved greatly. Thus, we record the number of every rule matching the packets during a period and calculate the ES values. Then we sort the rules by number in descending order. Certainly, Wrapset is also taken as one rule object. The rule allocation algorithm is shown in Algorithm 3.

Algorithm 3: Rule allocation

    Input Rule set-R, Statistic array-PacketList, Parallel level-T
    Output Rule allocation result
    1: ranges = devide PacketList into T parts
    2: for each range in ranges do
    3:    for each rule in R do
    4:      if rule's field is in range then
    5:        R_{tempt}.append(rule)
    6:    firewalls.append(R_{temp})
    7: for firewall in firewalls do
    8:    rearrange rules in firewall by PacketSort
    9: return firewalls

---

(2) Scaling Dynamically

The monitor module will record the status of the system. If the system efficiency gets low, CPFirewall will recalculate the ES values for the filed value and evaluate the load to decide the parallel level and redistribute the rules according to the rule allocation algorithm to change the number of the firewalls in the parallel system.

To ensure the correctness of the system during the reallocation time, firstly, we can add the new rules to the end of the single firewall. Then we update the classifier to distribute the packets according to the new allocation result. At last, we delete the old rules one by one. This algorithm is shown in Algorithm 4.

Algorithm 4: Scaling dynamically

Input Rule set-R, Statistic array-PacketList, Parallel level-T
Output Redistribution result
1: recalculate the new policies
2: for each newpolicy in newpolicies do
3:     get a firewall without newpolicy
4:       for each rule in newpolicy do
5:         firewall.append(rule)
6:         R_{temp}.append(rule)
7: update the packets distributor
8: for firewall in firewalls do
9:    for rule in firewalls do
10:      if rule is in oldpolicy
11:        firewall.remove(rule)
12: return firewalls

4.3   Classifier

As we can see from Section B, the classifier will distribute the packets only according to one of their filed values. The decision-making tree has only one layer as shown in Fig. 3, and just needs to compare once for each packet while the multi-layer classifier need to compare multiple times. Therefore, this can reduce the load of classifier greatly and improve the performance of the whole system significantly.

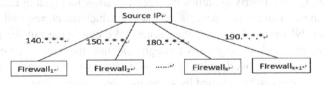

Fig. 3. Classifier mechanism

On the other hand, if the filed value of some packet is any value (denoted by '*'), it can make t copies and sent it to all the single firewalls to ensure the system can filter all the packets.

4.4 Monitor

In this component, the monitor can record the packet quantity information. Users can use this information to decide how to scale the system. At the same time, as described in Rule allocation section, if the packet's filed value is any value, the monitor can collect the result of all the single firewalls to decide the action for the packet.

As we can see, compared with traditional parallel firewall systems, CPFirewall simplifies the design of the classifier, and realize the features of load balance and dynamic scale. It not only improves the system performance but can also be applied to a cloud environment.

5 Experiment and Analysis

In this section, we test the CPFirewall system from two aspects. Firstly, we deploy a simple system to verify the system feasibility. In the experiment environment, we deploy the virtual network on the Openstack platform. The virtual router connects the Internet through a switch. We will use the virtual firewall to control the virtual network to access the external network.

Secondly, we test the system performance through simulation experiments.

5.1 Rule-Splitting Algorithm Experiment

Rule-splitting Algorithm correctness validation

In the anomaly detection section, we introduced the rule-splitting algorithm and building Wrapset. To verify the correctness, we list several rules as shown in Fig. 3. We get 11 segments through splitting the rules. We show the relationship between the rules and the segments in Fig. 10. The rule includes the segments which are below it. For example, r_1 includes these segments of s_1, s_3, s_4, s_6, s_{10}. We can find that every rule includes several segments from Fig. 4. These segments are disjoint with each other. We build the rule-segment matrix as shown in Table 1 to show this kind of relationship. In Table 1, 'A' means that the rule accepts the packets which the corresponding segment can filter, and 'N' means they deny the packets. We can figure out the relationship between every rule and the segments through scanning this table. For example, on segment s_{10}, r_0, r_1, r_3 have overlapping space. So, there exists an anomaly among them. Other anomalies also can be detected by scanning rule-segment table.

Then, we build the Wrapset (collide set) according to the RS table:

```
--------collide set---------
[{2}, {4}, {0, 1, 3}]
```

This shows that r_2 and r_4 have no overlapping filtering space with other segments, but r_0, r_1 and r_3 have overlapping filtering space.

```
-----------rule---------
TCP    10.10.82.0 - 10.10.82.255     0 - 65535     192.168.75.7 - 192.168.75.7    80 - 80
--segments--
s1    TCP    10.10.82.0 - 10.10.82.109     0 - 65535     192.168.75.7 - 192.168.75.7    80 - 80
s3    TCP    10.10.82.110 - 10.10.82.110   81 - 65535    192.168.75.7 - 192.168.75.7    80 - 80
s4    TCP    10.10.82.111 - 10.10.82.255   0 - 65535     192.168.75.7 - 192.168.75.7    80 - 80
s6    TCP    10.10.82.110 - 10.10.82.110   0 - 79        192.168.75.7 - 192.168.75.7    80 - 80
s10   TCP    10.10.82.110 - 10.10.82.110   80 - 80       192.168.75.7 - 192.168.75.7    80 - 80
-----------rule---------
TCP    10.10.82.110 - 10.10.82.110   80 - 80      192.168.75.0 - 192.168.75.255   0 - 25536
--segments--
s0    TCP    10.10.82.110 - 10.10.82.110   80 - 80      192.168.75.8 - 192.168.75.255   0 - 25536
s2    TCP    10.10.82.110 - 10.10.82.110   80 - 80      192.168.75.7 - 192.168.75.7     0 - 79
s8    TCP    10.10.82.110 - 10.10.82.110   80 - 80      192.168.75.0 - 192.168.75.6     0 - 25536
s9    TCP    10.10.82.110 - 10.10.82.110   80 - 80      192.168.75.7 - 192.168.75.7     81 - 25536
s10   TCP    10.10.82.110 - 10.10.82.110   80 - 80      192.168.75.7 - 192.168.75.7     80 - 80
-----------rule---------
TCP    10.10.84.0 - 10.10.84.255    0 - 65535    192.168.75.7 - 192.168.75.7    80 - 80
--segments--
s5    TCP    10.10.84.0 - 10.10.84.255    0 - 65535    192.168.75.7 - 192.168.75.7    80 - 80
-----------rule---------
TCP    10.10.82.110 - 10.10.82.110   80 - 80     192.168.75.0 - 192.168.75.255   0 - 25536
--segments--
s0    TCP    10.10.82.110 - 10.10.82.110   80 - 80     192.168.75.8 - 192.168.75.255   0 - 25536
s2    TCP    10.10.82.110 - 10.10.82.110   80 - 80     192.168.75.7 - 192.168.75.7     0 - 79
s8    TCP    10.10.82.110 - 10.10.82.110   80 - 80     192.168.75.0 - 192.168.75.6     0 - 25536
s9    TCP    10.10.82.110 - 10.10.82.110   80 - 80     192.168.75.7 - 192.168.75.7     81 - 25536
s10   TCP    10.10.82.110 - 10.10.82.110   80 - 80     192.168.75.7 - 192.168.75.7     80 - 80
-----------rule---------
TCP    10.10.83.110 - 10.10.83.110   0 - 65535   192.168.75.0 - 192.168.75.255   0 - 25536
--segments--
s7    TCP    10.10.83.110 - 10.10.83.110   0 - 65535   192.168.75.0 - 192.168.75.255   0 - 25536
```

Fig. 4. Rules and corresponding segments

Rule-splitting Algorithm performance analysis

In this part, firstly, we test the efficiency of this algorithm using different numbers of rules which are complex and have a lot of anomalies among them. We find that in this circumstance, the time needed increases exponentially with the rule number when the rule number exceeds 70. This is because the algorithm will split the rules which have overlapping filtering space with others into small segments. On this basis, to ensure these segments are disjoint with other segments, we need to split them further. In reality, there are almost no so many anomalies among the rules, so this is a limit test. Then we test it in ordinary circumstances and find that the processing time is basically proportional to the rule number. The experiment results are shown below (Figs. 5 and 6).

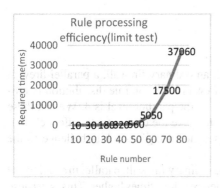

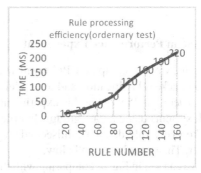

Fig. 5. Rule processing efficiency limit test   **Fig. 6.** Rule processing efficiency ordinary test

5.2   Rule Allocation Experiments

Validating the ES for the system stability

In this experiment, we test the ES's stabilization role for the system. We record the matching packet number of a rule every minute for half an hour. We can see the original data record fluctuates sharply. Its standard deviation is 30.6. We calculate the results using ES respectively when $\alpha = 0.1, 0.3, 0.5$, and the corresponding standard deviations are 17.7, 20.8, 23.6. So, when $\alpha = 0.1$, the data has the lowest fluctuation. Using ES can ensure the stability of the system to avoid rearranging rules in single firewalls frequently. The experiment result is shown in Fig. 9.

Validating the load balance feature

In this experiment, we get the load status of 20000 packets during a period in our system. We define parallel level as T. In this experiment, T takes five values: 4, 6, 8, 10, 12. And every single firewall's load is shown in Fig. 10 (Figs. 7 and 8).

As we can see from the result, no matter what the parallel level is, every single firewall's load in our parallel system almost are equal.

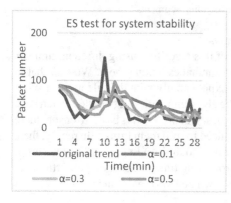

Fig. 7. ES test result   **Fig. 8.** Load balance test

5.3   System Performance Experiment

In this part, we will compare CPFirewall with an ordinary firewall, a parallel firewall based on packet classification and a parallel firewall based on rule distribution in efficiency. There are 25 rules in this experiment and the parallel level is 4. We test these firewall systems using 5 data groups. In order to make the experimental effect clearer, we make every rule delay a millisecond. If so, this also gives a bit challenge to the classifier. The result is shown below.

As we can see in Fig. 9, comparing with the ordinary firewall, parallel firewall based on packet classification's (Parallel FW 1) efficiency is four times higher. This is because in the parallel firewall based on packet classification, the number of packets every firewall needs to process is 1/4 of the ordinary firewall. For the parallel firewall system based on rule distribution (Parallel FW 2), every single firewall has about six rules.

According to probability theory, the average number of matching rules is 3. This is a great improvement over 12 in an ordinary firewall. We find that it saves 45 s compared to an ordinary firewall in processing 5000 packets. CPFirewall has combined the advantages of these two systems and has a great improvement in efficiency. In our simulation system, it just spends 16.3 s in processing 5000 data packets (every rule has 1 ms delay), while the other two parallel systems will spend 28.9 s and 63.6 s.

Next, we test the compare times between the packets and the rules when the packets pass through the firewall system. In the 5 experiments, the average compare times of the ordinary firewall and the parallel firewall system based on packet classification are almost equal, nearly 12 times. In the parallel firewall system based on rule distribution, because it needs to make copies of the packets, the average compare times is nearly 10. However, in CPFirewall, because rule redundancy is small in a single firewall, the average comparison times is near 2.4; much lower than other schemes.

In these experiments, we have verified the correctness of the algorithms and tested the performance of related mechanisms. Based on the integration of the upwards test results, we can draw a conclusion that CPFirewall is applicable to the cloud computing environment.

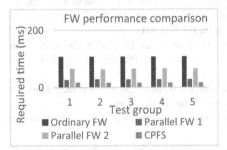

Fig. 9. FW performance test

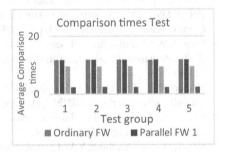

Fig. 10. Rule comparison times test

6 Conclusion and Future Work

In this paper, we design the CPFirewall, a novel parallel firewall scheme for FWaaS in the Cloud Environment, which matches the demands of the scalability and efficiency of the cloud environment. In order to realize these features, we present anomaly detection, load balance and dynamic scale mechanisms. At the same time, we also improve the performance and stability of the system through some algorithm. At last, we evaluate the flexibility and the performance of the system in experiments.

In the future, we will continue to improve our system. This includes strengthening the system monitoring ability to realize auto scale, optimizing algorithm to further improve system performance, adding more components to form a complete firewall system, solving the secure problem when migrating VMs. In summary, we believe Firewall as a Service is a key step to push the further development of network function virtualization in the near future.

Acknowledgment. This paper work is based on the Fudan-Hitachi Innovative Software Technology Joint Laboratory project-cloud virtualized resource management system. This work is also supported by 2014–2016 PuJiang Program of Shanghai under Grant No. 14PJ1431100 and 2015–2017 Shanghai Science and Technology Innovation Action Plan Project under Grant No. 15511107000. We would like to give our sincere thanks to them for all the support and advice.

References

1. Acharya, H.B., Gouda, M.G.: Firewall verification and redundancy checking are equivalent. In: INFOCOM, 2011 Proceedings IEEE, pp. 2123–2128. IEEE (2011)
2. Liu, C., Mao, Y., Van der Merwe, J., et al.: Cloud resource orchestration: s data-centric approach. In: Proceedings of the Biennial Conference on Innovative Data Systems Research (CIDR), pp. 1–8 (2011)
3. Lam, H.Y., Wang, D., Chao, H.J.: A traffic-aware top-n firewall approximation algorithm. In: 2011 IEEE Conference on Computer Communications Workshops (INFOCOM WKSHPS), pp. 1036–1041. IEEE (2011)
4. Al-Shaer, E., Hamed, H.: Design and implementation of firewall policy advisor tools. DePaul University, CTI, Technical Report (2002)
5. Al-Shaer, E.S., Hamed, H.H.: Discovery of policy anomalies in distributed firewalls. In: INFOCOM 2004, Twenty-third Annual Joint Conference of the IEEE Computer and Communications Societies, vol. 4, pp. 2605–2616. IEEE (2004)
6. Fulp, E.W.: Parallel firewall designs for high-speed networks. In: INFOCOM 2006, 25th IEEE International Conference on Computer Communications, Proceedings, pp. 1–4. IEEE (2006)
7. Hamed, H.H., El-Atawy, A., Al-Shaer, E.: Adaptive statistical optimization techniques for firewall packet filtering. In: INFOCOM 2006, vol. 6, pp. 1–12 (2006)
8. Chaure, R., Shandilya, S.K.: Firewall anamolies detection and removal techniques – a survey. Int. J. Emerg. Technol. **1**(1), 71–74 (2010)
9. Hajjat, M., Sun, X., Sung, Y.W.E., et al.: Cloudward bound: planning for beneficial migration of enterprise applications to the cloud. ACM SIGCOMM Comput. Commun. Rev. **40**(4), 243–254 (2010)
10. Khakpour, A.R., Liu, A.X.: First step toward cloud-based firewalling. In: 2012 IEEE 31st Symposium on Reliable Distributed Systems (SRDS), pp. 41–50. IEEE (2012)
11. Lee, S., Purohit, M., Saha, B.: Firewall placement in cloud data centers. In: Proceedings of the 4th annual Symposium on Cloud Computing, p. 52. ACM (2013)
12. Yu, S., Doss, R., Zhou, W., et al.: A general cloud firewall framework with dynamic resource allocation. In: 2013 IEEE International Conference on Communications (ICC), pp. 1941–1945. IEEE (2013)
13. Gardner, E.S.: Exponential smoothing: the state of the art. J. Forecast. **4**(1), 1–28 (1985)

Dependency Aware Business Process Analysis for Service Identification

Jiawei Li[1,2], Wenge Rong[1,3](✉), Chuantao Yin[2], and Zhang Xiong[1,3]

[1] State Key Laboratory of Software Development Environment,
Beihang University, Beijing, China
{jiawei.li,w.rong,xiongz}@buaa.edu.cn
[2] Sino-French Engineer School, Beihang University, Beijing, China
chuantao.yin@buaa.edu.cn
[3] School of Computer Science and Engineering, Beihang University, Beijing, China

Abstract. As a fundamental phrase in the life cycle in SOA, service identification has a huge impact in building up SOA based applications. Several service identification methods focus on the definitions of loosely coupled and a high cohesion inside services. There is a majority using business process as input. Because of the simplification of relation between process in most of the process modelling language, dependency between business process is ignored. However, dependency is an inevitable factor to performance of future system. In this paper, we proposed a procedure of dependency aware process analysis for service identification method to ensure not only the characteristics of SOA but also the dependency between services. With this procedure, we tried to have a group of services with visible dependency from analysing the business process and requirements.

Keywords: Service identification · SOA · Process fragment · Business process

1 Introduction

During the process of building business agility, service oriented architecture (SOA) is proposed and widely adopted for building flexible and efficient information systems [6,7,29], due to its advances in alignment of business-IT strategy [14]. Before fully utilising service oriented computing, there is an essential task assuring the alignment between business needs, which is normally referred as service identification [3]. Service identification is fundamental since its outcome will have an influence to the development of the system in the future [18].

There are three main strategies to identify services within SOA, i.e., bottom-up, meet-in-the-middle and top-down [2], among which top-down strategy is currently the most popular candidate to assure the alignment of business strategy and the services in SOA [10]. Within different kinds of top-down methods, process driven service identification is a typical approach since process is suitable for communication between business and the information technology [9].

© Springer International Publishing Switzerland 2015
L. Yao et al. (Eds.): APSCC 2015, LNCS 9464, pp. 137–152, 2015.
DOI: 10.1007/978-3-319-26979-5_10

Though process oriented service identification has been widely adopted, there are also several technical challenges and an important one is that the business process modelling usually ignores the dependency between process and the traceability of requirement [21].

The degree of dependency will influence the coupling and cohesion between software element and the dependency of requirement itself has a strong impact on future bugs [26]. Dependency is so critical that bottlenecks, blockages on the work flow or waiting occur due to a underestimate of dependency [23]. During the development phrase of services, business goal and business objectives are performances indicators to services [20]. Importantly, services with high adaptation to business changes demand a necessity to store the dependent relations between business requirements and the IT realization [22]. Therefore, using a business process as input is not enough to solve the problems of management and traceability of the services [9].

To solve this problem, a lot of efforts have been devoted in the literature and the methods are mainly studied from three perspectives, i.e., process modelling, goal dependency management, goal/process traceability. One method called User Requirement Notation (URN) tried to provide more powerful process modelling [21]. However, URN is an incomplete modelling language for business process and currently common modelling language is BPMN [5]. Some other solution tried to solve this problem by focusing on requirement dependency [26]. A method called i* [30] is proposed for traceability of non-functional requirement with all kinds of possible dependencies. Similarly, NFR is another goal oriented modelling method designed specially for non-functional requirements [19]. However, these methods have a strong focus in non-functional requirements. KAOS [16] is in other hand a goal oriented method concerning functional and non-functional requirements. Alternatively a method named GoalBPM proposed in [15] is an informal framework for goal/process traceability. However this traceability solution is dependent on an ambiguous definition of effect annotation.

In this research, a dependency aware process analysis method is proposed for service identification. Firstly the requirements will be modelled in the form of scenarios in the requirement acquisition phrase since the business process is another representation of requirements [4]. The scenario will be translated into process fragment [8] and each fragment represent a candidate service. Finally services will be grouped together according to the goal oriented model. By analysing the dependency between process fragment, this method identifies services in respect with the traceability succeeding from business goals and process the dependency relation coming from requirement analysis.

The rest of the paper is organised as follows. Section 2 will introduce the background about service identification and related methods from process oriented perspectives. In Sect. 3 we will present the proposed method and Sect. 4 will evaluate it via a case study and discussion. At last Sect. 5 concludes the paper and points out possible future work.

2   Related Work

To implement an efficient SOA based applications, one preliminary task is to obtain proper services [1]. In this phrase, there are three main strategies, i.e., top-down, meet-in-the-middle, bottom-up, among which the top-down strategy has been attached much importance due to its simpleness and capability of aligning IT-business strategy. A straightforward idea is to use business entity for service identification by analysing the connection between business entities using mathematical algorithm [18]. Every element of business for example a business process or a requirement is considered as a business entity. By using Particle Swarm Optimization (PSO) cooperated with metrics design, entities with strong connection will be grouped together but users have to research the definition of service.

A top-down strategy can have two types of inputs [11], i.e., use cases and business process [13]. Business process oriented top-down methods created services by analysing different business process or work flows as the minimum communications between tasks helps to group them [12]. A service means a group of tasks as such there is the less communication outside and the most centralization inside. Similarly Ma et al. tried to separate a business process by weighting different characteristics of SOA so that customers obtain a group of services with balanced characteristics according to their needs [17]. Because SOA enhances the flexibility and reusability of services according to the design principal of SOA [14], in order to balance the different contradictory characteristics the proposed matrix achieve the requirements of an information system. Another process driven method, P2S, analyses additionally the data transporting between tasks [1], which is suitable in solving complex process where the interoperation lies and it groups cooperating tasks together. Applying a new definition of business value for finding the services definitions, P2S provides a solution to combine data analysis with the design metrics.

However, all these process driven methods concentrate on decomposing the business process. Lack of the dependency between process and traceability of goals brings another challenge for further quality analysis [27]. One solution to find dependency of business process is using URN [21], which has a high quality in managing dependency but its design pattern is however incomplete in reality. To make it suitable for applications, there are three essential parts needed [21], i.e., a modelling language for business process in order to have a graphic representation; a goal oriented method for management of requirements; a method to relate the requirement engineering result with the business process. As a result a more sophisticated business process modelling language is proposed and named BPMN [30].

There are several methods to manage the goal oriented requirement traceability and its dependency. For example Koliadis et al. proposed GoalBPM framework [15] linking BPMN with KAOS (Knowledge Acquisition in Automated Specification of Software System) [16]. This framework serves to control the satisfaction of goals in business process development. An alternative efficient goal oriented method is i* [30]. Based on an analysis on the dependency relation between actors, i* form a self explained modelling languages for tasks in business process.

Another goal oriented requirement traceability method is NFR [19], which go further more on analysing the non-functional requirement and relations. By classifying goals in different layers, NFR built a goal oriented system similar to KAOS model. Instead of focusing on the logical hierarchy between goals, NFR manages to put non functional requirements as a soft goal in the dependency tree.

From the literature it can be concluded that the business process has a strong connection with the requirement engineering [4]. In other words, business process is another representation of requirements and it combines related and elicited requirements together. To model and verify a business process, we need a requirement engineering method [25,28]. Process fragment [8] is designed to specify an action needed for comparing a business process to it in order to control the business process. Matching a scenario of requirements to a process fragment helps to understand the logic inside a business process.

3   Dependency Aware Service Identification

From previous discussion, it is clear that dependency in service identification phrase is of much importance. In this research a dependency aware process analysis framework is proposed. In this research BPMN is employed for modelling business process. BPMN is a more powerful and XML oriented language which is a machine friendly language and propose a larger choices for gateway and for special cases and a graphic representation of business process for understanding easily [5].

Specially, a three stage service identification mechanism is adopted in this research. (1) In requirement acquisition phrase, we recognise requirements as scenarios using a traditionally popular method CREWS-SAVRE [24]. After having a library of requirements, we build up a goal oriented model according to KAOS [16]; (2) We will match scenarios with the business process for process fragment. (3) We will group different services candidate together according to the dependency tree, where requirements which have dependent relations and which locate to the same root of goal are highly recommended to group together.

3.1   Requirement Acquisition

In this research the requirements are defined as follow:

$$R = \{Id, D, S, Sc\} \tag{1}$$

where Id is a unique code for each requirement. D is an original semantic description of the requirement. S is the source of the requirement and the source should be a role instead of the physical person; Sc is a set of scenarios.

The goal central traceability map is built as KAOS defined in [16]. A branch of requirement dependency tree is defined as follow:

$$K = \{R, tr, Tt, Sr\} \tag{2}$$

where R is the requirement, tr is the trace between requirements. The trace direction is always coming from leaves to root. Tt is for specialising the type of the trace. Sr is the satisfaction relation which should be "satisfied", "weak" or "unsatisfied". A requirement with all its scenarios satisfied by the business process should be in state "satisfied". If only some scenarios are satisfied, the relation is "weak". Otherwise, the requirement is "unsatisfied".

According to the definition of scenario in [24], we define a scenario as a sequence of events which is one possible pathway through a use case.

$$Sce = \{ev_0, ..., ev_p\} \tag{3}$$

where two types of scenarios are further defined. Execution scenario is the scenario designed for execution. Forbidden scenario is a constraint and should not be executed.

In this research a scenario is formed by events. In [24] an event should save the state of system before or after an action showing a change. In order to simplify the comparison between scenarios and business process, we use only the information of changing state (event) but not the action needed in requirements. Therefore, event is defined as a set of data with its new state and the information of the changing source. Each data has a data object, a state of data and changing source.

$$ev = \{Dt_1, ..., Dt_q\} \tag{4}$$

$$Dt_l = \{Do, st, sc\} \tag{5}$$

As analysing the similarity between scenarios and business process, process fragment is a part of business process and can be located as follow.

$$PF = \{Id, T, E, F, A, G, L, \Delta\} \tag{6}$$

where Id is the indicator of the requirement. T is a set of tasks. E is a set of events. F is a set of flows and each flow is defined as $f = \{id, name, sourceRef, targetRef\}$. A is a suit of associations and there are two types of association, i.e., (1) Association $\{Id, name, sourceRef, targetRef, associationDirection\}$ and (2) DataAssociation $\{Id, name, sourceRef, targetRef, ioSpecification, DataSet\}$. If data association is linked to an input data, the $sourceRef$ is set as $itemAwareElement$. The $ioSpecification$ is $input$ and the $DataSet$ is an inputSet. If the data association is linked to an output data, the $targetRef$ should be $itemAwareElement$. The $ioSpecification$ is $output$ and the $DataSet$ is an outputSet.

G is a set of gateways and defined as $g = \{id, name, eventGatewayType\}$. L is a set of swim lanes and each lane in L is $lane = \{id, name, childLandSet, flowElementRef\}$. Δ is a set of Data and each data object belonging to Δ is defined as $data = \{id, name\}$. Input Data is a data object linked to a data association with $ioSpecification = input$. Output Data is a data object linked to a data association with $ioSpecification = output$. Message association is a data association connected at one end with a message. Simple data association is an association not linked with input data, output data or message. If an association is a simple association, this association serves to connect an internal task with a task outside of the current participant. Event shows changes in state of data or information before and after a task. We can now compare the difference to know a business fragment and the matching process will be illustrated in the next section.

3.2   Scenario Matching

The objective of this step is to locate a process fragment linking to scenario of a requirement in the business process. Business process (BP) has a similar definition as process fragment (PF):

$$BP = \{T, E, F, A, G, L, \Delta\} \tag{7}$$

But for a business process, it should have at least one start event and one end event. Normally BP belongs to PF but an instance of PF is not always a BP. In order to manage dependency in each business process, we define a relation matching matrix. This matrix serves to map the business process to the satisfied process fragment in a requirement. Satisfaction process fragment management matrix linked to the giving business process saves information of connected requirements. This matrix is defined as:

$$M_i := [Id, pf_1, ..., pf_m] \tag{8}$$

This matrix is a $1 \times (m + 1)$ matrix. where Id is the unique indicator of a requirement. If the business process satisfies the scenario of this requirement, the corresponding pf_j equals 1, otherwise 0. The size of this matrix is depending on the number of scenarios processed by the corresponding requirement. All the management matrix form a set $M[n]$. After searching for the corresponding requirements and scenario sets for each M_i, we have a set of requirements linked to the business process $R[n]$. For each chosen R_i we have a set of process fragment $PF_i[m]$ linked to them. For each R_i we will check each scenario Sce_i. If the sequence of process is found matching with the sequence of events in scenario, $pf_{ij+1} \in M_i$ is set as 1, otherwise 0.

Comparison of scenario and business process begins with the first event in Sce_i. According to the definition of PF, we can define a $\tau\{F, A, L, \Delta\}$. The elements belonging to F are sequence flows, condition flows or default flows. A condition flow or a default flow is considered as an special event. Inside this event, we recognise the condition as information inside the data. In other words, when we met a conditioned gateway, we should match the condition with the existing content of data in an event. Otherwise, a task can only have one in and one out, so the flow pointing to the task or the flow departing from the task do not influence the comparison of scenarios and business process. If a scenario has found a matching sequence of tasks, the flows in business process will be succeed by the process fragment. As the definitions of the relation between task and input, output data, we know that all the data are linked to a certain task by data association or association. In consequence, most of the comparison is laid in the difference between associations and lanes.

Events are normally positioned before a task or after one. In this research we supposed that our process fragment contains the task before the event but the task after the last event would not be counted. Firstly, we can identify if the belonging lane of a task is the same as the changing source of the event. A task's belonging lane should be the same to the changing source of the data in post event. When a event has more than one changing source, it is possible only if this event is after a gateway. The other case, an event has only one changing source to all the data, because a task can have only one in and one out. A piece of data can have in general four state: create, read, update, delete (CRUD) [1]. We group all the actions such as: rewrite, fill up, send, copy, etc. in the update state which means an operation done on data. Therefore given task τ and several flows F, associations A, swim lanes or collapsed pools L, some pieces of data D linked to the tasks, we can list the matching method as below for a satisfied scenario.

To match an event with an event in business process: (1) If a start event is an event with message mission, for the same message should at least have a data $data \in ev$ and $data.state =' R'$ with message being a part of the data. (2) If a end event is an event with message mission, for the same message should at least have a data $data \in ev$ and $data.state =' U'$ with message being a part of the data.

To match an event with a task: (1) $data_l \in ev$, $data_l.state = C$, $data_l.dataObject \in \tau.outputdata$ or $data_l.dataObject \in \tau.message$ with the message association pointed out and $data_l.dataObject$ does not be found before. (2) $data_l \in ev$, $data_l.state = U$, $data_l.dataObject \in \tau.inputdata$ or $data_l.dataObject \in \tau.message$ with the message association pointed to the task and $data_l.dataObject \in \tau.outputdata$ or $data_l.dataObject \in \tau.message$ with the message association pointed out. As updating is a complex operation on a piece of data, the detail definition of the same update action should be defined by the company itself. (3) $data_l \in ev$, $data_l.state = R$, $data_l.dataObject \in \tau.inputdata$ or $data_l.dataObject \in \tau.message$ with the message association pointed to the task. (4) $data_l \in ev$, $data_l.state = D$, $data_l.dataObject \in \tau.inputdata$ or $data_l.dataObject \in \tau.message$ with the message association pointed to the task. In this case we should be sure that this object will no longer be used any more.

When an event is found matching with a data state between two tasks, the task in front of the testing event will be examined if it situates in the supposing lane. If so, the task should be a part of the process fragment. As we are going through the business process for a scenario, we will have a result of satisfaction. For a scenario finding a process fragment fulfilling all the events of the scenario, this scenario is satisfied, otherwise it is unsatisfied. To build process fragment, we ignore the tasks or the gateways between two matched tasks and use a simple flow for connection. If the matched tasks have a parallel, inclusive or exclusive relation between them, the gateway relation should be heritage to the process fragment. After building up the matching process fragment, we obtain several matching matrix for the relation between business process and requirements.

3.3 Service Grouping

In this phrase, we already have business process linked to requirements by matching matrix. As we used the scenario comparison, the location of the requirement should group several task together or perhaps it is just located in one task. A matching scenario form a candidate service. For the candidate services answering the same requirement, we proposed to group together. If a task is identified being used by several requirements, we recommend to group services according to the minimum connection rule in respect of loosely coupled.

If two process fragments situated next to one another, we should go through the requirement dependency relation tree for minimizing the dependent relation between two service. The dependent requirement will only be analyse for one generation which means the leaf generation for the requirement. A resulting service dependency relation is defined as follow:

$$Rs = \{M_0, ..., M_x\} \tag{9}$$

It is a set of matching matrix for requirement with a satisfaction level at least weak. The dependency relation will be traced back to the dependency tree by the matching matrix. Each time, when a service changes we can trace back to the related requirements for a verification and according to the goal oriented model we obtain a list of possible impacted services. In the case that a change happened to a certain requirement, services linked to the requirement can be modified rapidly.

4 Case Study

In order to evaluate the service identification method, in this research a case study about booking process is employed to validate its capability. The booking process contains a basic hotel booking and an entertainment booking which is an alternative choice to customers. Each reservation should be paid for a confirmation of booking. The reservation process is shown in Fig. 1.

To deal with a reservation requirement, the employee of the sales department will show a detailed price table for customers. If the customers are not satisfied with the prices and decide not to reserve a room or a ticket, the process will end. If they continue to the next step, the customer can select from booking only for rooms, only for tickets or for both. After the booking process, customer will be satisfied or unsatisfied with the search result. If they need to look up an alternative choice, it will be easy to restart from the beginning. When the booking process is completed, the customers are required to pay for a reservation fare. Afterwards a booking confirmation will be sent to the customers.

4.1 Requirement Acquisition

This booking process is linked with several requirements. We have a list of main requirements shown in Fig. 2. The goal oriented model is built based on the model KAOS proposed in [16].

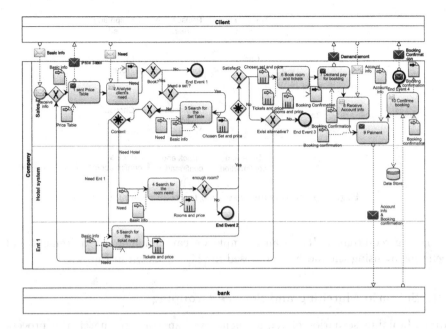

Fig. 1. Booking process

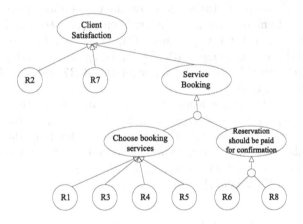

Fig. 2. Goal oriented model

(1) R1: Customers want to book hotels or entertainment tickets (2) R2: Customers can look up the price table (3) R3: Customers can have a booking confirmation feedback at the end of the booking. (4) R4: Marketing department propose different package of tickets for customers. (5) R5: Hotel wants to avoid over booking problem. (6) R6: The reservation fee should be pay before the booking process ended. (7) R7: If clients are not satisfied we can always try to go back to repose the price table. (8) R8: Booking fee will be charged to finally confirm the booking.

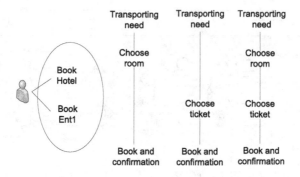

Fig. 3. Scenarios and use case for requirement R1

Taking requirement R1 as an example we can conclude a requirement and scenarios by using the method proposed in [24], as shown in Fig. 3

4.2   Scenario Matching and Service Grouping

After obtaining scenarios of requirement, we can start to match the process fragment, as shown in (Figs. 4, 5, 6, 7, 8, 9, 10 and 11). Therefore we can have a group of tasks as a candidate service $\{\tau_2, \tau_4, \tau_5, \tau_6\}$ for requirement R1. Similarly, we have $\{StartEvent, \tau_1, EndEvent_1\}$ for requirement R2, $\{EndEvent_6\}$ for requirement R3, $\{\tau_2, \tau_3, \tau_6\}$ for requirement R4, $\{\tau_4, EndEvent_2\}$ for requirement R5, $\{\tau_7, \tau_8, \tau_9, \tau_{10}, EndEvent_4\}$ for requirement R6, $\{StartEvent, EndEvent_3\}$ for requirement R7 and $\{\tau_{10}, EndEvent_4\}$ for requirement R8 (Table 1).

According to the dependency tree, we can group three main services:$\{StartEvent, \tau_1\}, \{\tau_2, EndEvent_1, \tau_3, EndEvent_2, \tau_4, \tau_5, EndEvent_3, \tau_6\}$ and $\{\tau_7, \tau_8, \tau_9, \tau_{10}, EndEvent_4\}$. For the reason of fulfilling the requirement 7, we can combine the first service and the second one for obtaining two services:$\{StartEvent, \tau_1, \tau_2, EndEvent_1, \tau_3, EndEvent_2, \tau_4, \tau_5, EndEvent_3, \tau_6\}$ and $\{\tau_7, \tau_8, \tau_9, \tau_{10}, EndEvent_4\}$.

Table 1. Scenarios of requirement R1

Scenario	Events
Scenario 1	Need(C) - Need(R) & RoomChosen(C) - RoomChosen(R) & Confirmation(C)
Scenario 2	Need(C) - Need(R) & Ent1Chosen(C) - Ent1Chosen(R) & Confirmation(C)
Scenario 3	Need(C) - Need(R) & RoomChosen(C) - Need(R) & Ent1Chosen(C) - RoomChosen(R) & Ent1Chosen(R) & Confirmation(C)

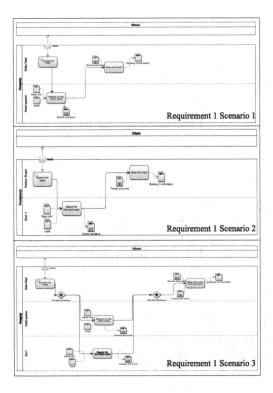

Fig. 4. Requirement R1

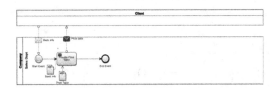

Fig. 5. Requirement R2

Fig. 6. Requirement R3

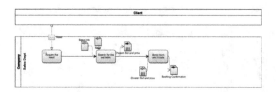

Fig. 7. Requirement R4

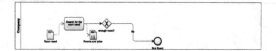

Fig. 8. Requirement R5

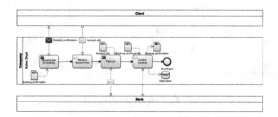

Fig. 9. Requirement R6

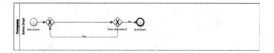

Fig. 10. Requirement R7

Fig. 11. Requirement R8

Table 2. Comparison between service identification methods

Method	Inputs	Method	Evaluation	Apply to this case?	For future governance in SOA?
Requirement center method	Business process in BPMN	top-down	case study and evaluations	Yes 3 services	Yes. No need to redo the calculation. Services are traceable linking to requirements and they change only when requirements change
[1]	Business process in BPMN	top-down	case study and evaluations	Yes 5 services	No. Should redo the calculation if business process changes
[12]	Petri Net	top-down	case study and evaluation	No	No. Should redo the calculation if Petri Net changes
[17]	Work flow	top-down	case study	No	No. Should redo the calculation
[18]	Business Entities	top-down	case study and evaluation	No	No. Should redo the calculation

4.3 Discussion

To evaluate the propose method, it is compared against other popular methods, as shown in Table 2. [18] method tried to cluster business entities but the definition of business entities can change from person to person. Once the business process changes, the standard of business entities will need to redefine. [17] method working on a work flow. The result can change according to the weight matrix at the same time, we should not fix the weight matrix. This matrix can serve for different business need. [12] is another process driven method working on analysing the Petri Net. But it mainly depends on a rule respecting the minimum communications between services. When a new communication build up, it might influence the belonging location of a task. The structure of services might also change. [1] is the only method testable to our case study. The result shows that R6 is separated. We recheck the reason should be a lack of connection between tasks 7 and 8 and tasks 9 and 10. Therefore, the business process should have enough details to enable the calculation.

The Dependency aware process analysis service identification method shows not only an advantage in service identification phrase but also in the continuity of life cycle of services. If the business process changes for a short period of time, all the other method have to redo the calculation. As the services identified from our method serving for certain requirements, there won't be a great change to the structure of services but the other method didn't assure this.

5   Conclusion and Future Work

In this paper we propose a dependency aware process driven goal oriented service identification method through (1) finding the business requirements in business process; (2) realising the dependency relation of requirements to business process and services in SOA; (3) proposing a service identification procedure. As shown in the case study, this method can give another proposition of standard to classified tasks to services. We consider the definition of cohesion and loosely coupling of SOA has a highly strong link to the requirement. A service serves to the less requirement possible then it is more specified. A service will be more independent if it does not have to cooperate with another service serving a same requirement.

As there still have variety of gateways and events in business process, the process fragment will be faced with more complex business process to match with. Also, we will study the case that a business process does not fully satisfy a requirement. For example, if the requirement is difficult to fulfil in the reason of limiting capability, the important part of scenario will be satisfied and the rest will be ignored. In this case, the requirement should still be able to be recognized in a business process.

Acknowledgement. This work was partially supported by the State Key Laboratory of Software Development Environment of China (No. SKLSDE-2015ZX-23), the National Natural Science Foundation of China (No. 61472021), the National High Technology Research and Development Program of China (No. 2013AA01A601), and the Fundamental Research Funds for the Central Universities.

References

1. Bianchini, D., Cappiello, C., Antonellis, V.D., Pernici, B.: Service identification in interorganizational process design. IEEE Trans. Serv. Comput. **7**(2), 265–278 (2014)
2. Bianchini, D., Pagliarecci, F., Spalazzi, L.: From service identification to service selection: an interleaved perspective. In: Agha, G., Danvy, O., Meseguer, J. (eds.) Formal Modeling: Actors, Open Systems, Biological Systems. LNCS, vol. 7000, pp. 223–240. Springer, Heidelberg (2011)
3. Börner, R., Goeken, M.: Identification of business services literature review and lessons learned. In: Proceedings of 15th Americas Conference on Information Systems (2009)
4. Castano, S., Antonellis, V.D., Melchiori, M.: A methodology and tool environment for process analysis and reengineering. Data Knowl. Eng. **31**(3), 253–278 (1999)
5. Chinosi, M., Trombetta, A.: BPMN: an introduction to the standard. Comput. Stand. Interfaces **34**(1), 124–134 (2012)
6. Choi, J., Nazareth, D.L., Jain, H.K.: The impact of SOA implementation on IT-business alignment: A system dynamics approach. ACM Trans. Manage. Inf. Syst. **4**(1), 3 (2013)
7. Dai, W.W., Vyatkin, V., Christensen, J.H., Dubinin, V.N.: Bridging service-oriented architecture and IEC 61499 for flexibility and interoperability. IEEE Trans. Industr. Inf. **11**(3), 771–781 (2015)
8. Daniel, F., Casati, F., D'Andrea, V., Mulo, E., Zdun, U., Dustdar, S., Strauch, S., Schumm, D., Leymann, F., Sebahi, S., Marchi, F.D., Hacid, M.: Business compliance governance in service-oriented architectures. In: Proceedings of IEEE 23rd International Conference on Advanced Information Networking and Applications, pp. 113–120 (2009)

9. Gu, Q., Lago, P.: Service identification methods: a systematic literature review. In: Di Nitto, E., Yahyapour, R. (eds.) ServiceWave 2010. LNCS, vol. 6481, pp. 37–50. Springer, Heidelberg (2010)

10. Huergo, R.S., Pires, P.F., Delicato, F.C., Costa, B., Cavalcante, E., Batista, T.: A systematic survey of service identification methods. Serv. Oriented Comput. Appl. 8(3), 199–219 (2014)

11. Inaganti, S., Behara, G.K.: Service identification: BPM and SOA handshake. BPtrends (2007)

12. Kim, Y., Doh, K.: Formal identification of right-grained services for service-oriented modeling. In: Proceedings of 10th International Conference on Web Information Systems Engineering, pp. 261–273 (2009)

13. Kim, Y., Doh, K.: Use-case driven service modelling with xml-based tailoring for SOA. Int. J. Web Grid Serv. 9(1), 35–53 (2013)

14. Kohlborn, T., Korthaus, A., Chan, T., Rosemann, M.: Identification and analysis of business and software services - A consolidated approach. IEEE Trans. Serv. Comput. 2(1), 50–64 (2009)

15. Koliadis, G., Ghose, A.K.: Relating business process models to goal-oriented requirements models in KAOS. In: Hoffmann, A., Kang, B.-H., Richards, D., Tsumoto, S. (eds.) PKAW 2006. LNCS (LNAI), vol. 4303, pp. 25–39. Springer, Heidelberg (2006)

16. van Lamsweerde, A.: Goal-oriented requirements engineering: A guided tour. In: Proceedings of 5th IEEE International Symposium on Requirements Engineering, p. 249 (2001)

17. Ma, Q., Zhou, N., Zhu, Y., Wang, H.: Evaluating service identification with design metrics on business process decomposition. In: Proceedings of 2009 IEEE International Conference on Services Computing, pp. 160–167 (2009)

18. Merabet, M., Benslimane, S.M.: A multi-objective hybrid particle swarm optimization-based service identification. In: Proceedings of 1st International Conference on Advanced Aspects of Software Engineering, pp. 52–62 (2014)

19. Mylopoulos, J., Chung, L., Nixon, B.A.: Representing and using nonfunctional requirements: A process-oriented approach. IEEE Trans. Softw. Eng. 18(6), 483–497 (1992)

20. Papazoglou, M.P., Traverso, P., Dustdar, S., Leymann, F.: Service-oriented computing: State of the art and research challenges. IEEE Comput. 40(11), 38–45 (2007)

21. Pourshahid, A., Amyot, D., Peyton, L., Ghanavati, S., Chen, P., Weiss, M., Forster, A.J.: Business process management with the user requirements notation. Electron. Commer. Res. 9(4), 269–316 (2009)

22. Stephan, B., Bauer, T., Reichert, M.: Bridging the gap between business process models and service composition specifications. In: Service Life Cycle Tools and Technologies: Methods, Trends and Advances, pp. 124–153 (2011)

23. Strode, D.E.: A dependency taxonomy for agile software development projects. Inf. Syst. Front. 1–24 (2015)

24. Sutcliffe, A.G., Maiden, N.A.M., Minocha, S., Manuel, D.: Supporting scenario-based requirements engineering. IEEE Trans. Softw. Eng. 24(12), 1072–1088 (1998)

25. Vanderfeesten, I.T.P., Reijers, H.A., van der Aalst, W.M.P.: Evaluating workflow process designs using cohesion and coupling metrics. Comput. Ind. 59(5), 420–437 (2008)

26. Wang, J., Wang, Q.: Analyzing and predicting software integration bugs using network analysis on requirements dependency network. Requirements Eng. 1–24 (2014)

27. Wetzstein, B., Leitner, P., Rosenberg, F., Dustdar, S., Leymann, F.: Identifying influential factors of business process performance using dependency analysis. Enterp. IS **5**(1), 79–98 (2011)
28. Xu, L.D., Viriyasitavat, W., Ruchikachorn, P., Martin, A.: Using propositional logic for requirements verification of service workflow. IEEE Trans. Ind. Inform. **8**(3), 639–646 (2012)
29. Yao, J., Tan, W., Nepal, S., Chen, S., Zhang, J., Roure, D.D., Goble, C.A.: Reputationnet: Reputation-based service recommendation for e-science. IEEE Trans. Serv. Comput. **8**(3), 439–452 (2015)
30. Yu, E.S.K.: Towards modeling and reasoning support for early-phase requirements engineering. In: Proceedings of 3rd IEEE International Symposium on Requirements Engineering, pp. 226–235 (1997)

Dynamic Allocation of Virtual Resources Based on Genetic Algorithm in the Cloud

Li Deng[1,2(✉)] and Li Yao[1]

[1] College of Computer Science and Technology,
Wuhan University of Science and Technology, Wuhan 430065, China
dengli@wust.edu.cn
[2] Hubei Province Key Laboratory of Intelligent Information Processing
and Real-time Industrial System, Wuhan 430065, China

Abstract. Cloud computing provides dynamic resource allocation using virtualization technology to greatly improve resource efficiency. However, current resource reallocation solution seldom considers the stability of VM placement pattern. Varied workloads of applications would lead to frequent resource reconfiguration requirements due to repeated occurrence of hot nodes. In this paper, a multi-objective genetic algorithm (MOGA) is presented to significantly improve the stability of VM placement pattern with less migration overhead. The group encoding scheme is employed in MOGA to express the mapping of physical nodes and virtual machines (VMs). Fitness function is designed based on the stability and migration overhead of group. Our simulation results demonstrate that, our MOGA is much more efficient than other algorithms for resource reallocation with good stability.

Keywords: Cloud computing · Resource allocation · Genetic algorithm

1 Introduction

Cloud computing [1] provides a huge resource pool shared by a large number of users. Virtualization technology enables dynamic resource configuration according to real demands of applications and greatly improves resource efficiency. Live migration of VMs is an important way to implement resource reallocation in the cloud.

Wrasse [2] is designed to handle generalized resource allocation in the cloud. It uses massive parallelism by orchestrating a large number of light-weight GPU threads to explore the search space in parallel. Server consolidation [3,4] has always been studied for green computing. Constraint programming is used to reduce the number of active physical nodes for energy efficiency while the Service Level Agreement (SLA) is guaranteed. Also, economic efficiency of cloud computing has been studied by many researchers [5,6]. Auction approaches are presented to balance the relationship between economic efficiency and computational efficiency.

© Springer International Publishing Switzerland 2015
L. Yao et al. (Eds.): APSCC 2015, LNCS 9464, pp. 153–164, 2015.
DOI: 10.1007/978-3-319-26979-5_11

However, current resource management methods seldom consider the stability of VM placement globally. Due to time-varying resource demands of applications, current mapping of VMs to physical nodes is not suitable for future workloads. New hot nodes would appear in the near future, which directly results in another resource reallocation. The stability of a VM placement pattern should be considered during dynamic resource configuration.

Resource allocation problem is a kind of combinatorial problem, known as NP-complete problem. In this paper, we present a multi-objective genetic algorithm (MOGA) to improve the stability of VM placement patterns. The group encoding scheme is used to exactly express the mapping of physical nodes and VMs. Fitness function is designed based on stabilization time and migration overhead of group. Our simulation results show that, our MOGA is efficient for resource reconfiguration with good stability.

The rest of the paper is organized as follows: Sect. 2 discusses related work about dynamic resource allocation. In Sect. 3, we give the description of problem formulation. Objectives and constraints of dynamic resource allocation are formulated. Section 4 introduces the details of our multi-objective genetic algorithm. Performance comparison of several algorithms are done in Sect. 5. Finally, we give our summary and future research directions in Sect. 6.

2 Related Work

Being completely different from traditional static resource configuration, cloud computing enables dynamic resource allocation in accord with time-varying workloads of applications. Resource efficiency are thus improved significantly. Many researchers have studied resource reallocation problems.

Dynamic resource allocation usually has the following objectives:

- *Green Computing*
  Server consolidation [3,4,7] is used to decrease the number of active physical nodes. Power efficiency is greatly improved. Constraints programming [3] and genetic algorithm [7] are respectively employed to find a solution using the minimum number of active nodes for green computing.
- *Resource Fairness*
  Resource in the cloud is shared among a large number of tenants. Resource fairness among numerous users is then studied [8,9]. A multi-resource allocation mechanism (called DRFH) [8] is presented to assure fair usage of resource among cloud users using heuristics.
- *Economic Efficiency*
  Resource in the cloud is usually rent in a pay-as-you-go model. Economic efficiency of cloud computing has been studied by many researchers [5,6]. Trading mechanisms for the demand response are designed to achieve the maximum social welfare with arbitrarily high probability.

In this paper, our work mainly focuses on the stability of VM placement pattern. Because workloads of applications are time-varying especially in mobile cloud computing, the stability becomes more important.

3   Problem Formulation

Due to dynamic workloads, resource demands of applications vary with time. Some nodes have frequent resource contention and become busy when workloads increase. These nodes are called *hot nodes*. Hot nodes should be alleviated by decreasing their workloads to assure service-level objectives (SLAs) of applications.

Live migration of virtual machine is an important method to alleviate hot nodes. It re-distributes VMs on a pool of nodes. When remapping VMs to nodes, we should consider future trends of application workloads to avoid "thrashing" – much more hot nodes arising in the future. So, stability is an important metric to choose new VM distribution on nodes. The stability of VM distribution mainly depends on the total workloads of each node.

Table 1 lists the definition of some symbols used in our discussion.

Table 1. Symbols and Definitions

Symbol	Definition
$\mathcal{M}$	The total number of physical nodes in the cloud
$\mathcal{N}$	The total number of virtual machines in the cloud
C_i	The amount of CPU resource that node i supplys
Mem_i	The amount of memory resource that node i supplys
$C\prime_j$	The amount of CPU resource that VM j requests
$Mem\prime_j$	The amount of memory resource that VM j requests
x_{ij}	Binary variable, if $x_{ij} = 1$, node i hosts VM j, or else, $x_{ij} = 0$
$\mathcal{D}_k$	The kth placement pattern of all VMs in the cloud
m_j	Binary variable, if $m_j = 1$, VM j migrates once, or else, $m_j = 0$
T_{node_i}	Stabilization time of a node $node_i$
$T_{\mathcal{D}_k}$	Stabilization time of a placement pattern $\mathcal{D}_k$

We give the following definitions:

Definition 3.1: A *placement pattern* $\mathcal{D}_k$ is the mode in which a group of applications (VMs) are distributed on physical nodes.

Definition 3.2: The node i is stable if and only if the node has enough resources for applications (VMs) residing on it during a certain period of time, no matther how the workloads of applications vary.

Definition 3.3: The placement pattern $\mathcal{D}_k$ is stable if and only if each node in the placement pattern is stable during a period of time.

Definition 3.4: *Stabilization time T* means the longest peroid in which a node or a placement pattern stays stable from a certain time. It is a straightforward metric to measure the stability of a node or a placement pattern.

The stabilization time of a placement pattern depends on that of each node in it, as shown in the following formula.

$$T_{\mathcal{D}_k} = \min\{T_{node_1}, T_{node_2}, \cdots, T_{node_{\mathcal{M}}}\}$$

Then, the problem of dynamic resource allocation is formulated as follows: having known dynamic workloads of VMs (including predicted future workloads), given a set of nodes, the objective of dynamic resource allocation is to find a placement solution of VMs on physical nodes with longest stabilization time and minimal number of VM migration:

$$\text{Objectives:} \qquad \max T_{\mathcal{D}_k}; \; \min \sum_{j=1}^{\mathcal{N}} m_j \qquad\qquad (1)$$

$$\text{Subject To:} \qquad \sum_{i=1}^{\mathcal{M}} x_{ij} = 1. \qquad\qquad j = 1, \cdots, \mathcal{N} \qquad (2)$$

$$C_i \geq \sum_{j=1}^{\mathcal{N}} x_{ij} C'_j. \qquad\qquad i = 1, \cdots, \mathcal{M} \qquad (3)$$

$$Mem_i \geq \sum_{j=1}^{\mathcal{N}} x_{ij} Mem'_j. \qquad\qquad i = 1, \cdots, \mathcal{M} \qquad (4)$$

$$x_{ij} \in \{0,1\}, m_i \in \{0,1\}. \qquad\qquad \begin{array}{l} i = 1, \cdots, \mathcal{M} \\ j = 1, \cdots, \mathcal{N} \end{array} \qquad (5)$$

We have two objectives: one is to make the new distribution of VMs with longest stabilization time ($\max T_{\mathcal{D}_k}$); the other is to only migrate the minimal number of VMs from current status to new status ($\min \sum_{j=1}^{\mathcal{N}} m_j$). The first objective means that, hot nodes would not appear in the new mapping in a short time. The second objective requests that migration overhead of VMs from old status to new status is minimal.

$$m_j = \begin{cases} 0, & \text{if the same node hosts VM } j \text{ both in old and new status} \\ 1, & \text{if different nodes host VM } j \text{ in old and new status} \end{cases}$$

In the above formulae, formula (2) indicates that each VM only resides one physical node. Formula (3) means that the total amount of CPU resource requested by VMs residing on a same node is not larger than the amount of resource supplied by the node. Formula (4) denotes that the total amount of memory requested by VMs is not larger than the amount of memory supplied by the node. Formula (5) explains that x_{ij} and m_i are binary variables.

4   Multi-objective Genetic Algorithm (MOGA)

As dynamic resource allocation problem is a kind of NP-complete problem, genetic algorithm can be used to tackle the problem in polynomial time. Genetic

algorithm is to find a solution to resource allocation problem using the evolution theory of biosphere. It uses computers to simulate species reproduction
procedure based on Mendel's laws of inheritance.

There are several key parts in genetic algorithm (GA) to consider: encoding,
initial population generation, fitness function, main operators, and termination
condition. Encoding is to express chromosomes, genes with elements of resource
allocation problem. Fitness function, which directs evolution procedure, should
be defined to quantify a certain attribute of each chromosome or individual.

Some details in genetic algorithm are discussed in the following.

4.1 Encoding

Encoding is important for genetic algorithm. There are three methods to express
bin packing problems in genetic algorithm: one gene per object, one gene per
bin, one gene per group (bin and objects in it) [10]. In this paper, the encoding
scheme based on group is employed because it can clearly express the relationship
between VMs and physical nodes.

Figure 1 lists examples of the encoding scheme using group. In Fig. 1(a), nine
VMs are deployed on three nodes. Accordingly, there are three genes in the
form of chromosome. Each gene includes one physical node and several VMs
residing on it. A chromosome or an individual signifies a possible solution – a
mapping between virtual machines and physical nodes. In Fig. 1(b), four nodes
host nine VMs altogether. So, the corresponding chromosome has four genes. The
chromosome in Fig. 1(a) has different length with the chromosome in Fig. 1(b).
The GA operators should handle different chromosomes with variable length in
the group encoding scheme.

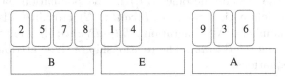

(a) Chromosome: BEA(B{2,5,7,8},E{1,4},A{9,3,6})

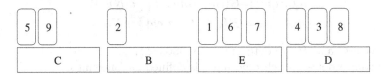

(b) Chromosome: CBED(C{5,9},B{2},E{1,6,7},D{4,3,8})

Fig. 1. Examples of group encoding scheme.

4.2   Initial Population Generation

A population is a set of chromosomes. Let the population size is $popSIZE$. Genetic algorithm usually starts from an initial population which is often generated randomly. Random generation provides wide search space to find a solution, but it takes much time to get an optimal global solution. First-fit heuristic is used to generate the first population. Note that, each individual should meet the constraints discussed in Sect. 3.

4.3   Fitness Function

Each chromosome is coupled with a fitness value, which signifies its certain attributes. The objective of genetic algorithm is to find the chromosome with optimal fitness value.

Fitness function is designed based on the objectives of dynamic resource allocation. Formula (1) in Sect. 3 lists the objectives. There are two objectives: longest stabilization time and minimal migration overhead. A fast multiobjective genetic algorithm (NSGA-II) [11] is used for fitness function. NSGA-II suits well for constrained multiobjective optimization in any evolutionary algorithm [11].

To simplify the problem, the objectives of dynamic resource allocation is re-defined as below:

$$\text{Objectives:}\qquad \min T_{\mathcal{D}_k}{}'; \ \min \sum_{j=1}^{N} m_j \qquad (6)$$

And,

$$T_{\mathcal{D}_k}{}' = MAXVALUE - T_{\mathcal{D}_k}$$

Each chromosome (the lth chromosome $ch[l]$) in the population has several attributes: ①, nondomination rank ($ch[l].rank$); ② crowding distance ($ch[l].distance$); ③ the number of VM migration ($ch[l].mNum$); ④ stabilization time $ch[l].staTime$; ⑤ time ($ch[l].time$). The relationship of these attributes is listed in the following equations:

$$ch[l].time = MAXVALUE - ch[l].staTime$$
$$ch[l].rank = G(ch[\ ].time, ch[\ ].mNum)$$
$$ch[l].distance = H(ch[\ ].time, ch[\ ].mNum)$$

Function G and H are determined by NSGA-II.
A partial order $\prec_n$ of chromosomes is defined as followings.

$ch[l] \prec_n ch[m]$,

if $ch[l].rank < ch[m].rank$

or $((ch[l].rank = ch[m].rank)$ and $(ch[l].distance > ch[m].distance))$

If $ch[l] \prec_n ch[m]$, it is believed that chromosome $ch[l]$ is closer to optimal global solution than $ch[m]$.

4.4   Three Main Operators

There are three main operators in genetic algorithm: crossover, mutation, and selection.

Crossover. Crossover is for two parents to produce offspring so that children can inherit much of meaningful information from parents. Using group encoding scheme, chromosomes may have different length. Crossover should be done on chromosomes with varied length.

As shown in Fig. 2, there are mainly four steps in operator crossover:

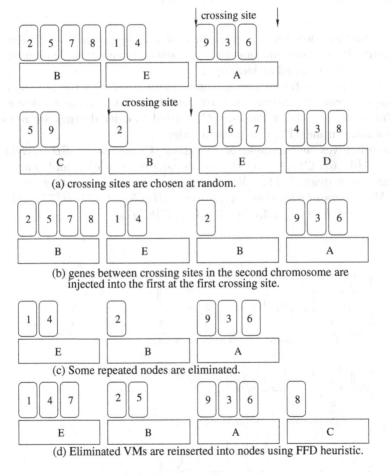

(a) crossing sites are chosen at random.

(b) genes between crossing sites in the second chromosome are injected into the first at the first crossing site.

(c) Some repeated nodes are eliminated.

(d) Eliminated VMs are reinserted into nodes using FFD heuristic.

Fig. 2. Examples of crossover.

(1) Crossing sites are chosen at random in both parents.
(2) Genes between crossing sites in the second chromosome are injected into the first, at the first crossing site. In Fig. 2(b), chromosome $BEA(B\{2,5,7,8\}$, $E\{1,4\}$, $A\{9,3,6\})$ becomes $BEBA(B\{2,5,7,8\}$, $E\{1,4\}$, $B\{2\}$, $A\{9,3,6\})$.
(3) Some genes with repeated nodes should be eliminated. Genes with repeated VMs are also removed. In Fig. 2(c), gene $B\{2,5,7,8\}$ has the same node as gene $B\{2\}$ and is deleted. Then, chromosome becomes $EBA(E\{1,4\}, B\{2\}, A\{9,3,6\})$.
(4) Some missing VMs are reinserted into genes using FFD (First Fit Decreasing, FFD) heuristic. In Fig. 2(d), missing VMs include VM 5, 7, and 8. These VMs are located on nodes again. Because node E, B, and A cannot host VM 8 due to limited resource, a new node C is added.

Mutation. Mutation may make an individual in the population different from his parents. It adds new information in an arbitrary way to widen search space and avoids being trapped at local optima.

Given a small mutation rate q_m, some chromosomes in the population are selected randomly to execute operator mutation. Mutation is to delete some genes at random in chromosomes. VMs related to deleted genes are re-located to other nodes using FFD in a random order.

Figure 3 gives an example of mutation. Chromosome $EBAC(E\{1,4,7\}$, $B\{2,5\}$, $A\{9,3,6\}$, $C\{8\})$ is selected to be done by mutation and gene $B\{2,5\}$ is chosen to be deleted. Then VM 2 and 5 are re-located to other genes using FFD. After operator mutation, chromosome $EBAC(E\{1,4,7\}$, $B\{2,5\}$, $A\{9,3,6\}$, $C\{8\})$ becomes $EAC(E\{1,4,7,5\}$, $A\{9,3,6\}$, $C\{8,2\})$.

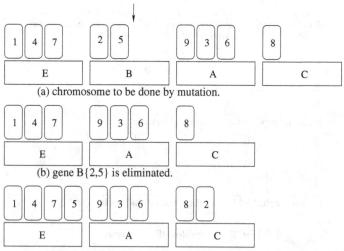

(a) chromosome to be done by mutation.

(b) gene $B\{2,5\}$ is eliminated.

(c) VM 2 and 5 in deleted gene $B\{2,5\}$ are re-located to other nodes.

Fig. 3. Examples of mutation.

Selection. For operator crossover and mutation, appropriate parents must be selected first in the population. Roulette wheel selection is used to choose parent chromosomes.

First, function $f(ch[l])$ is defined:

$$f(ch[l]) = k_{ch[l].rank} - ch[l].distance$$

The value of constant $k_{ch[l].rank}$ is obtained by experimental statistics. Chromosome $ch[l]$ is selected with the probability $p_{ch[l]}$ at random:

$$p_{ch[l]} = \frac{f(ch[l])}{\sum_i f_{ch[i]}}$$

4.5   Termination Condition

There are two termination conditions to stop iterations in genetic algorithm:

- stop iterations when maximum generation (MAX_GEN) is reached.
- stop iterations when a number of consecutive generations have nearly same fitness values.

5   Performance Evaluation

In this section, we evaluate the performance of our multi-objective genetic algorithm (MOGA) and respectively compare it with FFD-CPU, FFD-Mem, GA-S, and GA-N. Here, FFD-CPU indicates FFD algorithm based on CPU resource requests from applications or VMs. FFD-Mem denotes FFD algorithm according to memory size requested by VMs. GA-S is a genetic algorithm with only one objective – longest stabilization time of a placement pattern. GA-N denotes a genetic algorithm with another objective – minimum number of VM migration. Our simulation results show that, MOGA better balances the two objectives and achieves a win-win situation.

All the above algorithms are coded in Java and CloudSim [12] is used to simulate a cloud computing infrastructure. Our tests are done on a Dell optiplex with the fourth Gen Intel Core i5 CPU, 8 GB RAM, and 1TB hard drive.

In CloudSim, we simulate 12 physical nodes and 32 VMs. Dynamic resource requests (only CPU and memory) of these VMs are saved in matrices. Population size is set as 16 ($popSIZE = 16$). The value of constant MAX_GEN is 24 ($MAX\_GEN = 24$). Crossover rate (q_c) is 0.8 and mutation rate (q_m) is 0.1.

5.1   Evolutionary Process of MOGA

First, we observe the evolution process of multi-objective genetic algorithm from initial population to the maximum generation.

Figure 4 depicts the evolutional process of MOGA. Horizontal axis expresses the normalized value of stabilization time of each chromosome. We set the shortest

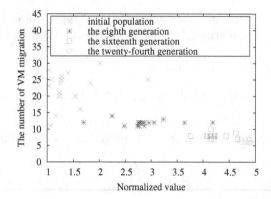

Fig. 4. Evolutional process of multi-objective genetic algorithm.

stabilization time as 1. Vertical axis shows the number of VM migration. The number of VM migration is just estimated roughly by comparing source node and destination node of each VM. Only the initial population, the eighth generation, the sixteenth generation and the twenty-four generation are listed in the figure.

From Fig. 4, we can find that, the reproduction process of individuals moves gradually towards the best solution (longer stabilization time and less number of VM migration). The process begins with quick changes. The initial population is quite different from the eighth generation. But the change becomes small in the later. The sixteen generation is close to the twenty-four generation. Evolutional process after the twenty-four generation is a little meaningful.

5.2   Stabilization Time and Migration Overhead

We also compare the performance of MOGA and other algorithms (FFD-CPU, FFD-Mem, GA-S, and GA-N). We mainly focus on stabilization time and migration overhead of a placement pattern. In simulation platform CloudSim, the

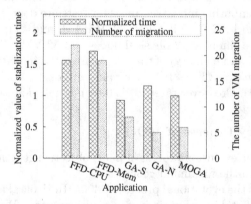

Fig. 5. Performance comparison of several algorithms.

above algorithms are tested based on the same workloads of applications. The simulation results are shown in Fig. 5.

As seen from Fig. 5, the solution obtained by MOGA are better than those found by the other algorithms. MOGA is the only algorithm being able to better balance two objectives and achieve win-win situation. Stability and migration overhead are two important metrics to evaluate a new VM placement pattern. They are also key elements of fitness function in MOGA.

In summary, compared with the other four algorithms, MOGA is much more efficient for dynamic resource allocation, especially for the stability of VM placement pattern.

6   Conclusion and Future Work

In this paper, a new problem of dynamic resource allocation for stability in cloud computing has been studied and a multi-objective genetic algorithm has been proposed to solve it. The group encoding scheme is employed to clearly express the mapping of VMs and physical nodes. Fitness function is designed based on stability and migration overhead of group. Our simulation results show that our MOGA is much more efficient than other algorithms for dynamic resource allocation with good stability.

In the future, we will continue to work on MOGA and further improve its performance. Firstly, fine parameter tuning in MOGA will be done, such as crossover rate, mutation rate. A new method to accurately compute VM migration overhead will be studied, considering migration dependence among VMs. Finally, the computing complexity of MOGA will be also analyzed.

Acknowledgment. This research was funded by Natural Science Foundation of Hubei Province (No. 2014CFB817), China.

References

1. Armbrust, M., Fox, A., Griffith, R., Joseph, A.D., Katz, R.H., Konwinski, A., Lee, G., Patterson, D.A., Rabkin, A., Stoica, I., Zaharia, M.: Above the Clouds: A Berkeley View of Cloud Computing. Technical report (2009)
2. Rai, A., Bhagwan, R., Guha, S.: Generalized resource allocation for the cloud. In: Proceedings of the 3rd Symposium on Cloud Computing (SOCC 2012). ACM, San Jose (2012)
3. Hermenier, F., Lorca, X., Menaud, J.M., Muller, G., Lawall, J.: Entropy: a consolidation manager for clusters. In: Proceedings of the ACM/Usenix International Conference on Virtual Execution Environments (VEE 2009), pp. 41–50 (2009)
4. Chen, L., Shen, H.: Consolidating complementary VMs with spatial/temporal-awareness in cloud datacenters. In: IEEE Conference on Computer Communications (INFOCOM 2014), pp. 1033–1041 (2014)
5. Zhang, L., Li, Z., Wu, C.: Dynamic resource provisioning in cloud computing: a randomized auction approach. In: IEEE Conference on Computer Communications (INFOCOM 2014), pp. 433–441 (2014)

6. Zhou, Z., Liu, F., Li, Z., Jin, H.: When smart grid meets geo-distributed cloud: an auction approach to datacenter demand response. In: IEEE Conference on Computer Communications (INFOCOM 2015) (2015)

7. Li, Q., Hao, Q.F., Xiao, L.M., Li, Z.J.: Adaptive management and multi-objective optimization for virtual machine placement in cloud computing. Chin. J. Comput. **34**(12), 2253–2264 (2011)

8. Wang, W., Li, B., Liang, B.: Dominant resource fairness in cloud computing systems with heterogeneous servers. In: IEEE Conference on Computer Communications (INFOCOM 2014), pp. 583–591 (2014)

9. Guo, J., Liu, F., Lui, J.C.S., Jin, H.: Fair network bandwidth allocation in IaaS datacenters via a cooperative game approach. IEEE/ACM Trans. Netw. (2015)

10. Falkenauer, E., Delchambre, A.: A genetic algorithm for bin packing and line balancing. In: IEEE International Conference on Robotics and Automation, pp. 1186–1192 (1992)

11. Deb, K., Pratap, A., Agarwal, S., Meyarivan, T.: A fast and elitist multiobjective genetic algorithm: NSGA-II. IEEE Trans. Evol. Comput. **6**(2), 182–197 (2002)

12. CloudSim: A framework for modeling and simulation of cloud computing infrastructures and services (2015). http://www.cloudbus.org/cloudsim/

Effective Mashup Service Clustering Method by Exploiting LDA Topic Model from Multiple Data Sources

Buqing Cao[1,2,3(✉)], Xiaoqing (Frank) Liu[2], Jianxun Liu[1],
and Mingdong Tang[1]

[1] School of Computer Science and Engineering,
Hunan University of Science and Technology, Xiangtan, China
{buqingcao,ljx529,tangmingdong}@gmail.com
[2] Computer Science and Computer Engineering Department,
University of Arkansas in Fayetteville, Fayetteville, USA
frankliu@uark.edu
[3] State Key Laboratory of Software Engineering,
Wuhan University, Wuhan, China

Abstract. Mashup is emerging as a promising software development method for allowing software developers to compose existing Web APIs to create new or value-added composite Web services. However, the rapid growth in the number of available Mashup services makes it difficult for software developers to select a suitable Mashup service to satisfy their requirements. Even though clustering based Mashup discovery technique shows a promise of improving the quality of Mashup service discovery, Mashup service clustering with high accuracy for discovery of Mashup services is still a challenge problem. In this paper, we propose a novel Mashup service clustering method for Mashup service discovery with high accuracy by exploiting LDA topic model built from multiple data sources. It enables to infer topic probability distribution of Mashup services, which serves as a basis of computation of similarity of Mashup services. K-means and Agnes algorithm are used to perform Mashup service clustering in terms of their similarities. Compared with other service clustering approaches, experimental results show that our approach achieves significant improvement in terms of precision, recall and F-measure rate, which will improve Mashup service discovery.

Keywords: Mashup service · LDA topic model · Multiple data source · Service clustering

1 Introduction

With the development of Web2.0 and its related technologies, many innovative Internet applications and software systems are developed, such as blog, wiki, network map, online shopping, and search systems. Service-Oriented Computing (SOC) and Web services play a key role in their construction. To satisfy users'

© Springer International Publishing Switzerland 2015
L. Yao et al. (Eds.): APSCC 2015, LNCS 9464, pp. 165–180, 2015.
DOI: 10.1007/978-3-319-26979-5_12

complex requirements, software developers need to compose different Web services on the Internet to create novel network applications and software systems. Recently, Mashup technology, which allows software developers to compose existing Web APIs to create new or value-added composite Web services, has emerged as a promising software development method [1]. It becomes popular since it has many advantages, such as easier programming and shorter development time [2]. Several online Mashup repositories have been established, such as ProgrammableWeb, myExperiment, and Biocatalogue. Programmableweb.com, has published more than 6215 Mashups and 13575 Web APIs as to June 2015. Several Mashup tools also have been developed, such as Microsoft Popfly, Google Mashup Editor and IBM Mashup Center. Typical Mashup applications in Mashup repositories include Map, Video and Image, Search and Shopping, News, Microblog Mashup, etc. The rapid growth in the number of available Mashup services, coupled with the myriad of functionally similar services, makes it difficult for software developers to select a suitable Mashup service to develop novel Internet applications and software systems.

Service clustering technology can effectively improve the quality of service discovery [3–10]. It can cluster Web services with similar functionalities and reduce service's searching space. Several methods [5, 6] first analyze users' requirements and service description documents, and then cluster Web services based on their functionality similarity. Other approaches [7–10] exploit user-contributed tags and perform service clustering by incorporating the similarity of WSDL document and tags information. Recently several researchers work on RESTful Web service clustering to optimize the searching and discovery of Mashup service [1, 11]. However, these existing methods have at least three problems.

- *Most of them focus only on SOAP/WSDL-based Web services which have a formal model and standard description of service capabilities. But the RESTful Web service only has an informal text description in natural language, which makes Mashup service clustering and discovery even harder.*
- *Most of them use only service description, which usually contain a limited collection of terms. Especially, the description of Mashup service is usually brief and other key information of specification of service functionalities were not used. For example, tags and Web APIs were not used for Mashup service clustering in many existing methods.*
- *Only classic clustering algorithms, such as K-Means algorithm, are usually used for Mashup service clustering in the existing methods. They have several shortcomings for Mashup service clustering, such as the size of clustering cannot be monitored, and convergence is to a local optimization. They produce poor service partition results and low accuracy of service clustering.*

Therefore, Mashup service clustering with high accuracy is still a challenging problem. In this paper, we propose a novel Mashup service clustering approach with a high accuracy. It exploits Latent Dirichlet Allocation (LDA) topic model based on multiple data sources to infer the topic probability distribution of Mashup services which is used to compute similarity degrees of Mashup services and combines K-means and Agnes algorithms to perform Mashup service clustering. The contributions of this paper are as follows:

- *We design an extended LDA topic model based on multiple data sources, called as DAT-LDA, which include content description of Mashup services, description of Web APIs, and user-contributed tagging data.*
- *We propose a novel Mashup service clustering approach with high accuracy, which use Jensen-Shannon (JS) distance to compute the similarity between Mashup service documents according to the topic probability distribution results of DAT-LDA, and then combine K-Means and Agnes algorithms to perform Mashup service clustering.*
- *We develop a real-world Mashup services dataset from ProgrammableWeb and conduct a set of experiments. Compared with other clustering approaches, the experimental results show that our Mashup service clustering approach achieves a significant improvement in terms of precision, recall and F-Measure rate, which can improve Mashup service discovery.*

The rest of this paper is organized as follows: Sect. 2 introduces related works. Section 3 presents our Mashup services clustering approach. Section 4 discusses the experimental results. Finally, we draw conclusions and discuss our future work in Sect. 5.

2 Related Work

To our best knowledge, most of the existing service clustering methods focused on exploiting description document of Web services capabilities to measure the similarity between Web services [12–14]. Liu et al. [3, 4] extracted content, context, host name and name from WSDL description text to perform Web services clustering. Sun et al. [5] combined the functionality and process similarities to achieve service clustering, which aggregated Web services with similar functionality and facilitated process-oriented Web services discovery. Platzer et al. [6] proposed Web services clustering using multidimensional angles as proximity measures.

User-added tags of Web services, as their functional descriptions have also been used to improve accuracy of service discovery. Wu et al. utilized tags to facilitate Web services clustering and discovery [8] by proposing a novel approach called WTCluster [7] and developing Titan System [9]. To further improve the precision of service clustering, LDA model was used for the topic extraction of service and service clustering. Cassar et al. [15] employed probabilistic latent semantic analysis and LDA to extract latent topics from OWL-S service description and then implement service clustering. Mustapha et al. [16] proposed a non-logic-based matchmaking approach that uses correlated topic model to extract topics from semantic service descriptions and model the correlation between the extracted topics. Based on the topic correlation, service description can be grouped into hierarchical clusters. These research works improves the accuracy of service clustering by using LDA to mine the implicit topic and deep semantics of service documents. However, they are applicable only to WSDL description documents of Web services.

Recently, with the rise of lightweight RESTful Web services (i.e. Web APIs and Mashup services), a few researchers have considered RESTful Web services clustering to optimize Mashup services searching and discovery. For example, Li et al. [11] designed a services clustering model named as DSCM based on probability and domain characteristics, and proposed a method of topic-oriented clustering of domain services based on the DSCM model. Xia et al. [1] proposed category-aware API clustering and distributed recommendation method for automatic Mashup creation.

3 Mashup Service Clustering Approach

In this section, we first describe Mashup service clustering framework in Sect. 3.1, then respectively discuss data preprocessing and the DAT-LDA topic model in Sects. 3.2 and 3.3, and finally present Mashup service clustering algorithm in Sect. 3.4.

3.1 Mashup Service Clustering Framework

Figure 1 shows the proposed Mashup service clustering framework, which consists of three major parts: data preprocessing, LDA topic model construction, and Mashup service clustering. In the first part, the Mashup description documents, their Web APIs and tags are crawled and parsed from the Internet and Mashup service documents are consolidated from multiple data source for Mashup service clustering. Specially, we extract the meaningful words from the Mashup services documents as feature words to construct a word-document matrix. In the second part, we design a DAT-LDA topic model to infer a word-topic matrix and a topic-document matrix by a series of trainings from the word-document matrix, and obtain a topic probability distribution of Mashup service documents. In the third part, we use the topic probability distribution results of Mashup service documents to calculate the JS distance and obtain the similarity among Mashup services documents. We then combine K-means and Agnes algorithms to perform Mashup service clustering according to the similarity. Finally, the clustered result of Mashup services can be applied to improve accuracy of Mashup service search engine.

3.2 Data Preprocessing

Firstly, we crawl the Mashup description texts, Web APIs, and tags of Mashup services from the Internet, and build Mashup service documents. Then, we extract the feature vectors representing Mashup service content information from the Mashup service documents. The main steps are as follows:

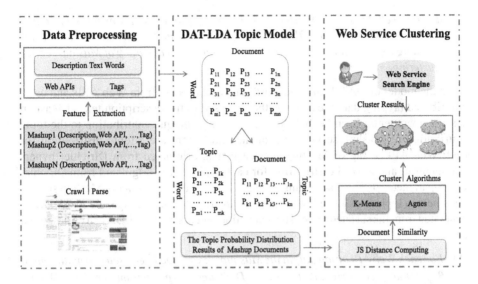

Fig. 1. Framework for Mashup service clustering

(1) *Build an Initial Vector. In this step, we use natural language processing toolkit NLTK[1] to divide sentences in Mashup service documents into words, and establish the initial feature vector.*

(2) *Remove Stop Words. We remove symbols and words, such as* +, −, *the, a, of, and, which are meaningless for the characterization and comparison of feature words. Meaningful feature words are typically nouns, verbs or adjectives.*

(3) *Extract Stemming. Feature words with a common stemming usually have the same meaning, for example, user, used, using, users, and useful all have the same stem use. We extract the stemming of all words by using the Porter Stemmer in the toolkit NLTK1, and produce a new feature vector.*

3.3 DAT-LDA

We construct an extended LDA Topic Model based on multiple data sources, called as DAT-LDA, including the content description of Mashup services, description of their Web APIs, the user-contributed tagging data. Main advantages of DAT-LDA are listed below.

(1) *It provides a generative probabilistic model of Mashup service description text to extract latent variables to improve accuracy of clustering of Mashup services.*

(2) *It calculates co-occurrence of words among Mashup service description texts and infers the topic distribution of individual Mashup service descriptions text, and obtains their topic vectors for Mashup service clustering.*

---

[1] http://www.nltk.org.

(3) *It takes Web APIs and the user-contributed tagging data of Mashup service into consideration. They have their topic distribution and contribute to the corresponding topic vectors, which will be used to improve accuracy of Mashup service clustering.*

In the model of DAT-LDA, the word selection process from Web APIs and user-contributed tagging data are same as the one used in the description text of Mashup service documents. Each Web API or tag has their unique contribution to the topic distribution of Mashup service document. Thus, there are a topic distribution for each Web API or tag of Mashup service that is originated from Dirichlet hyper-parameter α, and a word distribution for each topic that is originated from Dirichlet hyper-parameter β. Then, according to the selected Web API or tag in Mashup service document, a topic is extracted from the topic distribution and the specific word will be generated from the selected topic. The generation process of DAT-LDA can be described below

(1) *For each Web API or tag dat_d in the Mashup service document, select a multinomial θ_{dat_d} that is originated from Dirichlet hyper-parameter α. Where, $dat_d = 1, ..., DAT_d$, DAT_d is the number of Web APIs and tags in the Mashup service document.*

(2) *For each topic k, select a multinomial φ_k that is originated from Dirichlet hyper-parameter β. Where, $k = 1, ..., K$, K is the number of topics in the Mashup service document.*

(3) *For each Mashup service document d, $d = 1, ..., D$, D is the number of Mashup service documents.*
   (a) *We use datd to represent a Web API or tag vector for each Mashup document*
   (b) *For each word w_{di} in the Mashup service document d, $i = 1, ... M$, perform the following:*
      (i) *Extract a Web API or tag x_{di}, where x_{di} belongs to Uniform (dat_d) distribution;*
      (ii) *Extract a topic z_{di}, where z_{di} belongs to Multinomial $(\theta_{x_{di}})$ distribution; and*
      (iii) *Extract a word w_{di}, where w_{di} belongs to Multinomial $(\varphi_{z_{di}})$ distribution.*

The corresponding probabilistic graph model of DAT-LDA is shown in Fig. 2. Where, each topic is related with the distribution φ over words, which is independently originated from Dirichlet hyper- parameter β. x represents the corresponding Web API or tag for a given word that is selected from dat_d, and each Web API or tag has the distribution θ over topics that is independently originated from Dirichlet hyper- parameter α. The topic distribution of Web APIs or tags and the word distribution of topics are combined to produce the topic z_{di}. The word w_{di} is extracted from the selected topic.

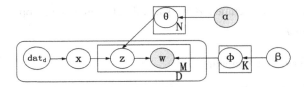

Fig. 2. The probabilistic graph model of DAT-LDA

In the above process, the posterior distribution of topics depends on the whole information of the Mashup description texts, their Web APIs, and their user-contributed tags. The parameter DAT-LDA is set as follows:

$$\theta_{dat}|\alpha \sim Dirichlet(\alpha) \tag{1}$$

$$\varphi_k|\beta \sim Dirichlet(\beta) \tag{2}$$

$$x_{di}|dat_d \sim Uniform(dat_d) \tag{3}$$

$$z_{di}|\theta_{x_{di}} \sim Multinomial(\theta_{x_{di}}) \tag{4}$$

$$w_{di}|\varphi_{z_{di}} \sim Multinomial(\varphi_{z_{di}}) \tag{5}$$

We use Gibbs sampling to approximately infer the parameters of DAT-LDA model. Gibbs sampling provides a simple and effective way to estimate the latent variables. It is a Markov Chain Monte Carlo algorithm by utilizing multivariate probability distribution to obtain a random sequence of samples. Each step of the Gibbs sampling is subject to the following distribution:

$$P(z_{di} = j, x_{di} = k|w_i = m, z_{-di}, x_{-di}, dat_d) \propto$$
$$\frac{m_{x_{di}z_{di}} + \alpha_{z_{di}}}{\sum_{v=1}^{V}(m_{x_{di}v} + \alpha_v)} * \frac{n_{z_{di}w_{di}} + \beta_{w_{di}}}{\sum_{v=1}^{V}(n_{z_{di}v} + \beta_v)} \tag{6}$$

Where, z_{-di} represents the topic assignments of all token words besides word w_{di}, x_{-di} represents the Web API or tag assignments of all token words besides word w_{di}, n_{zw} represents the total number of token word w which are assigned to topic z, m_{xz} represents the total number of token words in Web API and tag x which are assigned to topic z.

In Gibbs Sampling, we sample z_{di} and x_{di} by adjusting z_{-di} and x_{-di}. The other two latent variables, the word distribution of topic φ and the topic distribution of Web API and tag θ are respectively estimated from samples by the following formula:

$$\theta_{xz} = \frac{m_{x_{di}z_{di}} + \alpha_{z_{di}}}{\sum_{v=1}^{V}(m_{x_{di}v} + \alpha_v)} \tag{7}$$

$$\varphi_{zw} = \frac{n_{z_{di}w_{di}} + \beta_{w_{di}}}{\sum_{v=1}^{V}(n_{z_{di}v} + \beta_v)} \tag{8}$$

For Mashup service document d, we sum all θ_x to compute the topic distribution of d, where, $x \in dat_d$.

3.4 Mashup Service Clustering

The topic of Mashup service document is a simple mapping of their text vector space, and so the similarity of two Mashup service document MS_1 and MS_2 can be calculated by utilizing their corresponding topic probability distribution.

Since the topic is a mixture distribution of word vector, the following Kullback-Leibler (KL) distance can be introduced as the similarity measure standard.

$$D_{KL}(MS_1, MS_2) = \sum_{j=1}^{T} p_j \ln \frac{p_j}{q_j} \tag{9}$$

Where, j is a variable show the same topic in MS_1 and MS_2, T represents the total number of the same topic in MS_1 and MS_2, $1 \leq j \leq T$. p_j represents the topic probability of j in MS_1 and q_j represents the topic probability of j in MS_2, they can be gained in the Sect. 3.3. When $p_j = q_j$, $D_{KL}(MS_1, MS_2) = 0$. But KL distance is not symmetric, i.e. $D_{KL}(MS_1, MS_2) \neq D_{KL}(MS_1, MS_2)$. So, we often use the symmetrical version:

$$D_\lambda(MS_1, MS_2) = \lambda D_{KL}(MS_1, \lambda MS_1 + (1 - \lambda)MS_2) + \\ (1 - \lambda)D_{KL}(MS_2, \lambda MS_1 + (1 - \lambda)MS_2) \tag{10}$$

When $\lambda = 1/2$, the above formula will be transferred to JS distance, the value range of it is from 0 to 1. We use JS distance to compute the similarity between MS_1 and MS_2, which can be described as follows:

$$D_{js}(MS_1, MS_2) = \frac{1}{2}[D_{KL}(MS_1, \frac{MS_1 + MS_2}{2}) + D_{KL}(MS_2, \frac{MS_1 + MS_2}{2})] \tag{11}$$

Therefore, the similarities among all Mashup service documents can be achieved by the above formula (11), and then we can construct Mashup service document similarity matrix and perform Mashup service clustering. Different from the previous approaches [7], we combine K-Means and Agnes algorithms to achieve Mashup service clustering. The basic process can be described as follows:

(1) *Construct Mashup service document similarity matrix MSim, and then get average similarity set P between each Mashup service and all other Mashup services.*
(2) *Rank P, and find a part of Mashup services with the higher average similarity and other their similar Mashup services, and divide them into n' atom-clusters by using K-Means algorithm.*
(3) *Build the initial similarity matrix A among n' atom-clusters according to the similarity among Mashup clusters;*
(4) *Use Agnes algorithm to perform hierarchical clustering for the n' atom-clusters in A and merge a part of Mashup clusters with the higher similarity according to*

similarity threshold T among Mashup clusters, and form the updated similarity matrix B.

(5) Output Mashup service clustering result with K clusters until a termination condition is reached.

Where, *MSim* can be described as formula (12). $Sim(MS_i, MS_j)$ represents the similarity between MS_1 and MS_2, i.e. $Sim(MS_i, MS_j) = D_{js}(MS_i, MS_j)$. *n* is the total number of Mashup services, and *MSim* is a symmetric matrix, i.e. $Sim(MS_i, MS_j) = Sim(MS_j, MS_i)$, $Sim(MS_i, MS_i) = 0$.

$$MSim = \begin{bmatrix} Sim(MS_1, MS_1) & \cdots & Sim(MS_1, MS_n) \\ \vdots & \ddots & \vdots \\ Sim(MS_n, MS_1) & \cdots & Sim(MS_n, MS_n) \end{bmatrix} \tag{12}$$

Furthermore, *P* can be described as formula (13). Where, $AveSim(MS_i)$ represents the average similarity of MS_i, and it can be calculated by formula (14). The similarity between Mashup cluster C_i and C_j can be described as formula (15), and *X* is the number of Mashup services in C_i and *Y* is the number of Mashup services in C_j. Then, we can construct the initial similarity matrix *A* by element $Sim(C_i, C_j)$, which can be denoted by formula (16), and it is also a symmetric matrix, i.e. $n' <= n$, $Sim(C_i, C_j) = Sim(C_j, C_i)$, $Sim(C_i, C_i) = 0$. Meanwhile, we will update *A* by merging a part of Mashup clusters and form the updated similarity matrix *B*, the similarity between the merged cluster and other clusters is equal to the mean similarity of them. For example, the similarity between the merged cluster (C_2, C_4) and C_1 is equal to $(Sim(C_2, C_1) + Sim(C_4, C_1))/2$.

$$P = \{AveSim(MS_1), AveSim(MS_2), \ldots, AveSim(MS_n)\} \tag{13}$$

$$AveSim(MS_i) = \sum_{j=1}^{n} Sim(MS_i, MS_j)/n \tag{14}$$

$$Sim(C_i, C_j) = \sum_{x=1}^{X} \sum_{y=1}^{Y} Sim(MS_x, MS_y)/X * Y \tag{15}$$

$$A = \begin{bmatrix} Sim(C_1, C_1) & \cdots & Sim(C_1, C_{n'}) \\ \vdots & \ddots & \vdots \\ Sim(C_{n'}, C_2) & \cdots & Sim(C_{n'}, C_{n'}) \end{bmatrix} \tag{16}$$

To sum up, we combine K-Means and Agnes algorithms to achieve Mashup service clustering. The detailed algorithm can be shown as follows:

Algorithm 1. Mashup Service Clustering

Input: *Mashup Service Document Set MSDS, the similarity threshold T among Mashup clusters*

Output: *K Mashup Service Clusters*

//At first, get the topic probability distribution of MSDS by DAT-LDA, i.e. Document-Topic matrix//

1. ***Generate*** *Document-Topic matrix* **by** *DAT-LDA (MSDS) ;*
2. ***FOR*** *i=1 **TO** n*
3. ***FOR*** *j=1 **TO** n*
4. *//Compute the similarity between MS_i and MS_j by formula(11)//*
5. $Sim(MS_i, MS_j) = D_{js}(MS_i, MS_j)$ *;*
6. ***END FOR***
7. ***END FOR***
8. ***Construct*** *matrix **MSim** by formula(9);*
9. ***Construct*** *average similarity set **P** by formula(13) and (14);*
   // Secondly, obtain n' atom-clusters by applying K-Means algorithm //
10. ***Rank*** *P according to descending order;*
11. ***FOR*** *K=1 **TO** n'*
12. ***Select*** *Mashup service MS_i with the max average similarity in **P**;*
13. ***Divide*** *Mashup service MS_i into cluster C_k;*
14. ***Remove*** *Mashup service MS_i from **P** and **Update P**;*
15. ***FOR*** *j=1 **TO** n*
16. *{ IF(Sim(MS_j, MS_i)>AveSim(MS_i))*
17. ***Divide*** *MS $_j$ **into Cluster** C_k;*
18. ***Remove*** *Mashup service MS_j from **P** and **Update P**; }*
19. ***END FOR***
20. ***END FOR***
    *// Thirdly, use Agnes algorithm to perform hierarchical clustering for the **n'** atom-clusters //*
21. ***FOR*** *i=1 **TO** n'   **FOR** j=1 **TO** n'*
22. ***FOR*** *x=1 **TO** X   **FOR** y=1 **TO** Y*
23. *// X, Y represent Mashup service's number in C_i, C_j respectively //*
24. $$Sim(C_i, C_j) = \sum_{x=1}^{X} \sum_{y=1}^{Y} Sim(MS_x, MS_y) / X*Y \; ;$$
25. ***Construct*** *matrix A by formula(16);*
26. ***END FOR   END FOR***
27. ***END FOR   END FOR***
28. ***DO{ Find*** *two cluster C_i and C_j with the largest similarity in A;*
29. *// C_i and C_j represent the two clusters which will be merged //*
30. ***IF*** *(Sim(C_i,C_j)>T) **Merge** C_i and C_j into **a new cluster**;*
31. ***END IF***
32. ***Update*** *A   **and form** matrix B, the similarity between the merged new cluster and other cluster in B is equal to the mean similarity of them;*
33. *} **While** (! (all clusters are merged into a cluster || Sim(C_i, C_j)<T))*
34. ***Finally, output*** *K Mashup Service Clusters.*

---

4   Experiments

4.1   Experiment Settings

Experimental data is obtained from ProgrammableWeb[2], which provides detailed profile information of Mashup services. Figure 3 shows that the detailed profile information of

---

[2] http://www.programmableweb.com.

DUI Map Mashup service, including Mashup services' name, descriptive text, their Web API and tags information.

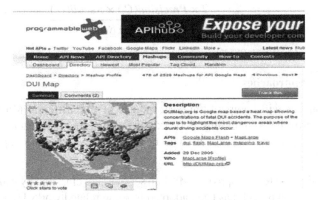

Fig. 3. The detailed profile information of DUI Map Mashup service

We crawl 6648 real Mashup services from the ProgrammableWeb as the data set. For each Mashup service, we get its Mashup service' name, descriptive text, its Web API and tags information. Then, we uniformly select 750 Mashup services from the data set to perform Web APIs clustering. In particular, we focus on 5 categories, which are "Deadpool", "Mapping" and "Shopping", "Search", and "Social". There are 119 Mashup services in "Deadpool" category, 355 Mashup services in "Mapping" category, 75 Mashup services in "Shopping" category, 95 Mashup services in "Search" category, and 58 Mashup services in "Social" category. In addition, we randomly choose 48 Mashup services as experimental noise.

In our proposed approach, when we use DAT-LDA to perform unsupervised learning, the values of two Dirichlet hyper-parameters α and β are all set to 0.01, and Gibbs sampling iterations is set to 2000 times.

4.2 Evaluation Metrics

In this paper, we choose the three metrics of precision, recall, and F-Measure from information retrieval to evaluate the performance of Mashup service clustering.

$$Precision(C_i) = \frac{SucDividedIn(C_i)}{SucDividedIn(C_i) + MisDividedIn(C_i)} \tag{17}$$

$$Recall(C_i) = \frac{SucDividedIn(C_i)}{SucDividedIn(C_i) + MisDividedOut(C_i)} \tag{18}$$

where $SucDividedIn(C_i)$ is the number of Mashup services successfully divided into cluster C_i, $MisDividedIn(C_i)$ is the number of Mashup services incorrectly divided into cluster C_i, $MisDividedOut(C_i)$ is the number of Mashup services which should be divided into C_i but are divided into another cluster.

Integrating precision and recall, F-Measure represents an overall assessment for Mashup service clustering results, and it can be calculated by the following formula:

$$F - Measure(C_i) = \frac{2 \times Precision(C_i) \times Recall(C_i)}{Precision(C_i) + Recall(C_i)} \tag{19}$$

4.3 Baseline Approaches

In this section, we will compare the performance of five different Mashup service clustering approaches, including two existing clustering approaches and three versions of our proposed approach. The details of these baseline approaches are described below.

- **K-Means**. For K-Means approach, Mashup services are clustered by using K-Means algorithm. This approach has been adopted in some related works [1, 7].
- **LDA**. For LDA approach, Mashup services are clustered by using LDA algorithm. This approach has been adopted in some related works [10, 11].
- **OD-LDA**. Only Consider Mashup Description Text based on LDA. Mashup services are clustered only by considering Mashup Description Text based on LDA.
- **CDA-LDA**. Combine Mashup Description Text and Web API based on LDA. Mashup services are clustered by combining Mashup Description Text and Web API based on LDA.
- **CDT-LDA**. Combine Mashup Description Text and tag based on LDA. Mashup services are clustered by combining Mashup Description Text and tag based on LDA.

4.4 Experiment Results

In this section, we discuss our experiment results. First, we determine the optimal topic number for our proposed approach to gain the best Mashup service clustering results. Then, we compare different approaches on precision, recall, and F-Measure metrics for evaluation of accuracy of Mashup service clustering. Finally, we will discuss the impact of the similarity threshold T in our approach on the service clustering results.

4.4.1   The Number of Optimal Topic

In our approach, we use the method described in the Literature [11] to measure Mashup service clustering results in the different topic numbers, and determine the optimal number of topic for five categories respectively. It is worth noting that we select the optimal topic number for all 5 categories instead of each single category. The reason for it is the number of Mashup services in each category is much less, which will lead to the unstable of Gibbs sampling process. The following Table 1 is the corresponding number of optimal topic for five categories.

Table 1. The number of optimal topic for five categories

Category	The optimal number of topic
Deadpool	5
Mapping	14
Shopping	3
Search	4
Social	2

4.4.2   Clustering Accuracy Comparison

Figure 4 shows the clustering accuracy comparison of five baseline approaches and our proposed approach, including the precision, recall, F-Measure and their average values.

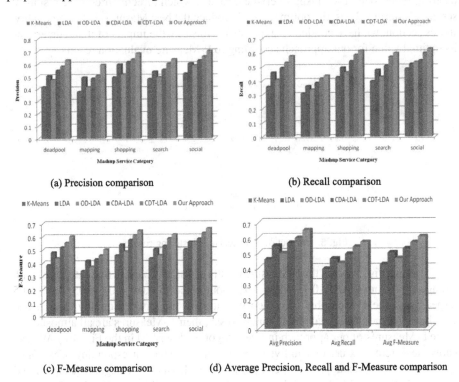

(a) Precision comparison

(b) Recall comparison

(c) F-Measure comparison

(d) Average Precision, Recall and F-Measure comparison

Fig. 4. The clustering accuracy comparison of different service clustering approaches

We can see from Fig. 4(a), (b) and (c) that the precision, recall, F-Measure and their average values of our proposed approach are significantly higher than the other five approaches. Specifically, it can be found that our approach outperforms the other two clustering approaches K-Means and LDA, which shows the advantage of our extended clustering algorithm by combining K-Means and Agnes algorithms. It can be also seen that the addition of multiple data source actually improves the performance of Mashup service clustering, as CDA-LDA and CDT-LDA outperforms LDA and OD-LDA, and our approach outperforms CDA-LDA and CDT-LDA. This observation illustrates that it is better for service clustering to simultaneously take multiple data source (the description text, Web APIs, tags of Mashup service) into consideration. Moreover, the performance of CDT-LDA is better than those of CDA-LDA, which represent the tag information is more important than the including Web APIs in Mashup service function description and clustering. LDA considers the domain characteristic of service, which makes the performance of LDA outperforming those of OD-LDA.

4.4.3   The Impact of T

In our approach, the parameter T aforementioned in the algorithm 1 of the Sect. 3.4 determines the size of K Mashup Service Clusters, which directly affects the results of service clustering. Thus, the optimal T will produce the best Mashup clustering quality. In this section, we have conducted experiments to study the impact of it and determine its optimal value for our approach according to F-Measure evaluation results.

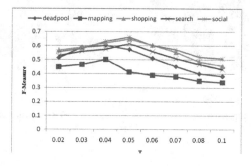

Fig. 5. Impact of T on service clustering results

The above Fig. 5 shows the F-Measure results of our approach in five categories with a range of T from 0.02 to 0.1. When T is about equal to 0.04, 0.04, 0.05, 0.05, 0.05 respectively, the F-Measures of category deadpool, mapping, shopping, search and social nearly reach their peak point, that is to say, their F-Measures are all almost close to the F-Measures of our approach in five categories described in the Fig. 4(c). We can know the different values of T in five categories achieve the corresponding peak points of F-Measure. For example, the peak points of F-Measure in category deadpool and mapping with more Mashup services respectively are 0.6088, 0.5021 when T is about equal to 0.04, while the peak point of F-Measure in category search and social with fewer Mashup services respectively are 0.6148, 0.6623 when T is about equal to 0.05. The reason for this is that category with more Mashup services needs a smaller T to merge more clusters and so produce optimal service clustering performance, and vice versa. Moreover, we can see from Fig. 5 that when T is more than 0.06, the F-Measures of our approach in all five categories constantly decrease and then almost close to those of K-Means described in the Fig. 4(c). This is because the merged clusters become less, even there are not any clusters for all n' atom-clusters in Algorithm 1 need to be merged when T becomes bigger (large than 0.06). Similarly, when T is less than 0.03, the F-Measures of our approach in all five categories constantly decrease. This is because merged clusters become more, and more noise Mashup services will be merged into clusters when T becomes smaller(less than 0.03), which will affect service clustering performance.

5   Conclusions

The rapid growth in the number of available Mashup services on Internet, coupled with the myriad of functionally similar services, makes it difficult for software developers to find suitable Mashup services to create novel Internet applications and software systems. In this paper, we propose a novel Mashup clustering approach with high accuracy for service discovery. Our approach firstly crawl and extract multiple data source of Mashup services, including the Mashup description documents, their Web APIs and tags, to form Mashup service documents. Then, we design the DAT-LDA topic model to get the topic probability distribution of Mashup service documents. Finally, we utilize JS distance to calculate the document similarity among Mashup service documents, and then integrate K-means and Agnes algorithms to perform Mashup service clustering. Our experiment results on real-world datasets show that our approach achieves a significant improvement on Mashup clustering accuracy in terms of precision, recall and F-Measure for service discovery.

Acknowledgements. The work was supported by National Natural Science Foundation of China under grant No. 61402168, 61402167 and 61272063, State Key Laboratory of Software Engineering (SKLSE) of China (Wuhan University) under grant No. SKLSE2014-10-10, and Scientific Research Fund of Hunan Provincial Education Department under grant 11C0689 and 11C0535.

References

1. Xia, B., Fan, Y., Tan, W., Huang, K., Zhang, J., Wu, C.: Category-aware API clustering and distributed recommendation for automatic Mashup creation. IEEE Trans. Serv. Comput. doi: 10.1109/TSC.2014.2379251 (preprinted)
2. Cao, B., Liu, J., Tang, M., Zheng, Z., Wang, G.: Mashup service recommendation based on usage history and service network. Int. J. Web Serv. Res. **10**(4), 82–101 (2013)
3. Liu, W., Liu, W.: Web service clustering using text mining techniques. Int. J. Agent-Oriented Softw. Eng. **3**(1), 6–26 (2009)
4. Liu, W., Liu, W.: Discovering homogeneous service communities through web service clustering. Serv.-Oriented Comput. Agents Semant. Eng. **5006**, 69–82 (2008)
5. Sun, P., Jiang, C.: Using service clustering to facilitate process-oriented semantic web service discovery. J. Comput. **31**(8), 1340–1353 (2008). (In Chinese)
6. Platzer, C., Rosenberg, F., Dustdar, S.: Web service clustering using multidimensional angles as proximity measures. ACM Trans. Internet Technol. **9**(3), 1–26 (2009)

7. Chen, L., Hu, L., Wu, J., Zheng, Z., Ying, J., Li, Y., Deng, S.: Wtcluster: utilizing tags for web service clustering. In: Proceedings of International Conference on Service-oriented Computing, pp. 204–218, Paphos, Cyprus (2011)

8. Wu, J., Chen, L., Zheng, Z., Lyu, R., Wu, Z.: Clustering Web services to facilitate service discovery. Knowl. Inf. Syst. **38**(1), 207–229 (2014)

9. Wu, J., Chen, L., Xie, Y., Zheng, Z.: Titan: a system for effective web service discovery. In: Proceedings of the 21st International Conference on World Wide Web, pp. 441–444. ACM, New York, USA (2012)

10. Chen, L., Wang, Y., Yu, Q., Zheng, Z., Wu, J.: WT-LDA: user tagging augmented LDA for web service clustering. In: Basu, S., Pautasso, C., Zhang, L., Fu, X. (eds.) ICSOC 2013. LNCS, vol. 8274, pp. 162–176. Springer, Heidelberg (2013)

11. Li, Z., Wang, J., Zhang, N., Li, Z., He, C., He, K.: A topic-oriented clustering approach for domain services. J. Comput. Res. Dev. **51**(2), 408–419 (2014). (In Chinese)

12. Yang, H., Chen, J., Meng, X., Qiu, B.: Dynamically traveling web service clustering based on spatial and temporal aspects. In: Hainaut, J.-L., Rundensteiner, E.A., Kirchberg, M., Bertolotto, M., Brochhausen, M., Chen, Y.-P.P., Cherfi, S.S.-S., Doerr, M., Han, H., Hartmann, S., Parsons, J., Poels, G., Rolland, C., Trujillo, J., Yu, E., Zimányie, E. (eds.) ER Workshops 2007. LNCS, vol. 4802, pp. 348–357. Springer, Heidelberg (2007)

13. Zhou, Z., Sellami, M., Gaaloul, W., Barhamgi, M., Defude, B.: Data providing services clustering and management for facilitating service discovery and replacement. IEEE Trans. Autom. Sci. Eng. **10**(4), 1131–1146 (2013)

14. Zhang, L., Cheng, S., Chang, C., Zhou, Q.: A pattern-recognition-based algorithm and case study for clustering and selecting business services. IEEE Trans. Syst. Man Cybern. Part A Syst. Hum. **42**(1), 102–114 (2012)

15. Cassar, G., Barnaghi, P., Moessner, K.: Probabilistic methods for service clustering. In: Proceedings of the 4th International Workshop on SMR2 Conjunction with the International Semantic Web Conference, pp. 4–20, Shanghai, China (2010)

16. Mustapha, A., Mohamed, Q., Zahi, J.: Leveraging formal concept analysis with topic correlation for service clustering and discovery. In: 2014 IEEE International Conference on Web Services, pp. 153–160, Alaska, USA, 27 June–2 July 2014

Efficient Search-Based Automatic Execution Replay for Virtual Machines

Tao Wang[✉], Jianhua Zhang, Wenbo Zhang, Jiwei Xu, and Jun Wei

Institute of Software, Chinese Academy of Sciences, Beijing 100190, China
wangtao@otcaix.iscas.ac.cn

Abstract. Execution replay of virtual machines is a useful method for debugging applications in the cloud computing environment. The traditional methods to reproduce a bug is recording every details during the system runtime. However, these methods will incur much overhead and affect the system performance, especially in a multicore processor system. In this paper, we present a virtualization-based execution replay method consisting of three steps. First, we only record some necessary events in the runtime and take a memory checkpoint in a regular interval. Second, we search for execution paths between every two adjacent checkpoints. Third, we reproduce the bugs according to these paths. We can decrease the logging overhead in the runtime by searching instead of logging. We have implemented the method and evaluate it on Xen. The experimental results demonstrate that our method can reduce the runtime overhead by 30 % effectively.

Keywords: Replay · Virtual machine · Xen · Cloud computing

1 Introduction

Virtualization is a core technology of cloud computing which is prevalent in recent years. It allows multiple operating systems to run on a single physical machine by partitioning this machine into multiple virtual machines. Each of these virtual machines is protected and isolated from the others, and performs as if it is an individual physical machine. Users of each virtual machine can execute their own applications without fearing causing a system crash by other users on the same physical machine. Because of the consolidation of virtual machines, virtualization also increases the utilization of physical machines and reduces the cost of system management. Thus, virtualization has been widely deployed and used by more and more companies. However, the systems deployed in the virtualized infrastructures often suffer from software failures because of the complex and dynamic cloud computing environment. And it is become more and more difficult to reproduce, diagnose and fix these software failures. In order to improve software quality, we needs a program execution replay system to facilitate the debugging process when the software failure happens. Besides software debugging, a program execution replay system can be used in many places.

© Springer International Publishing Switzerland 2015
L. Yao et al. (Eds.): APSCC 2015, LNCS 9464, pp. 181–194, 2015.
DOI: 10.1007/978-3-319-26979-5_13

For example, it can be used as a fault tolerance system or a dynamic analysis system. Computer forensics can also use system replay to analysis and obtain evidence.

Traditional replay systems are mainly implemented in the operating system level, but there are several reasons that we need implement it in a virtualized environment. First of all, virtualization has been widely deployed and used by more and more companies, so a replay system based on virtualization can be easily deployed widely. Second, implementing a replay system in a virtualized environment can record the entire operating system and software in it transparently. So it is more easily to replay a legacy program in a virtual machine. Finally, virtualization based replay system is essential in some situations, such as computer forensics. Because some failures are not only caused by software bugs, a deliberate hacker can intrude the operating system and eliminate the log files in the operating system. However, if we record the system execution path in virtual machine monitor (VMM) level, we can keep the log files safe and replay exactly what has happened inside the system in detail. In summary, the virtualization raise a great opportunity for implementing a replay system. Since the VMM can record the entire operating system and applications transparently, replaying a legacy system's execution process in a virtual machine is much more convenient.

The basic idea of replaying a system is recording and replaying. In the recording phase, all non-deterministic events, such as interrupts and keyboard inputs, are logged into files during the system runtime. And then we can deterministically replay the executions according to the log files in the following replaying phase. Uniprocessor replay is well studied, such as Hypervisor [1], Revirt [2], but multiprocessor replay is much more complicated because of the memory access races between processors. To address this issue, a typical method is to log the order of shared memory access [3], which incurs a significant overhead due to the serialization of memory accesses. Many studies focus on reducing the overhead of logging multiprocessors. One way is using modified hardware to facilitate the logging procedure. Although these methods are efficient, they cannot be used in commodity computers. Another way to reduce logging overhead is to record the necessary information to support replaying, such as PRES [4]. There are also some works reducing logging overhead through shift overhead from the recording phase to replay phase. But these works are all implemented in the operating system, and the log files are prone to be broken easily besides the high overhead.

In this paper, we present VMBR, a virtual machine based replay system in multiprocessors that has lower overhead both in recording and replaying phase. We separate the whole process into three phases including a record phase, a search phase and a replay phase. In the record phase, we log the nondeterministic input events, such as interrupts and keyboard inputs. We take memory checkpoints instead of logging the memory access orders in this phase. In the search phase, we search the execution paths between the two adjacent checkpoints. Although there may be more than one execution path satisfying the two adjacent checkpoints, it is sufficient for the debugging purpose. Furthermore, since each of the execution paths between the two adjacent checkpoints is independent, the searching process can be run in a parallel framework, such as MapReduce. After finding the execution path, we replay the system efficiently and debugging the system interactively.

The contributions of this paper are summarized as follows:

- We proposed a Xen based replay system which reduces overhead by searching execution paths instead of logging.
- Our system is implemented in the VMM level, and the business system logged runs as a guest virtual machine. So it is safe for the log files and pervasive for the legacy system.
- We have evaluated our system, and the lower overhead makes it reasonable to run in the deployment environment for a long time.

The rest of this paper is organized as follows. Section 2 and 3 describes the design and implementation of our proposed method. Section 4 reports on how our proposed method performs. Sections 5 and 6 discusses related work and give some future work.

2 Method

2.1 Method Overview

The basic idea of a replay system is record and replay. First a log process should be run with the business system, and the system's execution details will be logged to a file. Then in the replay phase, the logged file will be used to reproduce the execution history of the system. But the traditional replay system is a dilemma. As depicted in Fig. 1, when more details are logged in the record phase, it will cause more runtime overhead and decrease the service level. When less details are logged in the record phase, it will cause the replay phase less interactive and time consuming in the debugging process.

So we propose a method composed of three phases including a record phase, a search phase and a replay phase. Some of the logging operations are shifted to a separated search phase, as shown in Fig. 2. First, we only log the necessary event to the log file, and take a memory checkpoint every second instead of serializing all memory access orders. As there is more than one CPU or system changes memory frequently, serializing the memory access orders introduces large overhead. The reduced logging operations reduce significant overhead in the record phase. And then, we search execution paths that satisfies the logged information. For example, between the two checkpoints, an execution path should start at the first one and after the execution of some instructions, it comes to the end state. Every two adjacent checkpoints are independent, and can be searched separately. Thus it can be optimized by using some batch processing systems, such as MapReduce. We only need to search one execution path that satisfies the two adjacent checkpoints for the debugging purpose. After some interactive debugging with the programmer, we get some most suspected intervals, and a further search can be done in that interval to find out whether there are

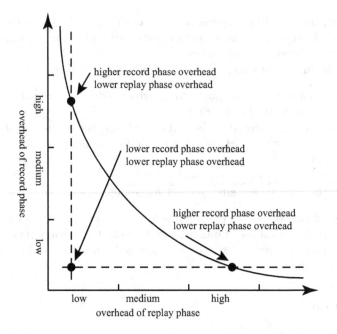

Fig. 1. The dilemma of record and replay phase.

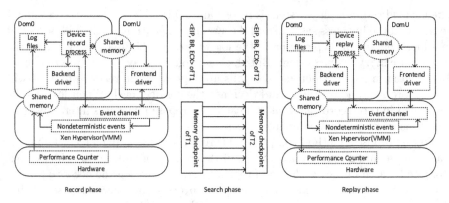

Fig. 2. Overview of the architecture.

abnormal status. After the search phase, we get the execution paths of the system. Thus, we can use this information to reproduce the execution history of the system more efficiently.

2.2 The Record Phase

The main purpose of the record phase is to log enough information to reproduce the execution path. There are two major kinds of information which needed to be logged. One is non-deterministic event, the other is memory access order.

```
            ...
loop: mov esi, mystr
      mov edi, mystr2
      cld
      mov ecx,6
rep movsb
      test ecx, ecx
      jne loop
      mov eax,ebx
            ...
```

Fig. 3. A fragment of instructions.

In a uniprocessor system, memory access order is deterministic. Replaying a virtual machine only requires logging the non-deterministic events (e.g. keyboard and mouse events, system interrupt) which change the virtual machine execution path in the runtime. Given the initial state and logged non-deterministic events, a uniprocessor system can re-execute the same execution path and output the same result as it did in the original execution. In a multi-processor environment, the current commercial off-the-self computers are mostly symmetric multiprocessor (SMP) systems, multiple CPUs of these computers can access the same memory address at the same time. So nearly every memory access is non-deterministic and needed to be logged. This will incur a large overhead and lower system performance in the runtime. For debugging, logging every execution detail to deterministically replay the system is not an efficient way. The most time consuming operation is to log the memory access orders under a multiprocessor environment. So, we take a memory checkpoint every second instead of logging every memory access orders. This will lower the overhead in the runtime significantly and make more computing resources used for key business computing.

All these events are logged with a time tag T composed of three CPU registers. The time T is defined as

$$< EIP, BR, ECX >,$$

where EIP is the extended instruction pointer used to locate a single line in the instructions as shown in Fig. 3. BR is a hardware performance counter called branch retired indicating the number of executed branch instructions. When there is a loop in the instructions, the register pair $< EIP, BR >$ can be used to locate the instructions and how many cycles it has been executed. When the REP prefixes is applied to an instruction, the instruction can be executed for many times depending on the value of register ECX. So the ECX should be added to the time vector T to indicate the number of an instruction has been repeated.

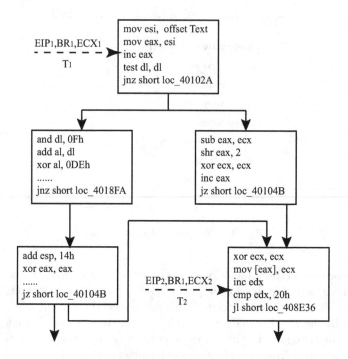

Fig. 4. An execution flow of a code fragment

2.3   The Search Phase

After the record phase, we get the system's runtime information consisted of the non-deterministic events and a sequence of memory checkpoints. In order to replay the whole system execution path, we need to search out the execution paths between every two adjacent memory checkpoints. Each of the adjacent memory checkpoints is independent, and can be searched simultaneously. The search process is divided into two steps, the path generation step and the path selection step.

As depicted in Algorithm 1, we generate all the possible execution paths between checkpoints C_1 and C_2. There are three kinds of input parameters C, T, N, where C represents the memory checkpoints (C_1, C_2), T represents the corresponding time (T_1, T_2) and N represents the VCPUs N_i. And time T_i is composed of EIP_i, BR_i, ECX_i of each VCPU. First we first calculate the number of branch instructions which executed between time T_1 and T_2 for each VCPU of a virtual machine.

$$nbranches = T_2.BR - T_1.BR,$$

where $T_i.BR$ is the number of branch instructions that a VCPU has been executed at Time T_i. Then we generate an execution path for this VCPU as shown in Fig. 4. The execution path starts from the instruction at $T_1.EIP$, and after executing *nbranches* branch instructions it ends at the instruction at $T_2.EIP$. In Fig. 4, there are two execution paths start and end at the same instructions. But they have different number of instructions in their execution paths, one is 3 and another is 2. So one of them can be eliminated by comparing the branch instructions in the execution path with *nbranches*. Finally we get same execution paths that satisfy the time T_1 and T_2 for current VCPU.

The second step of the search process is to select one execution path from the possible execution paths that we generate in the first step. The execution paths which we generated in the first step have satisfy the time T_1 and T_2, so in algorithm 2 we select an execution path which satisfies the memory checkpoints C_1 and C_2. Take a dual-core virtual machine for example, first we select an execution path of $VCPU_1$ and another execution path of $VCPU_2$. Then we intercross execute these two execution paths and verify if the end state of the memory satisfies the memory checkpoint C_2. And if the end state is the same as C_2, we return these execution paths as reasonable paths which can both satisfy the memory checkpoints and the time points.

The search process can be done offline. The programmer does no need to interact with the search process until an execution path has been searched out. So a longer search time is acceptable. There are also same methods to reduce the search space, such as using the output information. For example, we can log some I/O output at time T, which can be used to guide the search process to eliminate some search path that does not produce the same output. Since all the adjacent memory checkpoints are independent, we can assign different priority to different checkpoints. We can estimate a time T which most likely cause the system's error, and search around the time T instead of search the all execution paths from start to the end.

2.4 The Replay Phase

In the replay phase, a replay monitor reads the logged non-deterministic events and execution paths which have been searched out to replay the virtual machine. In order to replay the virtual machine according to the information we have logged, the non-deterministic events should be replayed exactly at the same point in the execution paths during replay. For example, the logged keyboard input events are simulated and sent to the virtual machine at the same time when they are logged in the virtual machine runtime. Since the checkpoints are independent from each other, the replay monitor can start from any checkpoints and execute forward. For debugging, the system may be abnormal only in a short period, so we can replay the system just before this period instead of the starting point over and over.

In order to check the virtual machine's state, such as registers and memory state, the replay process can be paused at any time. So a programmer can check the systems states, and decide if continuing current execution process or start a new execution process from another checkpoint.

Algorithm 1. Path-generation(C,T,N). Generate all possible execution path for each CPU, and filter them with time point $T < EIP, BR, ECX >$

Input:
    Two adjacent memory checkpoints C_1, C_2;
    Two corresponding time points T_1, T_2;
    Number of VCPUs of this virtual machine;
Output:
    Possible execution path for each VCPU which satisfy both the time points T_1 and T_2
1: **for** each $vcpu \in$ All the VCPUs of this guest virtual machine **do**
2:    $nbranches = T_2.BR - T_1.BR$
3:    Generate an execution path for current $vcpu$. The execution path starts from $T_1.EIP$, and executes $nbranches$ branches
4:    **if** Execute the instructions after the $path$ until reach to the next branch instruction. Check if this execution path can reach a state which satisfy $T_2.EIP$ and $T_2.ECX$ **then**
5:        We get a possible execution path for current $vcpu$ which satisfy T_1 and T_2
6:        Add this path to current $vcpu$ execution path set
7:    **else**
8:        This path cannot satisfy T_1 and T_2, ignore this path and search for another one.
9:    **end if**
10: **end for**

Algorithm 2. Path-selection(C,T,N,P). Select a possible execution path, which satisfy the memory state of both time point T_1 and T_2

Input:
    Two adjacent memory checkpoints C_1, C_2;
    Two corresponding time points T_1, T_2;
    Number of VCPUs of this virtual machine;
    The execution paths generated from algorithm 1;
Output:
    An execution path which satisfy both the time point T and memory checkpoints;
1: **for** each $path_1 \in$ the generated $paths$ of $vcpu_1$ of this guest virtual machine **do**
2:    **for** each $path_2 \in$ the generated $paths$ of $vcpu_2$ of this guest virtual machine **do**
3:        **for** each intercross execution of $path_1$ and $path_2$, we get a memory $state$ **do**
4:            **if** the state satisfy the memory checkpoints of T_2 **then**
5:                We get a execution path which satisfy both the memory checkpoints and time points
6:            **else**
7:                Ignore these path and search for another one.
8:            **end if**
9:        **end for**
10:    **end for**
11: **end for**

---

3   Implementation

We have implemented the proposed method based on SMP-Revirt. There are some challenges relevant to our method.

First, we take the memory checkpoints based on Xen's existing machinery for performing live migration. The key technology of live migration is a copying process, which copies the virtual machine's memory to another physical machine while the virtual machine still running on the original physical machine. During the migration, the memory is marked as read-only by the Xen VMM and all write operations are intercepted. The VMM marks the changed memory pages as dirty pages and transfers them to the new physical machine periodically. After several rounds of copying, the migration progress can copy the changed dirty pages out as fast as the virtual machine generates the dirty pages per round. Then the VMM suspends the guest virtual machine and copy the remaining dirty pages to the new physical machine. After that the virtual machine in the new physical machine is started and the virtual machine on the old physical machine is shut down. The downtime of this process is mostly under 100 ms which depends on the amount of dirty pages remaining when the guest virtual machine is suspended.

In order to track the guest virtual machine's dirty pages, Xen provides a mechanism called shadow page table which is a shadow version of the guest virtual machine's page table. The guest virtual machine's page tables are marked as read only, so the VMM can trap the guest access to its internal page tables and propagate to the shadow tables as appropriate. We use this technology to take an incremental memory checkpoint at intervals. At each interval, the VMM pauses the guest virtual machine and copies the dirty memory pages and CPU states to a ring buffer. Then the VMM resumes the guest virtual machine's execution on the current physical machine rather than on the remote physical machines. While the VMM writes data into the ring buffer, a process in dom0 reads data from the ring buffer and writes it to a log file.

Second, we record all nondeterministic events relevant to domU into a log file. There are two kinds of nondeterministic events. One kind is reading instructions which is used by the virtual machine to get the system's states, such as RDTSC and CPUID. Take RDTSC for example, we first set RETSC as a privileged instruction by setting the TSD bit of CR4. When RDTSC is set to privileged instruction, it can only be executed in privileged level 0, otherwise a general protection fault will be triggered. In the Xen virtualization system, only dom0 can be executed in privileged level 0, all domUs are executed in unprivileged level. So when a monitored domU executes a RDTSC instruction, a general page fault will be triggered. And this fault will be caught by Xen in do_general_protection interrupt handling function. In this function the VMM simulates this instruction for domU and logs an RDTSC event to the ring buffer. After returning from VMM, the domU continues to execute the next instruction. The logged RDTSC event in ring buffer will be written to the log file finally by dom0.

The other kind of nondeterministic events is the input event comes from the external devices, such as console input event. In a paravirtualized virtualization system, the devices are divided into two parts: the frontend and the backend.

The frontend of a virtualized device executes in the guest virtual machine, and the backend of the virtualized device executes in the VMM. These two parts send and retrieve data through event channel and shared memory page. So in order to log the console input event, we intercept the event channel and shared page between the frontend and the backend of the virtual console device. Every data send from the backend to the frontend of the virtual console device is copied out and sent to the log file.

Third, in order to locate every logged event to a particular time point, we retrieve the time $T < EIP, BR, ECX >$ and log it with every events. The EIP and ECX register is retrieved from the cpu_info structure. The amount of branch instructions executed until the current event cannot be read from register directly. First we enable the performance monitor function in the CPU. Then by setting the MSR_TBPU_ESCR0, MSR_MS_CCCR0 and MSR_MS_COUNTER0 registers, we enable the CPU to count branch instructions retired and select which event should be counted, such as unconditional jump, conditional jump, call, return and indirect jump. After that we count the branch retired number in each physical CPUs, but we cannot distinguish how many branches executed in dom0 and domUs. To solve this problem, we add performance counter variants in all virtual CPUs(VCPUs) of each domU. When one VCPU is scheduled, we save the current performance counter information to the last VCPU, and reset the performance counter. Then we can retrieve the current branch retired variant from the VCPU structure.

4   Evaluation

In this section, we present the experimental setup and evaluation results of our proposed method.

Our proposed method is implemented based on Xen hypervisor 3.0 which supports Pentium 4 CPU. So we conduct our experiments on a PC server which equipped with Intel Pentium 4 3.00 GHz CPU, 2 GB memory, and a 7200rpm 1 T disk. The Xen hypervisor 3.0 is installed on this server, and a CentOS operating system with the 2.6.11 kernel is setup as dom0. We create a para-virtualized virtual machine called domU, and install CentOS 4.8 with the 2.6.11 kernel. We change the domU's configuration file from one CPU to 2 CPUs, and test its overhead and the size of log files.

We compare our method with SMP-Revirt both in CPU utilization and log file size. We configure 2 CPUs for domU and generate the same workload in domU under our method and SMP-Revirt. SMP-Revirt uses CREW protocol to record the memory access order, and our method record the incremental memory states changes every second. The workload running in domU reads and writes the memory frequently, and generates some output in the console.

Figure 5 shows the comparison of overhead between our method and SMP-Revirt. We have normalized the SMP-Revirt's overhead to 1, our method will reduce 30 % CPU overhead, but slightly increase the size of log files. Because SMP-Revirt records all the memory access orders, it will spend more time to

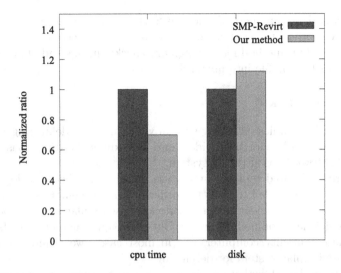

Fig. 5. The comparison of overhead between our method and SMP-Revirt.

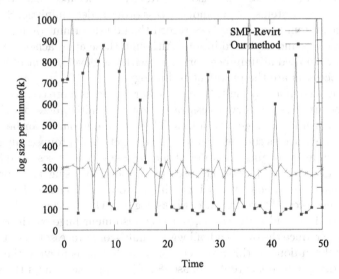

Fig. 6. The comparison of generated log size between our method and SMP-Revirt.

determine which CPU access memory first. And this procedure is the main overhead of SMP-Revirt. By only recording the memory state at intervals, our method can reduce the overhead dramatically. Now our method generates a little more log files. This is because we currently record the memory checkpoints in a memory page granularity. This causes a lot more redundant data logged into the log file. But it can be optimized to only log the changed content in a memory page instead of the whole memory page. Figure 6 shows the size of log files which generated by our method and the SMP-Revirt. Because of SMP-Revirt use a CREW protocol to log all memory access orders, it generates about 300k of log file steadily. Our method log the memory checkpoint instead of access order, so it generates the log file intermittently.

5 Related Work

There are a lot of studies on replay system which used for debugging and other purpose. These works can be divided into two kinds: hardware-based replay system and software-based replay system.

The hardware-based replay systems use modified hardware to log the execution information needed to infer the order of shared memory accesses. These approaches are more efficient than the software-based replay systems, because of their high efficiency modified hardware. But these approaches cannot be widely used on the commodity computers. So in most cases, we would like to use a software-based replay system for debugging.

The software-based replay approaches in a uniprocessor environment have been studied well, such as Hypervisor [1], Revirt [2]. Because there is only one processor in these systems, the memory access order is deterministic. So in order to replay the system, it only needs to record the non-deterministic input events, such as interrupts and keyboard inputs. Although these approaches are also efficient, they are limited at uniprocessors. And with the development of hardware, the multiprocessors are the prevalent technology.

Replaying a commodity multi-core processor system is more complex and less efficient. Because the memory access race between multi processors should be recorded and will incur huge slowdowns. These race outcomes can be recorded either by memory access content or order. LeBlanc et al. [5] proposes an approach to log the system by logging the order of shared memory accesses. And SCRIBE [6] logs the order of conflicting accesses to memory by using page protection in virtual machines. These works incur high overhead because of the intense memory access race. There are also works logging the system by logging the values returned by load instructions, but they also incur huge overhead [7].

In order to reduce the overhead of logging multi-core processor systems, many studies have been done in this area. One way to do this is moving the memory access race detection to the replay phase [8]. This approach traces the synchronization operations and use it to detect memory access race in the replay phase. Another way to reduce the recording overhead is relaxing the definition of replay. In some scenario, such as debugging, deterministic replay is not necessary, so we don't require all instructions return the same value as in the original run. For example, PRES only guarantee the replay phase can produce the same outputs as the original run instead of deterministic replay the whole system, because output determinism is sufficient for debugging [4]. Respec reduces recording overhead by speculative logging and externally deterministic replay, which also only guarantees the output deterministic [9].

There are also some studies focusing on reducing the logging overhead by shifting some work from the recording phase to the replay phase. These works also record a subset of the execution information, and use this information to search an equivalent execution path in the replay phase. Although these works can reduce the overhead of logging phase, they will incur large overhead in the replay phase, which makes the debugging process more time consuming.

Some works ensure that inter-thread communication is deterministic for a given input. These works don't need to log the memory access order by limiting the programs free of data races [10] or a particular language [11]. These works also incurs large overhead in the logging phase [12].

In summary, current replay systems which depend on hardware or uniprocessor cannot be widely used. The software-only multiprocessor deterministic replay systems are mostly have high overhead. And they are more vulnerable because of they are running in the same environment with the production system. The logged files can be easily destroyed by a software bug or malicious hackers. So we implement a replay system in a hypervisor which is transparent to virtual machines running on it. And we shift some work from the logging phase to a separate phase called search phase, which will reduce the logging overhead and not increase the replay overhead.

6 Conclusion and Future Work

In this paper we present a virtualization-based replaying system which consists of three phases. In the first phase, we record the virtual machine's execution information, such as input events and memory checkpoints. By logging the memory checkpoints instead of access orders, we reduce the record overhead effectively. In the second phase, we search for the execution paths between the checkpoints. After we have searched out an execution path, we can replay the system in the third phase. And the programmer can pause and debug the system interactively. We have evaluated our method on Xen. In the record phase, our method can reduce 30 % overhead than the SMP-Revirt, and the growth of record log is also acceptable.

In order to help a programmer to debug software bugs in a virtual machine, there are still some work need to do to facilitate the debugging process.

Since the generated log file is still so large for the programmer to inspect, it is hard to find the root cause of the bugs. And finding out all the execution path of the system is also a hard work. So we can help the programmer to analyze the log file and locate the suspect time point which the problem may happens around. Because the memory checkpoints are independent, we can search around that time point first to reduce the search time.

We log the memory with the page granularity now, there is many redundant data on these pages. So it can be optimized by logging the memory changes with only changed contents, which will reduce the size of generated log files. And this will help the system logging the execution information in a long period and using less disk space.

Acknowledgment. This work was supported in part by the National Natural Science Foundation of China under Project 61402450, and the Beijing Natural Science Foundation under Project 4154088.

References

1. Bressoud, T.C., Schneider, F.B.: Hypervisor-based fault tolerance. ACM Trans. Comput. Syst. **14**, 80–107 (1996)
2. Dunlap, G.W., King, S.T., Cinar, S., Basrai, M.A., Chen, P.M.: Revirt: enabling intrusion analysis through virtual-machine logging and replay. In: Proceedings of the 5th Symposium on Operating Systems Design and Implementation, ser. OSDI 2002, pp. 211–224. ACM, New York (2002)
3. Dunlap, G.W., Lucchetti, D.G., Fetterman, M.A., Chen, P.M.: Execution replay of multiprocessor virtual machines. In: Proceedings of the Fourth ACM SIGPLAN/SIGOPS International Conference on Virtual Execution Environments, ser. VEE 2008, pp. 121–130. ACM, New York (2008)
4. Park, S., Zhou, Y., Xiong, W., Yin, Z., Kaushik, R., Lee, K.H., Lu, S.: Pres: probabilistic replay with execution sketching on multiprocessors. In: Proceedings of the ACM SIGOPS 22nd Symposium on Operating Systems Principles, ser. SOSP 2009, pp. 177–192. ACM, New York (2009)
5. Leblanc, T., Mellor-Crummey, J.: Debugging parallel programs with instant replay. IEEE Trans. Comput. **36**(4), 471–482 (1987)
6. Laadan, O., Viennot, N., Nieh, J.: Transparent, lightweight application execution replay on commodity multiprocessor operating systems. In: Proceedings of the ACM SIGMETRICS International Conference on Measurement and Modeling of Computer Systems, ser. SIGMETRICS 2010, pp. 155–166. ACM, New York (2010)
7. Bhansali, S., Chen, W., De Jong, S., Edwards, A., Murray, R., Drinic, M., Mihocka, D., Chau, J.: Framework for instruction-level tracing and analysis of program executions. In: Proceedings of the 2nd International Conference on Virtual Execution Environments, pp. 154–163. ACM (2006)
8. Ronsse, M., De Bosschere, K.: Recplay: a fully integrated practical record/replay system. ACM Trans. Comput. Syst. **17**(2), 133–152 (1999)
9. Lee, D., Wester, B., Veeraraghavan, K., Narayanasamy, S., Chen, P.M., Flinn, J.: Respec: efficient online multiprocessor replay via speculation and external determinism. In: Proceedings of the Fifteenth Edition of ASPLOS on Architectural Support for Programming Languages and Operating Systems, ser. ASPLOS 2010, pp. 77–90. ACM, New York (2010)
10. Olszewski, M., Ansel, J., Amarasinghe, S.: Kendo: Efficient deterministic multithreading in software. SIGPLAN Not. **44**(3), 97–108 (2009)
11. Bocchino, R.L., Jr., Adve, V.S., Dig, D., Adve, S.V., Heumann, S., Komuravelli, R., Overbey, J., Simmons, P., Sung, H., Vakilian, M.: A type and effect system for deterministic parallel java. In: Proceedings of the 24th ACM SIGPLAN Conference on Object Oriented Programming Systems Languages and Applications, ser. OOPSLA 2009, pp. 97–116. ACM, New York (2009)
12. Berger, E.D., Yang, T., Liu, T., Novark, G.: Grace: safe multithreaded programming for C/C++. In: Proceedings of the 24th ACM SIGPLAN Conference on Object Oriented Programming Systems Languages and Applications, ser. OOPSLA 2009, pp. 81–96. ACM, New York (2009)

Leveraging Process Mining on Service Events Towards Service Composition

Yulai Li[✉], Hongming Cai, Chengxi Huang, and Fenglin Bu

School of Software, Shanghai Jiao Tong University, Shanghai, China
{tiffany1105,hmcai,hichens}@sjtu.edu.cn, bu-fl@cs.sjtu.edu.cn

Abstract. Service composition is a widely-used approach in the development of applications. However, well-designed service composition approaches always lacks the consideration of execution environment, and the approach designed for application execution is usually incomplete and lacking necessary business consideration. In order to improve the comprehensiveness covered both design and execution stages, a service composition approach based on process mining is proposed. First, a meta-model is designed to connect the information of execution environment and business requirement. Next, the scene model based on this meta-model is generated by leveraging process mining. Then the scene model is applied to do service composition, including service selection from the Service Registry. After that, BPEL instance is converted based on aggregated scene information so as to enable application execution. Finally, a cloud-based logistics platform is implemented to verify the approach, and the result shows that the approach has high requirement accuracy and execution effectiveness.

Keywords: Service composition · Service composition pattern · Process mining · Cloud computing

1 Introduction

At present, cloud computing is gradually transformed into a general technology because of its flexibility and scalability. With the popularization of cloud service, more and more people develop new web service by reusing existing services. Service composition is a popular reusing approach mainly because it can reduce the logic complexity.

As for the workflow-based service composition, the general approaches usually follow these 4 steps: (1) the business analysts do some analysis and discuss with related people such as consultants and end-users and to come out with the requirements of the project; (2) the developers design the workflow of the composite service according to the related part of the requirements; (3) the developers select the services, bind the services to the workflow and then test (4) the operation staffs deploy and monitor the composite service. The existing approaches usually differ in step 2 and step 3. There is a tradeoff between the business and the execution effectiveness for developers to choose the service composition approach. Focusing on business, well-designed service composition approach always lacks the consideration of execution information in production environment. Focusing on execution effectiveness, the approach considering execution

© Springer International Publishing Switzerland 2015
L. Yao et al. (Eds.): APSCC 2015, LNCS 9464, pp. 195–209, 2015.
DOI: 10.1007/978-3-319-26979-5_14

information is usually incomplete and lacking business support. Moreover, in step 1, the requirements of the composite services are usually from the business analysts, but their understanding of the requirements contains one-sided and subjective factors. Even well-designed approach cannot fully meet the practical needs. So we would like to propose a holistic service composition approach, considering both the practical business and execution effectiveness.

Process mining [1] is a process management technique that extracts information from event logs recorded by an information system to discover, analyze and enhance process models. The event logs are from the practical business. The processes discovered by process mining are usually the best practices during the execution period. Leveraging process mining technique to do service composition can just give consideration to the practical business and the execution effectiveness. The discovered processes with frequently used services can be regarded as composite web service patterns to help developers do service composition.

Similar to this idea, [2] proposed a mining algorithm based on statistical techniques to discover composite web service patterns from execution logs to better understand, control and eventually re-design the composite services. Also, the researchers in [3] proposed a novel approach to generate service composition pattern for cloud migration from a set of service composition solutions by a graph similarity analysis approach. And in [4], an event-based monitoring approach for service composition infrastructures is presented to provide a holistic monitoring approach by leveraging Complex Event Processing (CEP) techniques. The system is implemented for a WS-BPEL composition infrastructure. Its goal is to avoid fragmentation of monitoring data across different subsystems in large enterprise environments by connecting various event producers. Although these researches are all devoted to optimize the existing service composition based on the service composition patterns mined from history logs, but they do not use the standard and systematic process mining method which is proposed in [1].

Considering the above situations, in order to improve both execution effectiveness and comprehensiveness of existing service composition approaches, we propose a service composition approach covering both design and execution stages, which is based on process mining to take account of practical business, execution effectiveness and the completeness of the approach. Generally speaking, the main contributions in this paper can be summarized as follows:

- We propose an overall framework covering the whole lifecycle of service composition based on process mining (Sect. 3).
- According to the framework, we design a meta-model considering both the information of execution environment and practical business to support the holistic service composition approach which covers both the practical business and execution effectiveness. Then we propose the specific and detailed approach and strategy of service composition (Sect. 4).
- The proposed approach is applied to a collaborative data center of a logistics cloud service platform to discuss its feasibility and its adaptive scenario (Sect. 5).

2 Related Work

Nowadays, in the application area of service composition, the development approaches based on workflow standard such as BPEL occupy the vast majority of the service composition approaches. And in the research area, in recent years, a large amount of service composition researches are based on non-functional features of web services' QoS (Quality of Service) information.

Inspired by the concept of friction in physics, a service selection technique is proposed in [5] to select the best potential candidate service from a set of functionally equivalent ones. The proposed service composition approach is based on the QoS attributes of individual service nodes to minimize the waiting time for each request at the service nodes. The approach in [6] takes several aspects of web service into consideration, including QoS, user preference and the service relationship.

There are also some dynamic approaches such as context-aware dynamic service composition approach and AI planning techniques in addition. For example, [7] uses models at runtime to guide the dynamic evolution of context-aware web service compositions to cope with unexpected situations in the open world. The evolved model guides changes in the underlying WS-BPEL composition schema. As for AI planning, [8] explores how to use the domain-independent system named SHOP2 HTN (Hierarchical Task Network) planning system to do automatic composition.

With the popularization of cloud computing, multi-tenant and cloud-oriented service composition researches are growing. Based on granularity computing, [9] proposes a service granularity space for multi-tenant service composition, which provides a semantic basis for multi-tenant service composition.

In addition, in the service mining [10] area, i.e. the application of the process mining in service, the service governance researches are the majority. A methodology based on process mining is proposed in [11] to do business process analysis in healthcare environments to identify regular behavior, process variants and exceptional medical cases to conduct a hospital emergency service.

However, there are few approaches about service composition in the area of service mining, such as service composition analysis and optimization. The more similar researches are like [2–4], using data mining technique, not standard process mining technique. It lacks a comprehensive service composition approach based on standard process mining. Thus, the service composition approach based on process mining is quite novel and worthwhile for us researchers to discover its value and possibility.

3 Framework Overview

In this section, we will give an overall framework of our approach. As shown in Fig. 1, it is divided to 3 phases, process mining phase, service composition phase and service deployment phase. Among them, the result of process mining phase is used to support the other two phases.

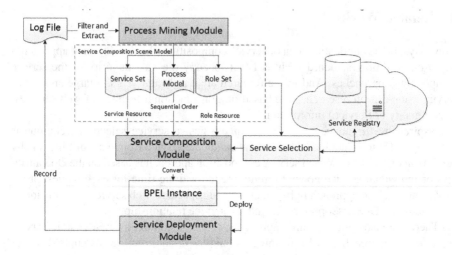

Fig. 1. An overall framework based on process mining

The process mining phase is to preprocess existing log files and to do process mining. Preprocessing existing log files includes analyzing the structure of existing log files, filtering errors or outdated information and extracting keys or other useful information. The process mining tool, ProM [12] is used in doing process mining. Besides mining process itself, other information also needs to be mined, including the relation between roles and the events in the process and the relation between services and the events in the process. These relations are aggregated to generate process model, service set and role set to support the latter 2 phases. Among them, process model is one kind of the presentation methods of service composition patterns.

The service composition phase is to collect the process model, service set and role set generated by the process mining phase, to use certain service selection strategy and do information mapping to generate composite service, according to the file format which service composition needs, such as the BPEL file and WSDL file. These two files need to be packaged with the related fine-grained component services and to be modified sometimes to do the deployment, entering the next phase.

The service deployment phase is to deploy the composite service generated by the service composition phase. This phase will record operations to the log file again to provide data support for the latter optimization and recomposition, being closed loops.

4   Detailed Approach

In this section, the framework mentioned above will be refined to introduce its specifics. Figure 2 shows all meta-models and their relationship involved in our approach, including Process Mining Model, Service Composition Scene Model, Business Process Execution Model and Service Management Model. Process Mining Model corresponds to process mining phase, Service Composition Scene Model corresponds to service composition phase, and Service Management Model corresponds to service deployment phase. Business Process Execution Model is the intermediate product between service composition phase and service deployment phase, which is generated after service composition phase and needs manual further development.

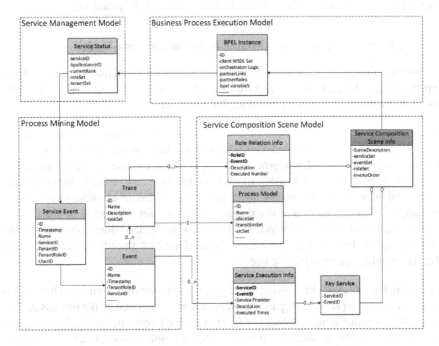

Fig. 2. Meta-models for this approach based on process mining

Process Mining Model shows the data information the process mining phase involves. Service Event is the data source of the process mining. Service Event records the information needed by process mining phase and the subsequent 2 phases, centering around events to record event name, the timestamp it occurred, ID of called service, the related tenant information, etc. Event is extracted and converted from Service Event, and a Trace consists of multiple Events.

Service Composition Scene Model shows the data information the service composition phase involves. Through the processing of process mining, Process Model, Role Relation Info and Service Execution Info are generated. Process Model is the discovered potential process, including the tasks in the process and their order. Role Relation Info represents the relations between tasks and the roles of tenants. Service Execution Info represents the relations between tasks and services. Service Composition Scene Info is aggregated from the information in Process Mining Model.

Business Process Execution Model shows the data information of the BPEL Instance converted from Service Composition Scene Model. It is a composite service instance in accordance with the BPEL standards. This approach is based on the history of the execution, while the current situation may have been changed. So it needs further develop by developers to meet the changes in current situation.

Service Management Model shows the data information the service deployment phase involves. Service Status represents the running status of a service, which will change with the deployment status of global services. Along with the records by Service Status, the logs in form of Service Event in next round will be produced.

4.1   Log Processing and Process Mining Based on the Meta-Model

The process mining phase mentioned in chapter 3 can be divided into these three parts: Preprocessing of Logs, Processing Event Logs and Leveraging Process Mining. In this section, these three parts will be introduced in more detail.

Preprocessing of Logs. Business process will change over time. For example, an existing task is cancelled, then the recent added logs will not record this task, but because the history logs have recorded it, process mining will also discover it. Therefore, before doing process mining, it is necessary to filter outdated data and noise in logs to extract the required information. At first, we integrate the distributed data sources and construct the mapping between models and databases. Then we set the period of time, some specific tenants and other conditions to filter noises. Finally, according to the meta-model of Service Event shown in Fig. 2, we build the database tables needed for extracting the information which process mining needs.

Processing Event Logs. Processing event logs is to convert the information for process mining we got from previous step into the input criterion required by the process mining tool, ProM, its required input format is XES (Extensible Event Stream). As Fig. 3 shows, XES file is a process instance that has integrated multiple Service Events. It contains multiple processes, which are called trace in XES standards, and every trace contains multiple events.

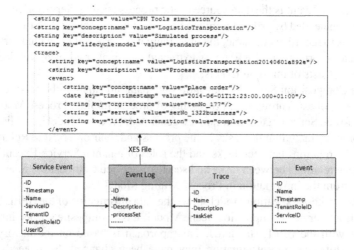

Fig. 3. Process event logs in XES format

Leveraging Process Mining. ProM is chosen to be our process mining platform. To do process mining, a certain process mining algorithm should be chosen in advance. In this paper, we just choose the α-algorithm. The algorithm candidates are all depended on the ProM plugins. After Event Logs Processing, the XES file is generated. It is legal input files of ProM, so it can be imported to ProM without extra processing.

We use ProM to analyze the interaction records among the business activities in the processes and through mining and reasoning to get the process model, i.e. process discovery. After process discovery, the process model is stored in the form of Petri Net, described in the form of the pnml file. The major information in pnml file are Place, Transition and Arc, which can be seen in Fig. 4. Place is condition, Transition is the event that may occur, and Arc describes which places are pre- and/or post conditions for which transitions.

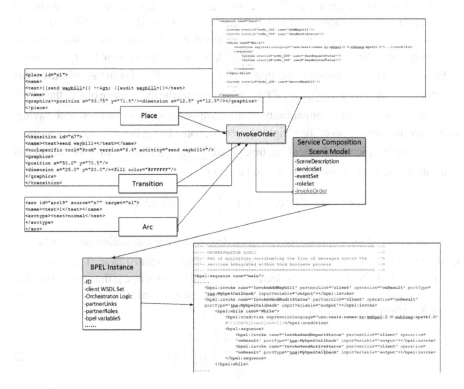

Fig. 4. Service composition phase

In addition, by means of statistics and analysis of the records of service calls of each event, we could get Service Execution Info. Similarly, by means of statistics and analysis of the relationship between the event and the tenant role the tenant uses at this time, we could get Role Relation Info.

4.2   Service Composition Based on Scene Model

According to the service composition phase which is mentioned in chapter 3, based on the scene model in Fig. 2, the Service Composition Method is designed. In this section, the method will be divided into two parts: Service Selection Strategy and Service Composition Method to describe it in more detail.

Service Selection Strategy. The same event may not invoke the fixed service every time, and one service may also be provided to multiple events, so we need a method to choose suitable services. That is the strategy we use to extract Key Service from Service Execution Info. We calculate the weight of the service for the event to measure its criticality to do service selection.

From Service Execution Info we can get the number of times the service s has been called for the event e, written as $ET(e, s)$. Because that we have done Preprocessing of Logs at first, which is mentioned in 4.1, the outdated data is filtered, so $ET(e, s)$ can be used to calculate the importance of service s to event e. Towards one certain event e_i, we first calculate the importance of its related s_j to this e_i, written as $w(e_i, s_j)$, as shown in Eq. (1). $ET(e_i, s_j)$ represents the importance of s_j to e_i, $E(s_j)$ is the set of the events which are related to s_j, and $\sum_{e \in E(s_j)} ET(e, s_j)$ represents the importance of s_j to the whole process, considering the compatibility of the component services and resource saving. u and v are the parameters of the weighted average, considering that $ET(e_i, s_j)$ does more contributions to the requirement accuracy and $\sum_{e \in E(s_j)} ET(e, s_j)$ may be calculated multiple times, so we pre-set u as 0.9 and v as 0.1.

$$w(e_i, s_j) = u \cdot ET(e_i, s_j) + v \cdot \sum_{e \in E(s_j)} ET(e, s_j) \tag{1}$$

Which s_j we should choose for e_i will be determined by calculating $priority(e_i, s_j)$, as shown in Eq. (2), $S(e_i)$ represents the set of services related to e_i. The closer that $priority$ is to 1, the higher service selection priority this service has. If the selected service is not available, the service with next priority will be selected. The selected service is called Key Service, as shown in Fig. 2 before.

$$priority(e_i, s_j) = \frac{w(e_i, s_j)}{\max\limits_{s \in S(e_i)} \{w(e_i, s)\}} \tag{2}$$

Service Composition Method. After the steps mentioned above, Role Relation Info, Process Model and Key Service, which is extracted from Service Execution Info, will be aggregated to build Service Composition Scene Info. Among them, according to the placeSet, transitionSet and arcSet in Process Model, centered around transition, i.e. event, we will get the relation and order of transitions, then it will be converted to be the invokeOrder in Service Composition Scene Info.

As shown in Fig. 4, by analyzing the source and target in arc, we could get the sequence of place and transition. The transitions then will be mapped to events and the places will be used to assist in analyzing the complicated invoke order, such as branch and loop. In it, the branch and loop in process model will be converted to be the conditions such as "if" and "while" in invokeOrder, in our custom xml format.

Then the Service Composition Scene Info will be generated to BPEL Instance according to the mapping relation in Table 1. After that, the BPEL variables and partnerLinks should do some extra operations, such as adding postfix to generate the attributes in BPEL Instance. For example, the partnerLink of the service named "Waybill-Service" is named as "waybillPL". Other attributes in BPEL Instance can all be found in Service Composition Scene Info. In this step, the client WSDL Set is in the format of files, a set of WSDL files of corresponding component services.

Table 1. The mapping relations from service composition scene info to BPEL instance

Attributes in BPEL instance	Attributes in SC scene model
Client WSDL Set	serviceSet
Orchestration logic	invokeOrder, eventSet and serviceSet
partnerRoles	roleSet
BPEL variables and partnerLinks	Generated by serviceSet and client WSDL Set

4.3 Further Development and Deployment Based on BPEL Instance

After service composition phase, the BPEL Instance is generated. Then the BPEL Instance need to do some further development and then deploy it in a certain server. This part is not our contribution, but a necessary part to maintain the closed loop of the service composition lifecycle.

Our approach is based on the history of the execution, so the generated BPEL Instance is inevitable to be a bit dated. While the current situation is likely to be changed. So it needs further develop by developers to meet the changes in current situation. That is the same as the traditional develop method. In our case study, we choose Eclipse BPEL Visual Designer to do further development.

The area of BPEL deployment is quite mature. In our case study, we just use Apache Ode (Apache Orchestration Director Engine) to do so. By using Apache Ode, we could intuitively see the status of all the composite services we have deployed. Also, it will record logs to support the service composition procedure in next cycle to maintain the closed loop of the service composition lifecycle.

5   Case Study and Discussion

5.1   Case Study

Our approach has been applied to a logistics cloud service platform. This platform supports the users from different companies to customize their functional services.

Every company is considered as a tenant. According to their business requirements, tenants could do service integration and service composition to register and customize their applications. When the custom application is running, all the data in each distributed node will be concentrated to be stored in this cloud platform, and then the service logs will be generated and stored as well. These logs are used as the input data of our service composition approach.

This case is about the waybill-related transportation process. The drivers and the related staff could upload the waybill, sending their status including location, car information and other information to the data center of the logistics cloud service platform through their devices, which are mostly the information sensing devices, using the technologies like radio frequency identification (RFID), infrared sensors, global positioning system (GPS) and laser scanner. This case is about how to develop the suitable waybill-related composite service. The initial input logs are the bills stored in the collaborative data center of this logistics cloud service platform. These initial logs are scattered in the data center in the form of electronic bills, a specific xml file format, which is shown in Fig. 5, containing the information of events, services, tenants and other supplementary information.

```
<Header>
<MessageReferenceNumber>FJWL_E-WAY_SEND_ARRIVE_STATUS</MessageReferenceNumber>
<MessageName>托运单到达状态上传</MessageName>
<MessageVersionNumber>v1.0</MessageVersionNumber>
<SenderCode>73360328-X</SenderCode>
<Role>tenNo_103</Role>
<MessageMakingDateTime>2013-4-26 14:59:25</MessageMakingDateTime>
</Header>

<Body>
<SupplierNo>73360328-X</SupplierNo>
<JobOrderNo>010a130422T057155</JobOrderNo>
<OrderNo>010a130422T057155</OrderNo>
<Type>at_transit</Type>
<SendTime>2013-4-26 14:16:00</SendTime>
<Content>无锡0x202013-4-26 14:16:000x20苏E670050x20无锡0x20接车员0x201</Content>
<Remark></Remark>
</Body>
```

Fig. 5. A raw service log file

About 200,000 electronic bills are gathered from the devices. After the preprocessing of logs, the Processing of event logs and the adopting of process mining, Process Model, Service Relation Info and Service Execution Info are gained. The discovered process model in the form of visualized Petri Net is shown in Figs. 6 and 7 shows a part of the pnml file of this mined process model, the real data source of Process Model.

Through service selection, a set of Key Services will be generated. After we import the data of process mining phase to service composition phase, the Service Composition Scene Model is generated. The invokeOrder in the Service Composition Scene Info is shown in Fig. 8. The Petri Net shows that there is a loop structure in it. In invokeOrder, the loop structure is converted into a "while" element.

According to the Service Composition Scene Info, we could get the general look at the composite service. Then we combine the service set with the event set, as Table 2 shows. The service is combined with the event according to the corresponding event ID. The similar phase is done to the role set as well.

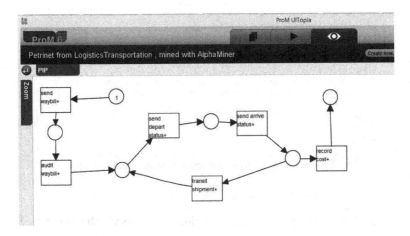

Fig. 6. A process model mined by our approach

```
<?xml version="1.0" encoding="iso-8859-1"?>
<pnml>
    <net id="net1" type="http://www.pnml.org/version-2009/grammar/pnmlcoremodel">
        <name>
            <text>Petrinet from LogisticsTransportation , mined with AlphaMiner</text>
        </name>
        <page id="n0">
            <name>
                <text />
            </name>
            <place id="n1">
                <name>
                    <text>{[send waybill+]} --&gt; {[audit waybill+]}</text>
                </name>
                <graphics>
                    <position x="93.75" y="71.5" />
                    <dimension x="12.5" y="12.5" />
                </graphics>
            </place>
            <place id="n2">
                <name>
                    <text>{[send depart status+]} --&gt; {[send arrive status+]}</text>
                </name>
                <graphics>
                    <position x="268.75" y="64.5" />
                    <dimension x="12.5" y="12.5" />
                </graphics>
            </place>
            <place id="n3">
                <name>
```

Fig. 7. The pnml file of the mined process model

Then we get the composite service, its BPEL instance is shown in Fig. 9. The "while" element in invokeOrder is converted to be "bpel:while" Control. The composite service contains the WSDL files of the key services, BPEL file and other essential files for deployment.

Then the composite service is registered in the service library and enters the service deployment phase. After long-term running, the execution of this service will leave behind service logs to be used for the new process mining phase of the next generation.

```
<sequence name="main">
    <invoke eventid="evNo_102" name="send waybill"/>
    <invoke eventid="evNo_103" name="audit waybill"/>
    <while name="While">
        <condition expressionLanguage="wsbpel:2.0:sublang:xpath1.0"></condition>
        <sequence>
            <invoke eventid="evNo_105" name="send depart status"/>
            <invoke eventid="evNo_106" name="send arrive status"/>
        </sequence>
    </while>
    <invoke eventid="evNo_108" name="record cost"/>
</sequence>
```

Fig. 8. The invokeOrder in the service composition scene info

Table 2. The service set combined with event set

Event ID and event name	Service operation	Service ID and service name
ev_102 send waybill	AddWaybill	sv_208 WaybillService
ev_103 audit waybill	SendAuditStatus	sv_202 SendAuditStatus-Service
ev_105 send depart status	SendDepartStatus	sv_203 SendDepartStatus-Service
ev_106 send arrive status	SendArriveStatus	sv_204 SendArriveStatus
ev_108 record cost	RecordWaybill	sv_208 WaybillService

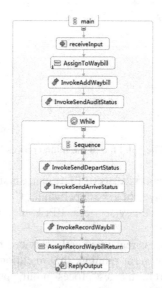

Fig. 9. A BPEL instance generated from the composite service

5.2 Discussion

Our work proposed a comprehensive service composition approach based on process mining, which takes into account both practical business, and execution effectiveness to help the developers do service composition.

The recent popular service composition approaches in service composition research area are QoS-based service composition approach [5, 6] and context-aware dynamic service composition approach [7]. From Table 3, we can know that our service composition approach in this paper focuses on different point from QoS-based service composition approach and context-aware dynamic service composition approach.

Table 3. Comparison between our approach and other service composition approaches

Features	Our approach	QoS-based	Context-aware
Main objective	Composition meets practical requirements	Good performance meets existed requirements	Dynamically changing situations adaption
Service selection criteria	Process and related info from process mining	QoS goals from SaaS providers	Context collected from the context-aware equipments
Requirement accuracy	High	Low	High
Non-functionality	General	High	Low
Time cost	Lower	Lower	High
Flexibility	General	General	High
Continuous optimization support	Yes	No	Yes
Comprehensiveness	High	Low	Low

In terms of the main objectives of these three approaches, our service composition approach is based on process mining and can select services according to the result of the process mining. Its objective is to meet practical requirements. While QoS-based approach needs definite requirements, on the premise of definite requirements and basic functionality, it selects services according to the given QoS objective. Its objective is to find the composition solutions with the best performance. But sometimes, in order to reduce the time and space cost, it may use the partial selection method instead of the global selection method like [13] does, thus it may lead the composition result to one-sided one. And context-aware dynamic service composition approach select services according to the context collected from the context-aware equipment. Its objective is to adapt to the dynamically changing situations.

The service selection criteria of these three approaches are different. The QoS-based approach selects services according to the QoS goals submitted by SaaS providers. The context-aware approach selects service according to the context information collected by the context-aware equipments.

In terms of other features, our approach provides high requirement accuracy and comprehensiveness to the whole workflow's service composition. That's mainly because our approach includes the business so that the service composition procedure could be brought forward. Also, the non-functionality is not bad because that the mined process and services are usually the best practices. Besides this, QoS-based approach provides high non-functionality and context-aware approach provides high flexibility at the cost of time. Among these three approaches, our approach and context-ware approach could optimize itself continuously.

In conclusion, our service composition approach based on process mining is outstanding in Comprehensiveness, also its Time cost and Flexibility is good. As these three approaches all have their own suitable adaptive situations, our approach is suitable for the situation with implicit requirements.

In addition, comparing with existing service composition approaches based on historical users usage information [2–4], our approach uses standard and systematic process mining method which is proposed in [1]. Therefore our approach and the process mining algorithm is decoupled. We can change mining algorithm with the evolvement of the process mining algorithm research.

6   Conclusion

In the area of Internet of Things, service composition is a widely-used popular approach for the development of applications. In this paper, in order to improve both execution effectiveness and comprehensiveness of existing service composition approaches, we propose a complete service composition approach based on process mining, considering both the practical business and execution information in production environment.

First, considering the information of execution environment and practical business, a meta-model for this approach is proposed. Next, the scene model based on this meta-model is generated by leveraging process mining. Then the scene model is applied to do service composition, including service selection from the Service Registry. After that, the composite service in form of BPEL instance is converted based on aggregated scene information, and the BPEL instance is deployed after the further development. During the execution of the composite service, new logs will be recorded, so that, the whole lifecycle of the service composition procedure is a closed loop. Finally, this approach is applied to a collaborative data center in a logistics cloud service platform to show its feasibility and to analyze its adaptive scenarios.

It is found that our approach is suitable for the situation without requirements or with indefinite requirements and without high performance demand of the result composite service. This approach is more comprehensive than the currently existing methods such as QoS-based and Context-aware methods lack.

In the future, we would focus on improving the flexibility of our approach, enlarging the supported service library, extending this approach to service management area, etc.

Acknowledgement. We would like to acknowledge the anonymous reviewers for their insightful and constructive comments and the support of the National Natural Science Foundation of China under No. 71171132 and No. 61373030.

References

1. Aalst, W.V.D.: Process Mining: Discovery, Conformance and Enhancement of Business Processes. Springer, Heidelberg (2011)
2. Gaaloul, W., Baïna, K., Godart, C.: Log-based Mining techniques applied to web service composition reengineering. In: Service Oriented Computing and Applications, vol. 2, no. 3, pp. 93–110. Springer, London (2008)
3. Wan, Z., Meng, F.J., Xu, J.M., Wang, P.: Service composition pattern generation for cloud migration: a graph similarity analysis approach. In: 21st IEEE International Conference on Web Services, pp. 321–328. IEEE Press (2014)
4. Moser, O., Rosenberg, F., Dustdar, S.: Event driven monitoring for service composition infrastructures. In: Chen, L., Triantafillou, P., Suel, T. (eds.) WISE 2010. LNCS, vol. 6488, pp. 38–51. Springer, Heidelberg (2010)
5. Ahmed, T., Srivastava, A.: Minimizing waiting time for service composition: a frictional approach. In: 20th IEEE International Conference on Web Services, pp. 268–275. IEEE Press (2013)
6. Cui, L., Li, J., Zheng, Y.: A dynamic web service composition method based on viterbi algorithm. In: 19th IEEE International Conference on Web Services, pp. 267–271. IEEE Press (2012)
7. Alferez, G.H., Pelechano, V.: Facing uncertainty in web service compositions. In: 20th IEEE International Conference on Web Services, pp. 219–226. IEEE Press (2013)
8. Sirin, E., Parsia, B., Wu, D., Hendler, J., Nau, D.: HTN planning for web service composition using SHOP2. Web Semant. Sci. Serv. Agents World Wide Web 1(4), 377–396 (2004). Elsevier
9. Cai, H., Cui, L., Shi, Y., Kong, L., Yan, Z.: Multi-tenant service composition based on granularity computing. In: 11th IEEE International Conference on Services Computing, pp. 669–676. IEEE Press (2014)
10. Aalst, W.V.D.: Service mining: using process mining to discover, check, and improve service behavior. In: IEEE Transactions on Services Computing, vol. 6, no. 4, pp. 525–535. IEEE Press (2013)
11. Rebug, Á., Ferreira, D.R.: Business process analysis in healthcare environments: a methodology based on process mining. Inf. Syst. 37(2), 99–116 (2012). Elsevier, Oxford
12. Weijters, A.J.M.M., Aalst, W.M.P., Mans, R., Rozinat, A., Song, M., Dongen, B., et al.: Process mining with ProM. In: the 19th Belgium-Netherlands Conference on Artificial Intelligence (2007)
13. Chen, Y., Huang, J., Lin, C.: Partial selection: an efficient approach for QoS-aware web service composition. In: 21st IEEE International Conference on Web Services, pp. 1–8. IEEE Press (2014)

RAID-6Plus: A Fast and Reliable Coding Scheme Aided by Multi-failure Degradation

Ming-Zhu Deng$^{(\boxtimes)}$, Yang Ou, Nong Xiao, Song-Ping Yu, Wei Chen, Zhi-Guang Chen, and Fang Liu

State Key Laboratory of High Performance Computing, College of Computer, National University of Defense Technology, Changsha 410073, China
dk_nudt@126.com

Abstract. Existing triple-failure-tolerant codes assume that failures are independent and instantaneous. Such assumptions overlook the underlying mechanism of multi-failure occurrences and ignored the effect of reconstruction window. These codes are not adapted to the occurrence pattern of failure in real-world applications. As a result, the third parity drive is almost idle as it set to handle the triple-failure scenario only with lower-level failure situations unattended. Furthermore, the problem of single failure rebuild deteriorates with the increasing disk capacity, and the system's reliability will decrease with user experience impaired. Aiming at these problems, a fast reconstructable coding scheme extended from RAID-6 has been developed in this study. RAID-6Plus maintains a smaller reconstruction window by recoding the third parity drive. Existing codes provide absolute reliability for triple failures via full combinations. As a contrast, RAID-6Plus employs short combinations which are able to greatly reuse overlapped elements during reconstruction to remake the third parity drive. The short combinations shorten the reconstruction window of single failure, which avoids multi-failure overlapping in the reconstruction window. The capability of multi-failure degradation provides RAID-6Plus with (1) a better system performance comparing to RTP and STAR and (2) an enhanced reliability comparing to RAID-6.

Keywords: Reconstruction window · Failure mode · Multi-failure degradation · Flexible reliability

1 Introduction

RAID-6 is conventionally the de facto default configuration of storage systems for its reliability. However, modern storage systems are more and more exposed to higher level of vulnerability in the big data era. Data reliability has unprecedentedly become a critical issue to be addressed. In terms of device level failures, Huang and Xu proposed STAR code [1] to protect a storage array from triple failures. Goel and Corbett extended RDP to RTP [2] with a faster coding algorithm for handling triple failures. Similar codes include RS codes and generalized EVENODD [3]. Jain et al. introduced redundancy between RAID groups by mirroring or replication at the

© Springer International Publishing Switzerland 2015
L. Yao et al. (Eds.): APSCC 2015, LNCS 9464, pp. 210–221, 2015.
DOI: 10.1007/978-3-319-26979-5_15

expense of a higher overhead [4]. Codes along this direction can generally provide satisfactory solutions to triple failures, but the following problems are still pending especially the lack of flexibility for enabling reliability:

- Existing codes assume that failures as independent and instantaneous and occurrences of failures conform to the exponential distribution [5, 6]. This ideal assumption does not apply in the fault pattern of modern storage systems [7]. Furthermore, those codes are not designed to support multi-failure degradation, which aims to convert a higher-level multi-failure into separate low-level multi-failures or single failures with a shorter reconstruction window.
- Existing codes largely ignore the pattern of failure occurrences in practice. For example, 99.75 % of recoveries are due to single disk failures [16], while triple whole-disk failures are rare. However, the third parity drive in RTP is set to handle the triple-failure scenario only with single failure rebuild unattended. The third parity drive is then almost wasted.
- Exiting codes focus on whole-device level failures while mixed-fault modes are more common in practice [8]. For example, when a fault consisting of two erasures and a sector error occurs, all of the three parity drives must be used. This directly overkills the effects of the third parity drive.
- As throughput is dwarfed by capacity [9], the reconstruction windows have grown exponentially from minutes to hours or even days in practice. This leads to a severe decrease of system performance and poor user experiences.

There exists a pressing need for a coding design to support multi-failure degradation and a smaller reconstruction window to deliver data reliability in a more flexible manner. In this study, we developed a fast reconstructable coding scheme with shorter reconstruction window, namely RAID-6Plus. The proposed coding scheme is an extension of RAID-6 [10] at the expense of three parity drives. RAID-6Plus employs short combinations which are able to effectively reuse overlapped elements during reconstruction to remake the third parity drive. This design shortens the reconstruction window of single failures by minimizing the total number of data reads. The possibility of multi-failure overlapping in the reconstruction window is therefore significantly diminished. The capability of multi-failure degradation provides RAID-6Plus with (1) a better system performance comparing to RTP and STAR and (2) an enhanced reliability comparing to RAID-6. RAID-6Plus balances reliability, performance and update penalty, which aims at practical uses in modern disk systems, flash-based systems, and even hybrid storage system on any array codes.

RAID-6Plus was then evaluated against the RTP and STAR codes. A mathematic analysis indicated that RAID-6Plus could achieve speedups of 33.4 %, 11.9 %, 47.7 % and 26.2 % comparing to RTP with conventional rebuild, RTP with optimal rebuild, STAR with conventional rebuild and STAR with optimal rebuild respectively. Furthermore, RAID-6Plus is orthogonal with some previous work on reconstruction speedup [10–13] and can be integrated together to further shorten reconstruction window.

The main contributions of this study are as follows:

- We developed a new XOR-based coding scheme with shorter reconstruction window and lower risk of multi-failure with no additional cost incurred comparing to RTP

and STAR, which enables a balanced solution with flexible reliability and better system performance.
- We design a novel redundancy scheme to minimize data reads in reconstruction scenarios. The redundancy scheme is generic and could be applied to other codes.
- To the best of our knowledge, RAID-6Plus is the first to extend given codes from the perspective of accelerating single failures instead of providing reliability boundary. The design is better adapted to the occurrence pattern of failures in practice.

The remainder of the paper is structured as follows. Section 2 motivates RAID-6Plus. Section 3 presents the design of RAID-6Plus. Finally, we evaluate the performance of RAID-6Plus in Sect. 4 with conclusions in Sect. 5.

2 Backgrounds and Motivation

In this section, we describe the different failure modes and recovery methods in RAID systems and introduce the concept of multi-failure degradation in reconstruction window. This motivates the need to design RAID-6Plus, which exploits multi-failure degradation mechanism.

2.1 Failure Modes in Disk and SSD

Multiple researches on the massive disks have all shown that mainstream disk drives have device failures or whole-disk failure [14] and sector failures. Device failure refers to the loss of the whole disk, which is mainly caused by hardware problems or manual misconducts. Sector failures could be caused by diverse reasons such as software glitches and so on [15]. All this failures directly cause data unavailability.

More specifically, newest findings by Ao ma reveal that despite infant mortality, multi disks tend to fail at similar age. That means that not only single failure happens at the device level, but also multiple devices may fail almost simultaneously, calling for higher reliability care than RAID-6. Other statistics show among all device failures, single failure accounts for the majority (99.75 %), and double failures is not eligible (roughly 8 %) and should be taken care of [16]. Meanwhile, often the cases are single failure coupled with other sector failures rather than multi whole-disk failures [8]. And sector errors are not rare and increase with time [14].

In terms of correlation between whole-disk failure and sector errors, reference [14] asserts whole-disk failure can be viewed as the consequence of accumulated sector error and uses the number of reallocated sectors to characterize probability of whole-disk failure. Further, the longer a functioning device undergoes, the higher probability of device failure could be. In other words, other failures could happen in ongoing process of failure recovery, aggravating system reliability and making it much more vulnerable [17]. In short, single failure is of vital importance to reduce window of system vulnerability.

Though there have not been any investigation published on massive SSD, SSD's failure modes must be similar while differ in some of its own vulnerability features, like its inborn limited endurance issues and wearing over time [18].

2.2 Growing Reconstruction Window for Single Failure

Basically, given the same rotational speed, reconstruction window for a single disk varies according to different disk capacity [19]. Therefore, through simple calculation, reconstruction windows has also grown hugely along with device capacity, leading to easier opportunity for potential data loss [17].

On the contrary, the shorter reconstruction window is, more failures with higher level can be avoided by degraded into low-level failures. Current reconstruction window is unbearably large. For example, it would normally take more than two hours to reconstruct the data of a 500 GB disk. This not only aggravates user experience for longer wait, but also puts the whole system into higher risks of data loss. Therefore, the urgent need to shorten reconstruction window is practical and would be welcome.

2.3 Motivation

Traditional codes with higher fault-tolerance, like RTP and STAR codes are originally designed for triple whole-device failures. Unfortunately, statistics have shown probability of single failure overwhelmingly accounts for most while triple whole-device failures is relatively rare, thus making current RAID-7 system with triple parity drives wasteful. Additionally, mixed-fault modes exhibited in modern storage systems, overkills the solution of those codes with higher reliability boundaries. In short, current RAID-7 system with triple parity coding is unpractical and needs to deliver more flexible reliability.

Further, reconstruction window is exponentially increasing with device capacity, thus worsening user experience and leaving larger space of system vulnerability and data loss. More notably, current coding schemes for triple-failure are unable to shorten reconstruction window.

Therefore, all these factors above motivate me to extend RAID-6 codes another way to shorten reconstruction window to provide flexible reliability and make it more practical.

3   System Design

In this section, we base on RDP code to illustrate our RAID-6Plus. First of all, we introduce the optimal way of RDP for single failure rebuild. Then, in RAID-6Plus, we keep the P disk and Q disk as it is, and present our new coding algorithm for the third parity drive X to reduce reconstruction window.

3.1 Optimal Reconstruction for Single Failure in RDP

In RDP, we have two kinds of parity drives, where P drive means slope 0 and Q for slope 1. whenever any single data disk fails, the conventional way to reconstruct a failed disk is merely using P drive while the optimal way is using equal number of parities from P and Q drives, maximizing the overlapping data, as shown in (Fig. 1) [10].

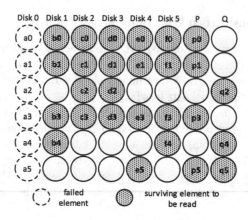

Fig. 1. The optimal reconstruction sequence of single failure in RDP. The hybrid use of equal numbers of P parities and Q parities maximizes the overlapping data for single failure rebuild.

Table 1. Feature comparison of three coding ways for X drive

Coding methods	Element example	Involved element	Merit	Shortcoming
Mirroring of single disk	a1	1	simple and can maximize reduction of reconstruction window for specific drive	Only cover replicated drive, not able to cover other drives, causing imbalance and fluctuation
Short combination	a1 + b1	2	In between	
Full combination	a1 + b1 + c1 + d1 + e1	5	Maximize fault-tolerance for the whole system	Unable to reduce reconstruction window

3.2 Redundancy Coding in RAID-6Plus

In order to provide solid and flexible reliability against multi-failure, RAID-6Plus uses three redundancy drives as other codes for triple-failure do, but in a new and more practical way. Actually, there is a reasonable compromise between reliability level and system performance in RAID-6Plus, where on the one hand, we maintain a basic reliability guarantee to accommodate non-negligible double-failure, while on the other hand, reconstruction window is shortened to degrade failure and reduce user wait.

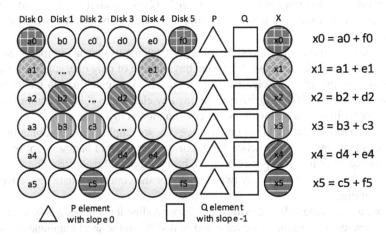

Fig. 2. Encoding for X drive in RAID-6Plus. Any new parity in the third parity drive X is the XOR of some two data elements. There are two data elements of each data drive to be covered in X drive. And those data pairs or tuples to construct X parity is specially chosen to satisfy fast reconstruction.

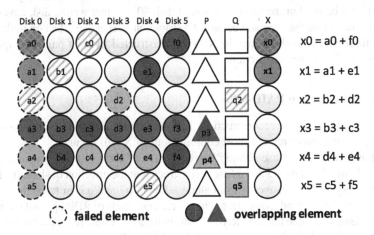

Fig. 3. Rebuilding single failure with X elements. a0 and a1are recovered by x0 and x1, a2 and a5 are rebuilt with the help of q2 and q5 while a3 and a4 are fixed with p3 and p4.

Of all the three redundancy drives, the first two are devoted to maintaining a lower bound of reliability, which we denote as "base reliability", given RAID-6 is widely deployed in diverse storage system for double-failure concern. Therefore, we have the exact P drive and Q drive in RAID-6Plus from RDP to keep a base reliability guarantee for double-failure.

The remaining redundant X drive is dedicated to reducing reconstruction window for single failure of any data drive, which is in charge of user access. Therefore, the key lies in the redundancy coding for X drive. Conventionally, there are three ways for X drive coding: (1) Mirroring of a single data disk; (2) Short combination; and (3) Full combination. The features of the three coding ways are so obvious that we summarize them in Table 1.

In essence, the three ways of coding for X parity elements vary in number of element involved in the encoding process. Mirroring has a length of only one, suggesting any X element is a replica of some element in other drives. Full combination has the same length of any element in P or Q drive, implying any element in X is the combination of many elements in data drives while Short combination is in between.

In order to reduce data reads as much as we can, we employ short combination with some 2 elements involved to encode for X parity. The thought behind is to find over-lapping elements as many as possible on the basis of optimal reconstruction for single data disk in RDP and combine any two of them for the concern of data disk coverage, as shown in Fig. 2.

We use the example in Fig. 2 to illustrate. In our coding for X drive, nearly all the data drives in covered. For example, we have and of disk #1 in the short combination, which will be used in the reconstruction of disk #1. In other words, disk coverage in some way guarantee the balance of effect distribution on multi disks.

We use the example in Fig. 3 to illustrate. In our coding for X parities, nearly all the data drives are covered. For example, we have a0 and a1 of disk #0 respectively in x0 and x1, which will be used in the reconstruction of disk #0. In other words, disk coverage in some way guarantees the even and balanced distribution of speedup effect on multi disks.

And those short combinations in X drive constructed by data pairs are specially chosen to satisfy fast reconstruction.

3.3   Reconstruction in RAID-6Plus

In terms of single erasure reconstruction, for example, if *disk#0* fails, reconstruction with the participation of related short combinations in X will occur.

As shown in Fig. 3, we use *x0* and *f0* to recover *a0*, *x1* and *e1* for *a1*. And respectively with the help of *p3* and *p4*, *a3* and *a4* are reconstructed in the direction of slope 0 while *a2* and *a5* are respectively recovered from slope −1 with *q2* and *q5*. In total, there are 22 elements needed to be read, comparing with 27 element reads of RDP optimal recovery.

In the views of double failures, RAID-6Plus maintains system reliability in two ways. First and foremost, RDP is actually included in RAID-6Plus, therefore there will be no data loss whenever any double failures happen.

Additionally, with shorter reconstruction window, some double-failure situations previous in traditional RAID-6 system could be converted to independent single failures, thus eliminating vulnerability undergone by the system. Also, the same way of using short combinations in X can be applied to save data reads for double-failure scenarios.

In fact, triple failures happen at an extremely tiny probability, therefore the third parity drive in traditional codes only exist for extreme cases. And according to Relia-bility Equation in [20], RAID-6Plus could convert a portion of triple-failure situations in traditional coding schemes to lower possibility, leaving unconvertible triple failures at a negligible level.

4   Performance Analysis

In this section, we analyze the performance of RAID-6Plus, with RTP and STAR codes. The analysis comes in parity computation complexity, update penalty, reconstruction cost, and storage ratio. Further, we propose Q metric to denote the induced performance improve per cost to validate our RAID-6Plus. The basic data matrix for STAR is $(p-1)p$ and $(p-1)(p-1)$ for RAID-6Plus and RTP.

4.1   Parity Computation Complexity

We compare parity computation complexity of different codes by XORs per element. The total XORs can be computed as the sum of XORs for different parity types. For example, the number of XORs for writing a stripe in RAID-6Plus is consisted of XORs for row parity, diagonal parity and X parity, making

$$
\begin{aligned}
A_{RAID-P} &= \frac{XOR_{row} + XOR_{dia} + XOR_x}{Blocknum} \\
&= \frac{(p-1)(p-2) + (p-1)(p-2) + p-1}{(p-1)(p-1)} \\
&= \frac{2p-3}{p-1} = 2 - \frac{1}{p-1}
\end{aligned}
\tag{1}
$$

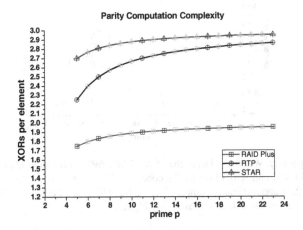

Fig. 4. Parity computation complexity. Obviously, RAID-6Plus uses least XORs per data block for parity computation and is fastest. The difference is more obvious when disk array size is smaller, which is the typical case of most storage systems.

Likely, we can get that of RTP is $A_{RTP} = 3 - \frac{3}{p-1}$. For $(p-1)p$ data matrix of STAR, similar computation pattern can be applied to get $A_{STAR} = 3 - \frac{1}{p-1}$. The results are shown in Fig. 4.

4.2 Update Penalty

Traditionally, redundancy incurs update penalty because parity must be consistent with data element involved. And update penalty is measured by number of introduced write. For example in RDP, any update on a single data block will result in two more writes to respectively update corresponding row parity and diagonal parity. We can get the update penalty of RTP in reference [2] $U_{RTP} = 5 - \frac{2}{p-1}$ and STAR is $U_{STAR} = 5 - \frac{4}{p}$ with the help of reference [21].

For RAID-6Plus, update penalty of all data elements requires

$$[(p-1)(p-1) - 2(p-1) - (p-2)] \times 2 + 2 \times (p-1) \times 3$$
$$+ (p-2) \times 1 = 2p^2 - 3p + 2 \tag{2}$$

Thus, the update penalty per block is

$$U_{RAID-P} = (2p^2 - 3p + 2)/(p-1)^2 = 2 + \frac{1}{p-1} + \frac{1}{(p-1)^2} \tag{3}$$

which is smaller than RTP and STAR, as shown in Fig. 5.

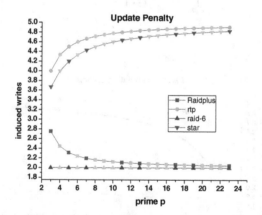

Fig. 5. Update penalty. When p is larger than five, RAID-6Plus introduces least update penalty than RTP and STAR, and it is close to RS-coded RAID6. For example, when p equals 11, RAID-6Plus has 2.11 update penalty compared with 2.82 of RTP and 4.636 of STAR. The different will be more obvious when disk array size grows bigger.

Significantly, RAID-6Plus bears least penalty when updating. The reason why RAID-6Plus has the least update penalty is RAID-6Plus employs short combinations while RTP and STAR use full combinations which involve more data elements to construct, therefore incurring more update penalty.

4.3 Reconstruction Cost

We hereby focus on single failure reconstruction scenario and use total data elements read to evaluate the cost.

For RTP and RDP share exactly same reconstruction sequences, they have the same reconstruction cost, which are respectively $(p-1)(p-1)$ with conventional recovery and $\frac{3}{4}(p-1)^2$ with optimal recovery. Similarly, STAR rebuilds a failed drive with the same cost of EVENODD, which is $(p-1)p$ conventionally and $\frac{(3p+1)(p-1)}{4}$ by optimal recovery.

When a drive fails in RAID-6Plus, two elements are recovered by two X parities and half of the rest elements are respectively rebuilt with P and Q parity as shown in Fig. 3. Therefore, we compute data elements read for rebuild in RAID-6Plus as

$$2 \times 2 + \frac{(p-1)-2}{2} \times (p-1) + \frac{(p-1)-2}{2} \times (p-1)$$
$$- \frac{(p-1)-2}{2} \times \frac{(p-1)-2}{2} - 2 = 2 + \frac{(p-3)(3p-1)}{4}. \tag{4}$$

We normalize the results based on STAR_conventional in Fig. 6. The secret of RAID-6Plus for single erasure rebuild is it uses short combinations in X drive to minimize elements needed and meanwhile reuse overlapping elements as much as possible on the basis of optimal reconstruction sequence of RDP.

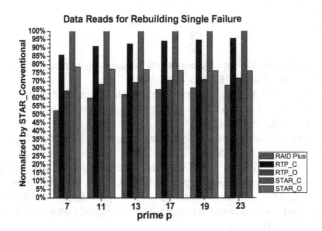

Fig. 6. Data reads for rebuilding single failure. Clearly, RAID-6Plus reads least data than any other coding schemes. For example, when p is 7, RAID-6Plus has a normalized speedup of 33.4%, 11.9%, 47.7% and 26.2% respectively over RTP_Conventional, RTP_Optimal, STAR_Conventional, STAR_Optimal.

4.4   Storage Overhead and Reliability

RTP, STAR are both MDS codes while RAID-6Plus is non-MDS. Regarding storage overheads, all the three codes have three drives dedicated to parities. In terms of absolute failures, RTP and STAR are strictly triple-failure tolerant while RAID-6Plus can tolerate only double failures at device level. But RAID-6Plus provides flexible and higher relative reliability by shorter single failure reconstruction. This effect are hard to be explicitly characterized because there are not any plausible reliability model for RAID systems.

5   Conclusions

Existing triple-failure-tolerant codes set the third parity drive to only handle the triple-failure scenario, which is very rare while single failure dominates in storage system breakdowns in real-world applications.

This study developed RAID-6Plus, a fast and reliable reconstructable coding scheme with shorter reconstruction window. RAID-6Plus remakes the third parity drive to shorten the reconstruction window to degrade multi-failures and to minimize short user wait. The design enables reuse of overlapped elements during reconstruction to balance the reliability and performance of the resulted coding scheme. The performance evaluation indicated that RAID-6Plus exhibited (1) a better system performance with no extra cost comparing to RTP and STAR and (2) an enhanced reliability comparing to RAID-6. The scheme held the potentials of practical uses in modern disk systems, flash-based systems, and even hybrid storage system on any array codes.

Acknowledgment.   We are grateful to our anonymous reviewers for their suggestions to improve this paper. This work is supported by the National Natural Science Foundation of China under Grant Nos. 61232003, 61332003, 61202121, 61402503, 61303073.

References

1. Huang, C., Xu, L.: STAR: an efficient coding scheme for correcting triple storage node failures. IEEE Trans. Comput. **57**, 889–901 (2008)
2. Goel, A., Corbett, P.: RAID triple parity. ACM SIGOPS Oper. Syst. Rev. **46**, 41–49 (2012)
3. Blaum, M., Bruck, J., Vardy, A.: MDS array codes with independent parity symbols. IEEE Trans. Inf. Theor. **42**, 529–542 (1996)
4. Jain, N., Dahlin, M., Tewari, R.: TAPER: tiered approach for eliminating redundancy in replica synchronization. In: FAST, pp. 21–21
5. Chen, P.M., Lee, E.K., Gibson, G.A., Katz, R.H., Patterson, D.A.: RAID: high-performance, reliable secondary storage. ACM Comput. Surv. (CSUR) **26**, 145–185 (1994)
6. Amer, A., Long, D.D., Thomas Schwarz, S.: Reliability challenges for storing exabytes. In: 2014 International Conference on Computing, Networking and Communications (ICNC), pp. 907–913. IEEE (2014)
7. Schroeder, B., Gibson, G.A.: Disk failures in the real world: what does an MTTF of 1, 000, 000 hours mean to you? In: FAST, pp. 1–16
8. Plank, J.S., Blaum, M.: Sector-disk (SD) erasure codes for mixed failure modes in RAID systems. ACM Trans. Storage (TOS) **10**, 4 (2014)

9. Leventhal, A.: Triple-parity RAID and beyond. Queue **7**, 30 (2009)
10. Xiang, L., Xu, Y., Lui, J., Chang, Q.: Optimal recovery of single disk failure in RDP code storage systems. ACM SIGMETRICS Perform. Eval. Rev. **38**, 119–130 (2010)
11. Xiang, L., Xu, Y., Lui, J., Chang, Q., Pan, Y., Li, R.: A hybrid approach to failed disk recovery using RAID-6 codes: algorithms and performance evaluation. ACM Trans. Storage (TOS) **7**, 11 (2011)
12. Zhu, Y., Lee, P.P., Xiang, L., Xu, Y., Gao, L.: A cost-based heterogeneous recovery scheme for distributed storage systems with RAID-6 codes, pp. 1–12. IEEE
13. Khan, O., Burns, R.C., Plank, J.S., Pierce, W., Huang, C.: Rethinking erasure codes for cloud file systems: minimizing I/O for recovery and degraded reads, p. 20
14. Ma, A., Douglis, F., Lu, G., Sawyer, D., Chandra, S., Hsu, W.: RAIDShield: characterizing, monitoring, and proactively protecting against disk failures. In: Proceedings of the 13th USENIX Conference on File and Storage Technologies, pp. 241–256. USENIX Association (2015)
15. Mingyuan, X., Mohit, S., Mario, B., David, A.P.: A tale of two erasure codes in HDFS. In: FAST, pp. 213–226 (2015)
16. Pinheiro, E., Weber, W.-D., Barroso, L.A.: Failure trends in a large disk drive population. In: FAST, pp. 17–23
17. Luo, X., Shu, J.: Load-balanced recovery schemes for single-disk failure in storage systems with any erasure code. In: 2013 42nd International Conference on Parallel Processing (ICPP), pp. 552–561. IEEE (2013)
18. Boboila, S., Desnoyers, P.: Write endurance in flash drives: measurements and analysis, pp. 9–9
19. Elerath, J.G., Schindler, J.: Beyond MTTDL: a closed-form RAID 6 reliability equation. ACM Trans. Storage (TOS) **10**, 7 (2014)
20. Corbett, P., English, B., Goel, A., Grcanac, T., Kleiman, S., Leong, J., Sankar, S.: Row-diagonal parity for double disk failure correction. In: Proceedings of the 3rd USENIX Conference on File and Storage Technologies, pp. 1–14
21. Rongdong, H., Guangming, L., Jingfei, J.: An efficient coding scheme for tolerating double disk failures. In: 2010 12th IEEE International Conference on High Performance Computing and Communications (HPCC), pp. 707–712 (2010)

The Searching Ranking Model Based on the Sharing and Recommending Mechanism of Social Network

Hongxiao Fei, Tianchi Mo, Yang Wang, Zequan Wu, Yihuan Liu, and Li Kuang[✉]

School of Software, Central South University, Changsha, China
{hxfei,wangyangcsusoft,wuzequancsusoft,3901130215, kuangli}@csu.edu.cn, motianchi@163.com

Abstract. The combination of social network and search engine is the trend of internet in coming years. Through introducing the widely utilized sharing and recommending mechanism in social network to search engine, this paper proposes a new searching ranking model. This model judges the quality of web pages and decide what extent do they meet users' personalized need through analyzing the records of users' social circle's behavior of sharing and recommending. Then, it can make search engine provide users with personalized results sequences. Both the experiment and the theoretical analysis show the proposed model can automatically help users to select the high quality search results, and provide users with better personalized service.

Keywords: Search engine · Personalized search · Ranking model · Social network · Sharing and recommending mechanism

1 Introduction

Search engine (SE) and social network are the two most active kinds of application systems in present internet. Google and Baidu are the most popular SEs, and both of them are belong to the second generation full-text search engine. The advantages of this kind of SE include fast speed, high recall ratio, timely updating, and total automation. However, with the sharp expansion of the online information quantity, some obvious defects of this kind of SE have been gradually exposed. Their precision is not very high, and many unrelated webpages, or the webpages that not accurately satisfied users' need are also contained in the result list, so, users have to spend lots of time in selecting the pages they really want. These SEs are also lack of proactivity, and not capable of providing precise personalized services [1]. Simultaneously, in the era of Web 2.0, social network, like Facebook and Twitter, has been rapidly ascending in recent years. Due to users spend much more scattered time in communicating on the social application, the social webs can accumulate massive data closely related to users' characteristic. It makes social networks able to supply more real-time and more personalized services to users. As a matter of fact, social network is becoming the new major source of information after the time of SE appearance [2]. In the future, it may

© Springer International Publishing Switzerland 2015
L. Yao et al. (Eds.): APSCC 2015, LNCS 9464, pp. 222–234, 2015.
DOI: 10.1007/978-3-319-26979-5_16

overstep SE and become the most important traffic entrance of other websites [3]. Nevertheless, there is randomness and uncertainty when people obtain knowledge online, and their social relations have boundaries, that cause the information given by the social web cannot cover whole searching requirement of single user.

SE and social network both have pros and cons. The combination of them is the trend of cyber application. Literature [4] points out that the proactivity, personality and sociality of social application can make up the defects of modern SE. And SE's ability to gather resource from whole internet makes it capable of providing much more information than social webs. Therefore, in next decade, SE may comprehensively combine with social network. The new mixed application will possibly become the most critical online system in daily life.

So, in this paper, to make SE become more intelligent, accurate and personalized, we introduce the common sharing and recommending mechanism in developed social nets, and try to transplant it into search systems. A new ranking model named Sharing and Recommending Result Rank (SRRR) is proposed in this purpose. SRRR is designed to help users select search results. It can return different webpage lists to different users, which is the real personalized searching based on social information.

The rest of the paper is organized as follows. Section 2 reviews the related work about the combination of social web and SE. Section 3 introduces the proposed SRRR model. Theoretical, empirical and experimental analysis is presented in Sect. 4. And finally, the conclusion and the future work are given in Sect. 5.

2 Related Work

Actually, industry does not have consensus on the definition of "social search". Literature [3, 5, 6] deem social search to be the search function in the social application, which means when users submit the keywords, systems return the related contents (like tweets, photos, blogs and so on) with order. Meanwhile, other scholars regard the core of social search as the full-text S, which is faced with the whole internet. In their opinion, social search means using the intelligent data of social web to comprehensively optimize the second generation SE, rather than searching the social website itself [1, 2].

Academic research about the combination of SE and social application began with the deep exploration of bookmarking sites [7, 8]. Bao et al. put forward the social similarity rank model (SSR) and the social page rank model (SPR) [7]. Both SSR and SPR are based on the del.icio.us, and the hypothesis that people search pages through annotations. SSR is an iterative algorithm to identify the similarity of annotations through checking the people who use them, while SPR extends the page rank algorithm to the network consist in pages and annotations. Lo et al. pose the expert vote rank (EVR) and recommendation page rank (RPR) [8]. EVR is based on users' prestige in del.icio.us, and RPR uses the weighted directional recommendation graph to do the ranking. After above approaches, some algorithms and models have been designed for the search function in social platforms. Xing et al. propose that RDF data can be used to solve the heterogeneity problem in social network searching [5], but they do not design an effective rank model. And Xu et al. put forward a ranking model involving the page publishing time, the relevance of content and key word, the page publishers' prestige and the relation between publisher and searcher [9]. However, this model is confined in Twitter and only covers the data of two months. As for the optimization of full-text search engine with some social mechanisms, there is few study achievements so far. And some companies have developed productions, but their effects need to be checked by time. Facebook gets a patent called "curated search", while they also cooperating with Bing, the SE developed by Microsoft [3]. Google also designs Google+, hoping to build their own social platform. In academic area, Akiyama et al. also proposed a method through extending page rank. They apply the page rank to a new class of nets with two kinds of vertices, pages and people who are reading these pages right now. It is very creative, but it is too hard to implement.

As far as we can see, academia is now focusing much more on the site search of certain social platform, and the improvement of page rank with shallow social information. However, until today, there is no effective approach to incorporate the sociality into the full-text search engine on a high level. As mentioned above, the functional limitation makes social web cannot totally replace full-text search engine. Frequent searching behaviors need to be supported by the data from whole World Wide Web. For this purpose, the design of SRRR aims at using the superiorities of social network to prove the effect of SE. The application environment will not be confined in certain social platform.

3   Using Social Sharing and Recommending Mechanism to Optimize Search Ranking

Virtually, as literature [3] pointing out that what threatens the SE is not the social network itself, but the intelligent data of the social web. Users spend lots of time on interacting with social applications and other users, so these applications can accumulate massive data related to users' information. Therefore, social applications get better know about their user. Without users' active expression, we can use the data to build interest vectors [1]. Generally, social nets provide the completed mechanism of sharing and recommending, large quantity of information can be diffused rapidly along the links of web. If we can utilize the features hiding in the actions of sharing and recommending guiding search ranking, then the SE will become more personalized due to the personality of everyone's social circle.

3.1   Relevant Conceptions of SRRR

In existing social application, the conceptions related to SRRR include social relation, social distance, and sharing and recommending mechanism.

1. Social Relation: User u and v having social relation on a social platform means an explicit record of the edge between 2 vertices in the database. There are 2 kinds of relation, unidirectional and bidirectional. Unidirectional relation means if there is a link from u to v, the link from v to u is not necessary. In Twitter and Sina micro

blog, "Follow" is a kind of unidirectional relation. We denote unidirectional relation as $u \rightarrow v$. Meanwhile, in Facebook and Tencent QQ, the social relation is bidirectional, that means if there is a link from u to v, the link from v to u also must exits. It is denoted as $u \leftarrow\!\!\rightarrow v$. The proposed model only takes unidirectional relation into consideration, because $u \leftarrow\!\!\rightarrow v$ is equal to $u \rightarrow v \&\& u \leftarrow v$. The conception of social relation maps the social network as a directed graph.

2. Social Distance: In existing social nets, information is one-step-visible [5], which means u can read contents posted by v iff $u \rightarrow v$. Social distance is defined for Information transmission. If $u \rightarrow v$, the social distance is 1, denoted as $dis(v, u) = 1$. If $dis(v, u) = d$, $w \rightarrow u$, w not $\rightarrow v$, and $\forall w \rightarrow x \&\& x \neq u$, $dis(v, x) \geq dis(v, u)$, then $dis(v, w) = d+1$. Due to the huge number of users in social web, ordinary algorithms are incapable of finishing the calculation in short time. The method proposed by literature [10] can be taken to curtail the run time.

3. Sharing and recommending mechanism: Most of the social applications provide the service of sharing and recommending. In different positions of different webpages, we can see the sharing button. And in social applications themselves, users also can use "Forward" function to share the contents they like. If $u \rightarrow v$, then, when v has retransmitted something u will be noticed and can read the content v shares. But to users who do not follow v, they will not directly know v's behavior.

Recommending, also known as "Like", is a kind of behavior visible to whole site. The application will record users' positive evaluation to certain content. When other users request a set of data, the content with more recommendation is nearer to the top of result list.

3.2 Process of SRRR

Nowadays, the second generation full-text SE works with following procedure: collecting webpages, analyzing webpages, building the inverted index, and handling the queries [11]. The proposed model will influence all of these steps. The thought of SRRR comes from users' daily behavior of sharing and recommending. Normally, only the pages/resources with high quality and exactly satisfying users' requirement can get users' "Forward" and "Like". And the behavioral basis of sharing and recommending is user and his social circle having analogous interests. Therefore, SRRR is based on following hypothesizes:

- Hypothesis 1: The shorter the social distance between two users is, the more similar the requirement they share [1]. Namely, user and his social circle have similar interests.
- Hypothesis 2: Page's quality is proportional to the times it has been shared. Simultaneously, if the user who shared the page has high prestige in social web, this certain sharing behavior should have higher importance. The more a content has been retransmitted, the nearer it is to the top of the result list in searching.
- Hypothesis 3: Page's quality is proportional to the times it has been recommended. Meanwhile, to certain webpage, if its recommenders have shorter social distance to the searcher, and the number of them is larger, the page's position should be nearer to the top of return list.

Through calculating the share ranking factor (SRF) and the Recommending Ranking Factor (RRF), SRRR combines the attitude to webpages of searcher's social circle with other elements of search ranking, and helps users to make decision when selecting search result.

1. SRF Calculation: Before calculating SRF, the prestige of users in social network needs to be identified. We call it prestige factor (*PF*). Page rank algorithm [9, 12] is taken to solve this problem. The primary idea of this algorithm is: the more users follow u, the larger *u*'s *PF* is. Meanwhile, if u has large *PF* and $u \rightarrow v$, v also has large *PF*. Equation (1) is utilized to calculate *PF*:

$$pf(u) = (1 - d) + d \sum_{v \in f_u} \frac{pf(v)}{N_v} \qquad (1)$$

In Eq. (1), *pf(u)* denote *u*'s *PF*, $d \in (0, 1)$ is attenuation factor, usually, $d = 0.85$. And fu is a set of users, $\forall v \in f, v \rightarrow u$. N_v is the number of users followed by v. Application system can measure users' *PF* when its load is light, like 12 p.m $\sim$ 4 a.m every day. This is a preprocessing step, so it does not cut down the system performance. When user *u* submits the search request with keyword *k*, system firstly retrieves out pages, the set of webpages whose keyword is *k* (existing SE has this step, too). Then, Eq. (2) is taken to calculate SRF of every $p \in pages$ for *u*:

$$srf(u, k, p) = \sum_{v \in (bf_u \cap Share(k,p))} \frac{pf(v)}{N_u} \qquad (2)$$

In Eq. (2), bf_u is a user set. $\forall v \in bf_u, u \rightarrow v$. Share (*k, p*) is also a user set. It has two kinds of elements. One includes the users who directly share *p* into social net without using the SE, and after analysis the system reckon *k* as the keyword of *p*. The other contains the users who share *p* when they searching keyword *k* through SE. N_u is the number of users followed by *u*.

2. RRF Calculation: As the definition in Sect. 3.1 (3), recommending, or "like" is visible to full site. User *u* can recommend the webpage *p* he likes to the whole social web. According to Hypothesis 1 and Hypothesis 3, proposed model computes RRF with following way:

Build a pattern *M*, which is a triplet (*u, k, url*). It means when *u* searching keyword *k*, he recommended one of the results whose web address is *url*. Or *u* directly recommended the webpage whose address is *url* to the system, and system regards *k* as its keyword. When user *u* submits the search request with keyword *k*, same as SRF calculation, system figures out pages, then, uses Equation to measure RRF of every $p \in pages$ for *u*:

$$rrf(u,k,p) = \sum_{v \in Rec(u,k,p)} \frac{threshold - dis(u,v) + 1}{num(u, dis(u,v))} \qquad (3)$$

Rec(u, k, p) is a user set. $\forall v \in Rec(u, k, p)$, there must be a record in *M* values *(v, k, url of p)*, and *dis(v, u)* ≤ *threshold*. In Eq. (3), threshold is a constant number, which denotes the upper bound of social distance. According to six degrees of separation [6], we set *threshold* = 6. Users with longer social distance will be ignored. Meanwhile, suppose *dis(v, u)* = *d*, *num(u, dis(v, u))* is the number of users whose social distance to *u* is also *d*.

3. Search Ranking Based on SRF and RRF: After SRF and RRF are obtained, combine them with existing ranking algorithm by weighting. Final step to get ranking value needs the following equation:

$$srrr(u,k,p) = \left(\sum_{i=1}^{n} c_i f_i \right) + c_s srf(u,k,p) + c_r rrf(u,k,p) \qquad (4)$$

In Eq. (4), *srrr(u, k, p)* is the ranking value of webpage *p* when user *u* search keyword *k*. The webpage with bigger SRRR value can get closer position to the top of the result list. Meanwhile, $c_1, \ldots, c_n$, c_s and c_r are weighting factors, $f_1, \ldots, f_n$ is the n ranking factors (like page rank, timing factor, bounce rate and so on) already utilized by current SEs. To any f_i ($1 \le i \le n$), if pages with bigger f_i can be nearer to the top, then, c_i is positive, otherwise, c_i is negative. Because both SRF and RRF are the larger the better, c_s and c_r are positive. In this case, SRF and RRF are capable of affecting the ranking result as importantly as other factors like page rank value. If searching system needs to be influenced more by social information, the value of c_s and c_r should be amplified. Different users have different social circle, and the attitude and emotion tendency of user's social circle can help user to make personalized decision. So, it can be concluded that with sharing and recommending mechanism, SRRR is able to make the ranking of full-text SE more in line with users' personalized requirement.

4 Analysis of SRRR

The rationality of SRRR will be analyzed in theoretical and empirical aspect, and then, a simulation experiment will be presented.

4.1 Theoretical and Empirical Analysis

What has to be mentioned is most of the existing social search approaches are focused on the adaptation of page rank. It is not very reasonable. Firstly, page rank algorithm was designed for static internet in late 1990s. Today's internet is dynamic. Page rank evaluates webpages through links between them, which only based on the assessment of developers, not scanners or readers. To some web sites that are not good at link optimization, even though they may have content with high quality, their page rank value may be very low. This problem also bothers the new born websites. Meanwhile, some sites with high SEO capability might cheat SE by using spam links, that makes users get many irrelevant and inaccurate information when doing their search [6]. No matter how to change the page rank algorithm, above defects will not be fixed as long as the information construction of internet is based on hyperlink. Under this frame, page rank value is an inherent property to the webpages or sites.

Secondly, although page rank is not perfect, it is still a very effective ranking method. Most of SEs takes it as a core ranking factor all over the world. Modification of page rank may change the kernel systems of SEs, which requires huge workload. In addition, developing a system with both search service and social service may be not a good idea. It is hard for new application to gather users in the area of social web and SE today. And all of these thoughts are contradictive to the open closed principle of software engineering. Finally, page rank is only one of many ranking factors, its weight is not very high. Take Google as an example, when sorting the search results, there are dozens of factors influencing the final order, including page rank value, page view value, bounce rate, response rate and so on. Just combining the social information with page rank cannot make the social data affect ranking comprehensively.

Additionally, considering the personalized searching, the page rank value of website is not very meaningful. As mentioned, page rank value is an inherent property of webpages. It is based on the hyperlinks, and relatively static. It does not change with different users and keywords. In other words, page rank is not a personalized algorithm. So, when the personalized SE is being designed, we need find a new approach to utilize the social data independent to the page rank model.

The proposed model, SRRR, does not have the defects mentioned above. Through analyzing Eq. (2), it can be found that to different users, one webpage has different SRF. The number of user u's friends does not have effect on SRF calculation and only the behavior of users directly followed by u will be involved into the calculation. At the same time, SRF and the prestige of sharers have positive correlation. So, Eq. (2) is in line with the Hypothesis 1 and Hypothesis 2, which make SRF represent users' personalized requirement objectively. And Eq. (3) takes the number of recommenders and social distance into consideration, which meet the requirement of Hypothesis 1 and 3, reflecting users' need through analyzing the recommending behavior in social circle. Like SRF, the same page has unequal RRF according to various people. More importantly, Eq. (4) defines that SRRR model is a complementary optimization to existing SE. When using SRRR, the modification of internal structure for existing systems is not necessary. What is necessary is only adding two factors with proper weight. So, SRRR has high usability, acceptable to the software engineering. Because we consider the social factors as independent ones, when the social network is not mature (if the SE enterprises choose building social web by themselves), the weight of SRF and RRF can be lowed down to make sure the old ranking result from existing models will not be disturbed. In fact, if there is no "Forward" or "Like" behavior, the output of SRRR will be same with the existing algorithm. Further more, with the development of social net, the weight of SRF and RRF can be amplified, then, social behavior can affect search ranking globally. With SRRR, new webpages, and the sites with good content and poor SEO can get better rank rapidly, and the search optimization with social information and personalized search will be truly implemented.

It is worth mentioning that when exploring the way to combine social network with SE, what has been found very hard is how to protect users' privacy. On this aspect, SRRR only uses the public behavior of people instead of gathering and analyzing users' sensitive information deeply. Meanwhile, when designing SRRR, we fully consider about the different visibility of sharing and recommending. We use information from full site to calculate RRF, however, when calculating SRF, only users' friends will be involved. So, SRRR will not cause privacy problems.

4.2   Experimental Verification

Unfortunately, there is no developed platform with both search function and social service. So, it is impossible to get real data for experiment. To finish the experimental verification for SRRR, we build a tiny dataset to stimulate the situation. It is supposed there is a social-SE. Some users in this system form a network as Fig. 1 shows.

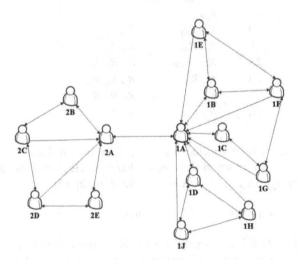

Fig. 1. Users' social network.

In this web, users with ID starting by 1 is a software develop team (referred as Circle 1), and users with ID from 2A to 2E are some fruit peasants (referred as Circle 2). Peasant might ask the team to produce an APP to promote their apple, so, there might be links between two circles (2A $\longleftrightarrow$ 1A). When these 14 people search word "apple" by Google or Bing, they will get the results with same following order:

$$R_{original} = \{R_{f1}, R_{p1}, R_{p2}, R_{p3}, R_{f2}, R_{p4}, R_{f3}, R_{f4}, R_{f5}, R_{f6}\}$$

In $R_{original}$, left members have the position nearer to the top than right ones on the output page of SE. Briefly, these results (webpages) are divided into two categories. R_{fi} pages are relevant to apple cultivation, while R_{pi} pages are related to Apple Inc. (iPhone, iPad, et al.). Index i means this page get the ith position in its category.

Obviously, Circle 1 wants R_{pi} more, while Circle 2 is more concerned about P_f pages. Now, in social-SE, suppose users will share or recommend the result pages they like or feel helpful, like Table 1 shows. We will exclude user 1G and 2C.

Table 1. Records of sharing and commending

User ID	Pages	
	Sharing pages	Commending pages
1A	R_{p1} R_{p3}	R_{p1} R_{p3} R_{f4}
1B	R_{p2} R_{p3}	R_{p3}
1C	R_{p1}	R_{p1} R_{p3}
1D	R_{p3} R_{p4}	R_{p3} R_{p4} R_{f6} R_{f1}
1E	R_{p2} R_{f6} R_{p4} R_{p1}	R_{p2} R_{p3}
1F	R_{p3}	R_{p3}
1H	R_{p1} R_{p3}	R_{p1}
1J	R_{p4} R_{p3} R_{p1}	R_{p3} R_{p1}
2A	R_{f4}	R_{f4} R_{f1} R_{f2}
2B	R_{f4} R_{f5}	R_{f4} R_{f5} R_{p2}
2D	R_{f4}	R_{f4} R_{f6} R_{p3}
2E	R_{f1} R_{p3}	R_{f1} R_{f3}

From Table 1, it can be concluded that R_{p3} is popular in Circle 1, while Circle 2 prefers R_{f4}. Some noise is set in the dataset, like that 1E shared R_{f6} and 2D recommended R_{p3}. Briefly, the Eq. (4) will be simplified as Eq. (5):

$$srr(u, k, p) = c_1 f_1 + c_s srf(u, k, p) + c_r rrf(u, k, p) \qquad (5)$$

In Eq. (5), $f_1 = 11-i$, i means page p is on the ith position in $R_{original}$. Here, f_1 is a comprehensive factor that covers all ranking factor in existing SE. Weight c_1 is given a small value to eliminate the error caused by the simplification. We use Java to implement SRRR, and set $d = 0.85$ for Eq. (1), $threshold = 6$ for Eq. (3) and $c_1 = 0.1$, $c_s = 0.7$, $c_r = 0.2$.

When taking 1G and 2C into consideration, SRRR values of different webpages to different users for searching "apple" are shown in Table 2.

According to Eqs. (4) or (5), SRRR value is a function of user, keyword and webpage, and in Table 2, it can be found to different users, one page has different SRRR value when the keyword is "apple". So, it proves that SRRR is a personalized rank model.

If the results are arrayed by SRRR value in descending order, the output lists of user 1G and 2C will like Tables 3 and 4 present:

Table 2. SRRR value

Webpage	User ID				
	1A	*1B*	*1C*	*1D*	*1E*
R_{f1}	1.7429	1.7000	1.6600	2.8933	1.7000
R_{f2}	0.9000	0.8500	0.8500	0.9333	0.8500
R_{f3}	0.5429	0.6000	0.5600	0.5600	0.6000
R_{f4}	3.1422	1.7500	2.2454	1.7933	1.7500
R_{f5}	0.3429	0.4000	0.3600	0.3600	0.4000
R_{f6}	1.2000	1.1685	0.9500	2.100	0.9500
R_{p1}	3.0514	2.4726	4.0116	3.1794	2.2542
R_{p2}	1.4006	1.8685	1.7200	1.2800	2.8794
R_{p3}	4.3173	5.0861	5.4616	5.0928	5.3154
R_{p4}	0.9739	0.9685	0.7500	2.1210	0.7500
Webpage	User ID				
	1F	*1G*	*1H*	*1J*	*2A*
R_{f1}	1.8667	1.7000	1.8933	1.8933	3.2229
R_{f2}	0.9333	0.8500	0.9333	0.9333	2.0000
R_{f3}	0.6000	0.6000	0.5600	0.5600	0.8000
R_{f4}	1.901	1.7500	1.7933	1.7933	3.1229
R_{f5}	0.4000	0.4000	0.3600	0.3600	0.8229
R_{f6}	1.1972	0.9500	1.1000	1.1000	0.7000
R_{p1}	2.4930	2.6251	3.9585	3.9585	2.1242
R_{p2}	1.9359	2.6500	1.2800	1.2800	1.5200
R_{p3}	5.0345	5.0861	4.1036	5.1036	3.0671
R_{p4}	0.9972	0.7500	1.3528	1.1318	0.7000
Webpage	User ID				
	2B	*2C*	*2D*	*2E*	
R_{f1}	2.5396	2.1667	2.516	3.2667	
R_{f2}	1.2000	1.0000	1.000	1.2000	
R_{f3}	0.7333	0.9000	0.8000	1.8000	
R_{f4}	4.3050	4.1151	3.9922	3.2383	
R_{f5}	1.6000	0.8229	0.7000	0.5333	
R_{f6}	1.6396	2.1667	2.3930	1.3000	
R_{p1}	1.7400	1.9067	1.9067	1.7400	
R_{p2}	2.4400	1.4400	1.5400	1.3733	
R_{p3}	2.6467	2.8800	4.1029	2.9133	
R_{p4}	0.7667	0.7667	0.7667	0.7667	

Because 1G belong to Circle 1, so, as Table 3 shows, R_{p3} and R_{p2}, the webpages with much sharing and recommendation in this circle, are put on the top of 1G's list. As well as R_{f4} and R_{f1} in 2C's list. Namely, with SRRR, SE can choose the results through analyzing behaviors of users' social circle, find the good pages, and put them on the head of the list. In summary, compared with existing SE, the simulating system with SRRR is more personalized, meanwhile, it is capable to save users' time to find the best pages they really want.

Table 3. Search result list of user 1G

No.	Search result of user G1	
	Webpage	*SRRR value*
1	R_{p3}	5.0861
2	R_{p2}	2.6500
3	R_{p1}	2.6251
4	R_{f4}	1.7500
5	R_{f1}	1.7000
6	R_{f6}	0.9500
7	R_{f2}	0.8500
8	R_{p4}	0.7500
9	R_{f3}	0.6000
10	R_{f5}	0.4000

Table 4. Search result list of user 2C

No.	Search result of user 2C	
	Webpage	*SRRR value*
1	R_{f4}	4.1151
2	R_{p3}	2.8800
3	R_{f1}	2.1667
4	R_{f6}	2.1667
5	R_{p1}	1.9067
6	R_{p2}	1.4400
7	R_{f2}	1.0000
8	R_{f3}	0.9000
9	R_{f5}	0.8229
10	R_{p4}	0.7667

5   Conclusions and Future Work

This paper briefly introduces the research status of the combination of social network and search engine, and analyzes the advantages and deficiencies of existing study. Then, we propose a new model called sharing and recommending result rank (SRRR). Through mining the behavior of users' social circle, especially the sharing and recommending action, SRRR can gain the evaluation of users' social circle to the content, which make it able to help people choose the search result. To different users, SRRR can provide different search result ranking. Theoretical, empirical and experimental analysis is in line with the expectation. So, it is proved that SRRR is an effective approach with proactivity, personality and sociality, just as literature [4] anticipated.

Like we mentioned in 4.1, it is very hard to build a new and fully developed social network in existing search engine. And until now, an application system with both full-text search function and social service has not been developed. So, in future research, we hope find a way to combine existing social web and SE, like Google and Twitter, or Baidu and Sina Micro Blog. The data from developed social net can make SE more intelligent, while the full-text SE can provide much more information to the social web. The approach to integrating their capability of allocating network flow will be explored, and SRRR might be the starting point.

Future work also include adapting and ascertaining the best range of SRRR parameters, analyzing users' interests, improving the effect of SRF and RRF calculation, and taking the specialization of users (expert users) into consideration. Further more, social search engine is based on the hypothesis that users will actively interact with the system. So, how to encourage users sharing and recommending contents, and how to prevent spam behavior (like malicious "Forward" or "like") is also need to be researched.

Acknowledgement. The research is supported by "National Natural Science Foundation of China" (No. 61202095, No. 61073186) and "National Undergraduates' Innovation and Entrepreneurship Training Program of China" (NO. 201310533018).

References

1. Cao, J., Tang, Y., Lou, B.: Social search engine research. In: 2010 3rd IEEE International Conference on Computer Science and Information Technology (ICCSIT), vol. 7, pp. 308–309 (2010)
2. Akiyama, T., Kawai, Y., Matsui, Y.: A proposal for social search system design. In: The 11th IEEE/IPSJ International Symposium on Applications and the Internet (SAINT), pp. 110–117 (2011)
3. Li, F.: The game between search engines and social networks. Commun. CCF 7(9), 54–57 (2011)
4. Elowitz, B.: Search and social: how the two will soon become one[EB/OL]. http://techcrunch.com/2012/06/17/search-and-social-how-two-will-soon-become-one/. 17 June 2012
5. Xing, C., Zhang, W., Zhang, X., et.al.: SNSearch: a search engine for social network search [EB/OL]. http://www.paper.edu.cn/releasepaper/content/201301-172. 5 January 2013
6. Li, J.: The development trend of next generation search engine: social search. Sci. Inf. **13**, 14 (2010)
7. Bao, S., Wu, X., Fei, B.: Optimizing web search using social annotations. In: Proceeding WWW 2007 Proceedings of the 16th International Conference on World Wide Web, pp. 501–510 (2007)
8. Lo, C.-H., Peng, W.-C., Chiang, M.-F.: Ranking web pages from user perspectives of social bookmarking sites. In: 2008 IEEE/WIC/ACM International Conference on Web Intelligence and Intelligent Agent Technology - Workshops WI-IAT Workshops, vol. 1, pp. 155–161 (2008)
9. Xu, J., Kang, M., Dong, G.: The research of ranking algorithm for real-time search engine based on social networks. Sci. Technol. Eng. **11**(28), 6879–6882 (2011)
10. Shi, L.: Research and Application of Graph Query Algorithm Based on Friends Relationship in Social Network. Nanjing University of Science and Technology, Nanjing (2012)

11. Pan, X., Hua, G., Liang, B.: Stepping into Search Engine (V2). Publishing House of Electronics Industry, Beijing (2011)
12. Page, L., Brin, S., Motwani, R., et al.: The page rank citation ranking: bringing order to the web[EB/OL]. http://ilpubs.stanford.edu:8090/422/1/1999-66.pdf. 29 January 1998
13. Vu, T., Baid, A.: Ask, don't search: a social help engine for online social network mobile users. In: 2012 35th IEEE Sarnoff Symposium (SARNOFF), pp. 1–5 (2012)
14. Li, D.: Cloud Computing Technology Development Report 2013]. Science Press, Beijing (2012)

WebCDN: A Peer-to-Peer Web Browser CDN Based WebRTC

Kai Shuang[✉], Xin Cai, Peng Xu, and Qiannan Jia

State Key Laboratory of Networking and Switching Technology,
Beijing University of Posts and Telecommunications, Beijing, China
shuangk@bupt.edu.cn

Abstract. Over the past decade, though web contents increase rapidly, the architecture of the web services still remains the same. As the growth of users, web servers must provide huge network bandwidth and computing power. This paper realizes the WebCDN system, which achieves a content distribution network by web users. Through the WebRTC and HTML5 technology, the web site needn't require their users installing anything to use this service. This paper describes the design and implementation of the system in terms of both server and client side in detail. Through simulating the web resources and user behavior, the experiment result shows that the WebCDN system greatly reduces the network traffic with acceptable service latency.

Keywords: Web content delivery network · WebRTC · WebSocket · Local storage

1 Introduction

Over the past decade, the web contents become rich and varied. Image, flash and video can come in any website. The average size of a single web page is 1.6 MB (in 2013), which is fifteen times larger as 93 KB (the beginning of the Internet in 2003). Unfortunately, the architecture of the web services doesn't change as the page size does. Although, some new technology (just like LVS [1]) can make the website burn more visits by using more servers. Primarily the servers must serve every request, response the page content, and need huge network bandwidth and computing power. So the website's expense will increase as linear.

Just as Moore's law dictate, the hardware price becomes cheaper and the memory size and CPU speed of personal PCs become better. When a user browses the web, the free processing power will be a huge waste. Recently, some professors have focused on this problem and have done some research in this field to allow browser users to distribute the resources of web pages. For examples, in paper Web2Peer [2], Flower-CDN [3] and Buddy Web [4], they all present a P2P solution. To use their system, users have to install some browser plugins or PC client software, which will extremely affect user experience. Hence, all of the above systems can't be applied widely.

L. Yao et al. (Eds.): APSCC 2015, LNCS 9464, pp. 235–243, 2015.
DOI: 10.1007/978-3-319-26979-5_17

In this paper, we build a WebCDN system embedded into a Web Server. We use the browser technology HTML5 [5] and WebRTC [6] supported by a lot of browsers, without installing anything. The goal of this system is to eliminate the bottleneck of increasing users by distributing some static resources requests to other users and handle them in a P2P mode. For a P2P system, the more active users it has, the better service it will provide. We give sufficient consideration to the design of WebCDN from two aspects: user experience and system expansibility. In this paper, we describe the implementation of the system from both the server and client side in detail. In the last, we conduct a simulation experiment for evaluating the practicality.

2   Development of WebRTC and HTML5

The first draft of HTML5 has been published in 2008 and the fifth revision has been completed in October 2014. Now, most of the browsers have supported it. Local storage is one of the new features, which is different with session storage. The storage will be saved persistently whenever a user closes the browser, and it also removes the cookie restrictions of 4 KB memory space. The stored data are shared in a domain, thus even if the user browses other pages, we can also easily access the stored data. The HTML5 technology provides guarantee of storage for the WebCDN system.

WebRTC (Web Real-Time Communication) is an API definition drafted by the W3C that supports browser-to-browser applications for voice calling, video chatting, and P2P file sharing without the need of either internal or external plugins [7]. In 2011, Google releases the open source project of WebRTC. Currently, Firefox and Chrome browsers have provided friendly support both in PC and android endpoint. A new report from Disruptive Analysis shows that, by 2019, there will be more than six billion WebRTC-enabled devices worldwide [8]. The WebRTC project is developed by C++, which guarantees high efficiency. It's embedded into the browser and developers can use its JavaScript API easily. Two browser users complete the P2P connection by exchanging their SDP [9] information through the offer/answer messages. WebRTC also supports an ICE [10] framework to complete the NAT [11] traversal, which greatly promotes the application scenario. In our system, we mainly use the API of Datachannel for P2P file sharing. WebRTC technology provides guarantee of data transition for the WebCDN system.

3   Design of WebCDN Architecture

The WebCDN will be embedded into a Web Server and build a content delivery network for distributing static web resources, such as images, videos and files, which saves some hottest resources in the user browser after the user requests them. The system mainly contains two components: the WebCDN client by javascript code and the WebCDN server by java code. The WebCDN system is deployed independently and the web server will share the web resources with it (Fig. 1).

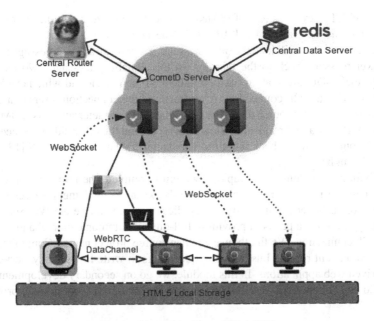

Fig. 1. The architecture of WebCDN

3.1 Design of WebCDN Client

The WebCDN client consists of two modules, browser local storage and data transfer, which are both developed by JavaScript. We mainly use two technologies of web browser, HTML5 and WebRTC, which are supported by most browsers and can be used by users without installing any plugins or software.

Browser Local Storage Module. To store the resources into the user's browser persistently, we use the HTML5 Local Storage JavaScript API. It's a key/value storage system and the API is very easy to use, just like "localStorage.setItem('key', value)". The key (resource's ID) generated by the server will be globally unique. The value is indicated by Uint8Array, which is a binary mode and uses less memory than common JavaScript array.

The resources stored on the user's disk can't be unlimited. So there will be a storage threshold in the WebCDN system. We use the FIFO (first in first out) algorithm to keep the threshold and the operation will be triggered after the user completes a request.

Data Transfer Module. We will build WebRTC connection with different users for different resources. Requesting the resource in turn will lead to an intolerable latency. Fortunately, HTML5 has provided Web Workers which makes JavaScript become multithreaded. In this module, we will run five workers to deal with the resource request concurrently.

The WebRTC provides an Offer/Answer model to realize the build of connection and this module is realized by JSEP [12]. The connection initiator sends an offer message. The receiver will load the offer message, generate an answer message and send the answer message back to the initiator. In the signalling message, two users will exchange their SDP information (ip, port and so on) and session-id which can be used to identify a unique P2P connection. After finishing the connection, users can call the send-function to send resource data. The send-function parameter supports DOMString, Blob, ArrayBuffer and ArrayBufferView data type. And the other side will receive the data in the onmessage callback function. The data is transmitted by DTLS [13], which is secure enough.

For building the connection rapidly, we must complete the transmission of offer/answer message in a real time. In other words, we need implement Server Push. In consideration of better expansibility and applicability, we choose the WebSocket [14] technology, which is a protocol providing full-duplex communication channels over a single TCP connection. We finally use an open source project named CometD that is a scalable web event routing bus allowing developers to write low-latency, server-side, event-driven web applications. In this module, we go on secondary development based on the CometD client. Also it will help us solve the javascript cross-domain problem.

3.2 Design of WebCDN Server

The WebCDN server consists of two modules, CometD management module and central data service module. The main function of server is locating the position of resources and routing the signalling messages for the foundation of WebRTC connection.

CometD Management Module. In the WebCDN system, we need to satisfy a real-time connection between server and user for exchanging offer/answer message rapidly. So we use the open source project CometD to manage the WebSocket connection. For one CometD server, the numbers of WebSocket connection is limited.

We develop a central router server to solve the problem above. The server can dynamically distribute a user to a CometD server and update the global routing hashtable in the central data server. And it also have TCP connections with all CometD servers. When one user J wants to contact the user K in another CometD server, J's CometD server will firstly query K's server in the central data server and then forward the message to K's server over TCP. By this design, we can support millions of active users through extending the amounts of CometD server.

Data Service Module. Data server provides all kinds of data operation for CometD servers and the central router server in this module. All the data structure is a key/value model and the dates should be cached in the memory for fast access. So Redis is a very appropriate choice, which supports distributed deployment. In this system, we deploy multiple Redis servers to store data. The data structure is below.

The global routing hashtable is a mapping between user_id and CometD server, which is updated and queried by the central router server.

The user-resource hashtable is a mapping between user_id and the resources list stored by the user. This hashtable will be update when the user visits our system, includes adding new stored resources and removing the expired ones.

The resource-location hashtable is a mapping between resource_id and the locations list. The location consists of the resource url in the Web Server and the online users who keep the resource. When a user changes his online or offline state, this hashtable will be updated. When a user requests a resource, the CometD will query this hashtable and return the resource url and two user_ids randomly.

The resource-count hashtable is a mapping between resource_id and its request numbers in one day. In this server, there's also a thread executed per five minutes, which calculates the top $x\%$ resource_id for all CometD servers. The WebCDN will only make user store the hot resources. The value of x will affect the update frequency of user's local storage and hit rate of the resource.

4 Workflow of WebRTC

In this part, we will describe two most important workflows in the WebCDN system, the client initial workflow and the resource request workflow.

4.1 Client Initial Workflow

In this workflow, every user needs a unique user_id to identify himself. When a new user visits our system, he will be allocated a user_id and the system will store the id in his cookie. When an old user comes, he will carry his user_id and the Router server will update his status (Fig. 2).

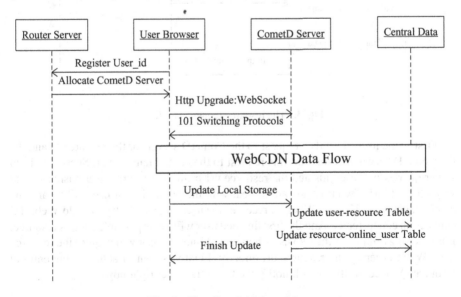

Fig. 2. The client initial workflow

When a user visits our site, firstly, he sends a normal HTTP request to the Web Server. The Web server will ask the Route Server to allocate a CometD Server for the user and response with the resource list and ServerID. Then, the user sends to the CometD server the upgrade request for building a WebSocket connection. Next, the user will complete the resource request workflow. After the user gets all resources, he will check his local storage whether the total storage has exceeded the threshold and then send the update message. If any resources are removed from the user, the CometD server will synchronize the user-resource table and resource-oneline_user table in the central data server, which will guarantee the consistency of data between the server and the user.

4.2 Resource Request Workflow

In this workflow, the user requests a resource from the WebCDN system. The integral process includes finding the online users with the resource, building WebRTC connection with other users and achieving the resource (Fig. 3).

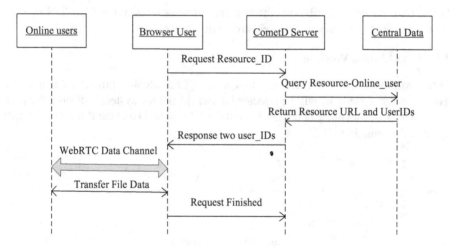

Fig. 3. The resource request workflow

Firstly, the user M sends a request to the CometD server for the resource Q and the the CometD server will forward this request to the central data server. Secondly, Data server queries two user_ids and the resource url from the resource-user hashtable and sends them to M. Two user_ids are to deal with the possibility of user offline and the complex network. Thirdly, When M receives the user list, he will try to build WebRTC connection concurrently. And the first finished user will be responsible for this resource. If no user is available or the two users both get offline, the user will acquire the resource from Web Server by the resource url directly. Finally, When M achieves the entirely resource Q successfully, he will notify the CometD server right now.

5   Experiment and Evaluation

The WebCDN system is used for the web application with large scale users, which will greatly reduce the server's load and bandwidth. The percent of flow served by WebCDN is the most important evaluation criterion. As the connection of WebRTC and the probability of user's offline, fetching resources from our WebCDN network will bring extra latency. So we also need to prove that the latency shouldn't affect the user's experience. We implement WebCDN by java(server) and javascript(client) and acquire some valuable data through testing. Then we complete the simulation experiment and evaluate the performance of WebCDN based on the test date.

5.1   Latency of WebRTC

The latency of WebRTC comes from two parts, the connection establishment and the data transmission. We go on the experiment in our lab's LAN. In the process of WebRTC connection establishment, the average exchange data is 0.78 KB between two users by the CometD Server and the average latency is 123 ms. The transmission latency of 50 KB data by the DataChannel is about 17 ms, which linearly increases with the size of file. The total average latency will greatly decrease by using the Web Worker technology. The process of establishment takes up more time. So serving bigger resource will have better cost performance. In our system, the resources will be transmitted through WebCDN network only when they're more than 20 KB.

5.2   Simulation Experiment

For evaluating the system, we have done a simulation experiment. In the experiment, we simulate the users' behaviors and web page resources. The user behavior includes the visiting moment and the average time on page. The visiting moment concentrates upon between 9:00 ~ 11:00 and 19:00 ~ 21:00. The users can be divided into two kinds, light users and heavy users. For light users, they will come once every one to three days, browse one to three web pages every time and stay ten to thirty seconds for one page. For heavy users, they will come one or two times every day, browse three to six web pages every time and stay thirty to sixty seconds for one page. The page size is random distribution from 800 KB to 1.2 MB and we assume that the resources follow Gaussian distributions from 1 KB to 100 KB. The resources increase by 10 % every day and the requested probability decreases with the publish time, which means that the new resources are requested more frequently. The size of local storage is 10 MB in our simulation experiment and the stored resources are in the top 20 %. For solving the data cold start, all users have some resources in their local storage randomly. Based on the conditions above, we have accomplished the simulation experiment for three weeks. The results are presented in the Figs. 4 and 5. By the analysis of the results, the WebCDN system can reduce 48.39 % network traffic of web server, while bringing 147.85 ms extra latency for each request.

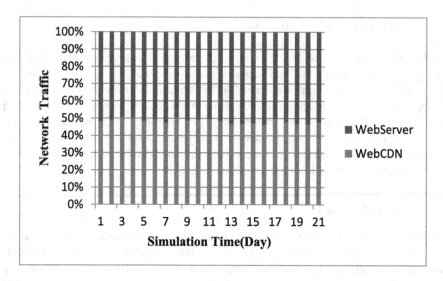

Fig. 4. The network traffic between WebCDN and Web Server

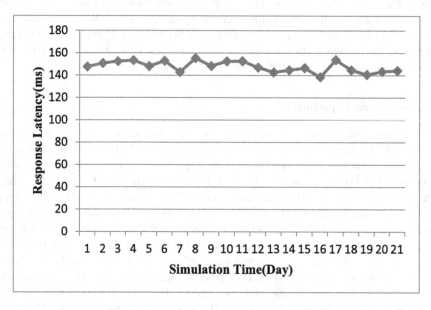

Fig. 5. The extra average latency of WebCDN

6   Conclusion

In this paper, we design and implement a WebCDN system, which can distribute resources by the P2P model. It's a new architecture that can reduce the Web servers' network traffic. We use the Local Storage technology to store resources in users' browsers and use the WebRTC technology to accomplish data transmission. Without any plugins, the system can be easily promoted. At last, we simulate the web resources and user behaviors. The experiment result shows that the WebCDN greatly reduces the network traffic with acceptable service latency. Further, we are thinking about combining some files from the same web page as one resources, which will solve the problem of small files and reduce the amount of resources in the WebCDN system.

Acknowledgements. National Key Basic Research Program of China (973 Program) under grant 2011CB302506, 2009CB320504; National Natural Science Foundation under grant 61170274.

References

1. WIKI LVS. http://en.wikipedia.org/wiki/Linux_Virtual_Server
2. Ribeiro, H.B., Lung, L.C., Santin, A.O., Brisola, N.L.: Web2Peer: implementing a peer-to-peer web browser for publishing and searching web pages on internet. In: Advanced Information Networking and Applications (AINA), Sedona, AZ, pp. 421–428 (2007)
3. Dick, M.El, Pacitti, E., Kemme, B.: Flower-CDN: a hybrid P2P overlay for efficient query processing in CDN. In: Extending Database Technology - EDBT, pp. 427–438 (2009)
4. Wang, X.Y., Ng, W.S., Ooi, B.C., Tan, K.-L., Zhou, A.Y.: BuddyWeb: a P2P-Based collaborative web caching system. In: Networking, pp. 247–251 (2002). (in Chinese)
5. Hickson, I., Berjon, R., Faulkner, S., Leithead, T., Navara, E.D., O'Connor, E., Pfeiffer, S.: HTML5: A vocabulary and associated APIs for HTML and XHTML (2014)
6. Bergkvist, A., Burnett, D.C., Jennings, C., Narayanan, A.: W3C Editor's Draft-2013. WebRTC 1.0: Real-time Communication Between Browsers (2013)
7. How WebRTC Is Revolutionizing Telephony. http://blogs.trilogy-lte.com
8. Disruptive Analysis. http://www.disruptive-analysis.com/webrtc2014.htm
9. Rosenberg, J., Schulzrinne, H.: RFC 3264. An Offer/Answer Model with Session Description Protocol (SDP), USA, IETF (2002)
10. Rosenberg, J.: RFC 5245. Interactive Connectivity Establishment (ICE): A Protocol for Network Address Translator (NAT) Traversal for Offer/Answer Protocols, USA, IETF (2010)
11. Egevang, K., Francis, P.: RFC 1631. The IP Network Address Translator (NAT) [S], USA, IETF (1994)
12. Uberti, J.: Draft-ietf-rtcweb-jsep-08-2014. Javascript Session Establishment Protocol, USA, IETF, (2014)
13. Rescorla, E., Modadugu, N.: RFC 4347. Datagram Transport Layer Security, USA, IETF (2006)
14. Fette, I., Melnikov, A.: RFC 6455. The WebSocket Protocol, USA, IETF (2011)

Short Papers

DDS: A Deadline Driven Workflow Scheduling Algorithm for Hybrid Amazon Instances

Zitai Ma, Jian Cao[✉], and Shiyou Qian

Shanghai Jiao Tong University, Shanghai, China
mazt1024@gmail.com, {cao-jian,qshiyou}@sjtu.edu.cn

Abstract. Workflows can orchestrate multiple applications that need resources to execute. The cloud computing has emerged as an on-demand resource provisioning paradigm, which can support workflow execution. In recent years, Amazon offers a new service option, i.e., EC2 spot instances, whose price is on average more than 75 % lower than the one of on-demand instances. Therefore, we can make use of spot instances to execute workflows in a cost-efficient way. However, the spot instances is cut off when their price increases and exceeds the customer's bid, which will make the task failed and the execution time becomes unpredictable. We propose a deadline driven scheduling (DDS) algorithm which is able to use both on-demand and spot instances to reduce the cost while the deadline of workflows can also be guaranteed with a high probability. Especially, we use an attribute, called global weight, to represent the interdependency relations of tasks and schedule the tasks whose interdependent tasks need longer time first to reduce the whole execution time. The experimental results demonstrate that DDS algorithm is effective in reducing cost while satisfying the deadline constraints of workflows.

Keywords: Workflow · Scheduling algorithm · Spot instance · Cloud computing

1 Introduction

Cloud Computing has become a hot and popular topic for researchers and companies in recent years. It has been found that more and more applications are deployed in the cloud instead of grid [1] or other traditional computation platforms for the lower cost, higher availability, larger scalability and other attractive features of cloud.

Workflows are able to orchestrate multiple applications so that they have been extensively applied in scientific research and business areas. Generally, time and cost are two contradicting objectives. In order to balance the two objectives, an effective scheduling algorithm is needed for workflows to be deployed in the cloud environment.

There have been some works focusing on workflow scheduling problem in the era of grid computing [2, 3]. On-demand instances are typical resources provided in clouds, with a stable and fixed price for each size and capacity (small, medium, large). Customers are charged with a fixed cost according to their running time. Some workflow scheduling algorithms have been proposed to optimize the workflow execution cost with on-demand resources [4, 5].

© Springer International Publishing Switzerland 2015
L. Yao et al. (Eds.): APSCC 2015, LNCS 9464, pp. 247–257, 2015.
DOI: 10.1007/978-3-319-26979-5_18

Recently, Amazon provides EC2 spot instances, which allows customers to purchase compute resources at hourly rates that are lower than the on-demand ones for most of the time. Therefore, we can make use of spot instances to execute workflows in a cost-efficient way [6, 7]. However, spot instances are cut off when their spot price increases and exceeds the customer's bid which will lead to the failure of task execution within a workflow so that the execution time of the whole workflow becomes unpredictable. Therefore, in the algorithms [6, 7], fault tolerance strategies are utilized to avoid aborting the whole workflow when the spot instances are cut off. Obviously, fault tolerance strategies, such as redundancy and check point, incur additional overhead.

In this paper, we propose a new deadline driven scheduling (DDS) algorithm which considers both on-demand and spot instances for workflow execution. The target of DDS is to reduce the execution cost while the deadline specified by customers can also be met with a very high probability. The main contributions of this paper are as follows: (1) we proposed a deadline driven scheduling (DDS) algorithm. (2) We used an attribute, called global weight, to represent the interdependency relations of tasks and schedule the tasks whose interdependent tasks need longer time first to reduce the whole execution time. (3) We developed a simulator to mimic the workflow scheduling process in the cloud, and conducted a series of experiments to verify the effectiveness of DDS algorithm in terms of monetary cost.

The rest of the paper is organized as follows: In Sect. 2, we introduce the background. In Sect. 3, we give the simulator model. Section 4 details the DDS algorithm. Section 5 presents the experimental results. The paper is concluded in Sect. 6.

2 Background

2.1 Related Work

To date, best-effort based and QoS constraint based are two major types of workflow scheduling algorithms which have been applied in the grid or cloud platforms [4, 5, 8–10]. Heuristics or meta-heuristics algorithms that belong to the best-effort based algorithms are widely used. For example, Zhong et al. [8] proposed a scheduling algorithm based on Genetic Algorithm for optimal scheduling plans. In addition, some meta-heuristics like particle swarm optimization (PSO) and ant colony optimization (ACO) have also been employed in the cloud workflow scheduling field [9, 10].

Some algorithms are designed based on deadline and budget constraints. For example, a deadline assignment method is proposed for tasks to meet the deadline constraint in [11]. Kllapi et al. consider the tradeoff between execution time and monetary cost [12]. Although the aforementioned studies have shown their effectiveness in reducing monetary cost, most of them only consider scheduling workflows with fixed pricing models, such as on-demand instances.

After Amazon launched spot instances in the market, it has been shown that using spot instances to run scientific applications can save money [13, 14]. Workflow scheduling algorithms considering spot instances are introduced in [6, 7]. These algorithms are based on different bidding strategies and fault tolerance techniques like checkpoint and duplication. Although these techniques can ensure the completion of workflows, the customer's deadline requirement is not guaranteed.

In addition, there are also some studies using spot instances while considering workflow deadline constraints like [15]. However, in [15], the author used a priority queue to select the minimal task for scheduling which belongs to batch scheduling category and ignored the interdependency relations in the workflow. In DDS, we use an attribute, called global weight, to represent the interdependency relations of tasks and schedule the tasks whose interdependent tasks need longer time first to reduce the whole execution time. When it comes to a batch of tasks that have the same global weight in parallel structures, another selection strategy would be active in DDS. DDS is able to enhance the performance in both pipeline and parallel structures by a wise scheduling strategy.

Hence, we will focus on three points: spot instances, deadline constraint and wise scheduling sequence to improve the performance of DDS algorithm.

2.2 Instances in Cloud

Amazon EC2 offers a wide range of instance types that is optimized for different applications. We consider general purpose (T_2, M_3) types only. Table 1 shows some configuration information about different capacities and the fixed prices of on-demand instances in USA East zone with Linux/Unix systems on May 22nd, 2015. Unlike the fixed price for on-demand instances, spot instances have dynamic prices which fluctuate periodically depending on the supply and demand relationships in the cloud.

Table 1. Configuration information with different types.

Type	vCpu	Memory (GiB)	Price ($/hour)
T_2.small	1	2	0.026
T_2.medium	2	4	0.052
M_3.medium	1	3.75	0.070
M_3.large	2	7.5	0.140
M_3.xlarge	4	15	0.280
M_3.2xlarge	8	30	0.560
M_1.small	1	1.7	0.044

2.3 Workflows

Workflows are usually represented as a directed acyclic graph (DAG). Given a workflow, let T represent the set of nodes T_i ($1 < i < n$), and E represent the edges in the workflow. (T_i, T_j) is one of edges in E, showing AND/OR data dependencies between node T_i and node T_j. T_i is called the parent task node, and T_j is the child task node. The tuple *(T, E)* represents a workflow.

In this paper, we focus on scientific workflow applications. Juve et al. [16] have shown and explained the characteristics of various scientific workflows, such as Montage, Cybershake, Sipht and LIGO. We use the above four different workflows which have the basic patterns in the workflows, such as pipeline, parallel, data aggregation and distribution structures.

In/out node: a task node that has no parent task nodes is called an in node, denoted as T_{in} while the node which has no children is called an out node, denoted as T_{out}.

Tasksize: The size of a task, denoted as *Tasksize,* is in terms of millions instructions.

Cost: Cost is the sum of costs of the instances that are used during the workflow execution.

Deadline: Deadline is a parameter of a workflow which specifies when the workflow should be completed. *Deadline* is one of the customer's QoS requirements.

3   System Model

We have designed a simulator to mimic the process of workflow execution in the cloud. The architecture is shown in Fig. 1. First, a user submits a workflow into the system. Then, the workflow parser handles the workflow's input files and conducts the pre-schedule steps in DDS algorithm: deadline assignment and global weight calculation. The workflow parser analyzes and processes the AND/OR data dependencies and task information in the workflow. After these steps are completed, the workflow parser sends the processed workflow to the workflow engine which is in charge of group partition and maintains a waiting list for unscheduled tasks. Only tasks whose parent tasks are all completed can be added to the waiting list. A task selected from the waiting list is submitted into the searching and mapping engine which is responsible for looking through the available instances in the cloud and finding a suitable instance that can

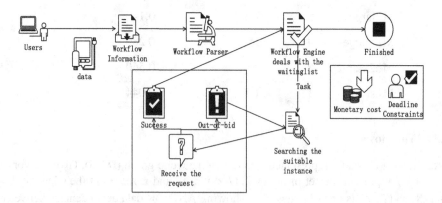

Fig. 1. The architecture of simulator

minimize the monetary cost. The key pair of (*VM, task*) will run in the cloud. If a spot instance terminates halfway, a strategy coping with out-of-bid failures will be active to guarantee the task to be finished. After all the tasks in a workflow have been done, the scheduling process is completed. The details about our DDS algorithm will be introduced in the next section.

4   Deadline Driven Scheduling Algorithm

For the workflow scheduling problem in the cloud, finding a suitable instance types with a minimal price for tasks in a workflow is an NP-Complete problem [17]. Inspired by a batch of workflow scheduling algorithms in grid computing [18, 19], especially the idea of hybrid [19], we have proposed DDS algorithm that makes special changes to adapt the price dynamics and out-of-bid failures in the cloud.

Next, we will introduce DDS algorithm in 4 subsections. Sections 4.1 and 4.2 present the necessary steps that should be done before executing a workflow. The following section presents the details about the real-time property of DDS algorithm for the execution of workflows.

4.1   Deadline Assignment

According to the user's deadline requirement, we attempt to minimize the cost under the deadline constraint. DDS algorithm puts deadline constraint at a very preliminary and important stage, which is a leading thread throughout our scheduling algorithm.

For each task, DDS does the deadline assignment by a depth first search (DFS) procedure and calculates the deadline. At first, the deadline of T_{out} is set, whose value is equal to the deadline constraint of the workflow. If there are two more T_{out} nodes in a workflow graph, we estimate the deadline according to each T_{out} and choose the earliest

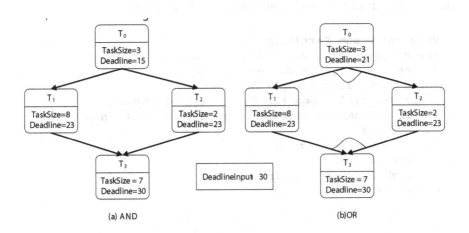

Fig. 2. Deadline Assignment example

one as the latest time to complete this task. Further, the deadlines of the tasks that are in the current task's parent task list are calculated. Figure 2 shows an example. Suppose the fastest instance executing a task in the cloud with 1 task size needs 1 h, without considering the data transfer time.

4.2 Global Weight Calculation and Group Partition

The next step is to figure out the global weight of tasks and do the group partition. The global weight is evaluated in a breadth-first manner, calculating from bottom to top in the DAG. First, Eq. (1) is used to calculate the average execution time of task T_i as the local weight of T_i, denoted as w_i. $w_{i,j}$ represents the average data transfer time from task T_i to task T_j. Equations (3) and (4) show how to calculate the global weight for T_{out} and other nodes [20], where $child(T_i)$ denotes all the child nodes of T_i.

$$w_i = \frac{\sum_{v \in VM_i} Time\,(T_i, v)}{|VM_i|} \tag{1}$$

$$w_{i,j} = \frac{\sum_{v_i \in VM_i, v_j \in VM_j} Time\,(v_i, v_j)}{|VM_i|\,|VM_j|} \tag{2}$$

$$GW\,(T_i) = w_i \tag{3}$$

$$GW\,(T_i) = w_i + max_{T_j \in child(T_i)}\,(w_{i,j} + GW\,(T_j)) \tag{4}$$

According to the global weight that we have computed and the interdependency relations between task nodes, tasks are divided into several groups so that nodes in the same group are interdependent and nodes between groups may have data dependencies. The following program shows the procedure of group partition.

```
Program: Group Partition
Input: Queue of task in decreasing order
create a task group Gk where k = 0;
While Queue is not empty do
  current_task   pop the first task of Queue
    If any parent task of current_task in Gk then
    k   k+1
      create a new group Gk
    end if
    push current_task into Gk
end while
```

4.3   Mapping Instances and Out-of-Bid Strategy

After generating a number of groups in terms of the global weight, DDS algorithm schedules groups one by one. For each group, tasks are sorted according to their local priority which is set in terms of task size. Usually, the bigger the task size, the higher the local priority. Tasks are accepted as input in the sorted order, mapping to suitable instances. If out-of-bid happens, the spot instances are terminated and DDS algorithm immediately goes through the available instances and finds an instance with the lowest cost meeting the deadline constraint of the task. If there is no suitable instance, an on-demand instance will be added to the instance pool. However, if there is no available schedule plan for this task that satisfies the deadline requirement, the schedule is failed. The flow chart is shown as part of Fig. 1.

5   Experiments

5.1   Instance Resource Pool

Resource Pool. In our simulation, we have considered 5 instance types (m3.medium, m3.large, m3.xlarge, m3.2xlarge and m1.small), two on-demand instances and one spot instance for each type. And more on-demand instances will be added if necessary.

Instance Price. There have been some studies focusing on how to calibrate the spot price of Amazon EC2 [21]. We have obtained the price history from February 22nd, 2015 to April 24th, 2015 for the experiments. From the data we got, we can see that the spot price is lower than the on-demand price for most of the time. Occasionally, it is possible that the spot price is larger than the on-demand one.

5.2   Bidding Strategies and Baseline Algorithms

Bidding Strategies. There are three bidding strategies in the experiments. (1) *naïve bidding strategy (gaussian)*: use the last spot price plus a Gaussian increment as the bid price. (2) *average bidding strategy (average)*: use the average price of the last 4 history prices as the bid price. (3) *KNN bidding strategy*: search the history prices, and predict a bid price with KNN method.

Baseline Algorithms. In order to test the effectiveness of our DDS algorithm, we have developed 5 algorithms as baselines. (1) *Basic Scheduling Algorithm (BS)*: BS algorithm uses task mapping only. Tasks are scheduled by adding them into the waiting list. (2) *On-Demand only (ODO)*: ODO algorithm only considers on-demand instances for scheduling. (3) *Spot Instance only (SIO)*: SIO algorithm only considers spot instances and assumes that there is no possibility that out-of-bid takes place. (4) *Local Sorted scheduling (LSS)*: every time the engine queries a task from the waiting list, the waiting list will be sorted and LSS selects the task with the largest local priority.

5.3   Experimental Results and Analyses

Deadline with Success Rate and Cost. The deadline of a workflow is an important parameter in workflow scheduling. In the experiments, we set the execution time of the longest path in the workflow with the fastest instances as the standard time. And we use the integer multiples from 1 to 10 of this standard time as input deadline requirements.

Table 2 shows the success rate with different algorithms in a range of deadline inputs. Since the stable of on-demand instances and the assumption that no out-of-bid happens in SIO, the success rates of ODO and SIO are both 100 %. Other algorithms with one deadline is simulated 100 times and the scheduling success times are recorded. Figure 3(a) shows the mean execution cost with different algorithms.

Table 2. The success rate with different algorithms in a series of K.

Algorithm\K	4	5	6	7	8	9	10
BS	60 %	69 %	86 %	93 %	98 %	100 %	100 %
LSS	40 %	58 %	88 %	96 %	98 %	100 %	100 %
DDS	64 %	78 %	92 %	99 %	100 %	100 %	100 %

Different Bidding Strategies. We evaluate the effect of bidding strategies on monetary cost. In this experiment, we use three strategies: (1) Gaussian (2) average

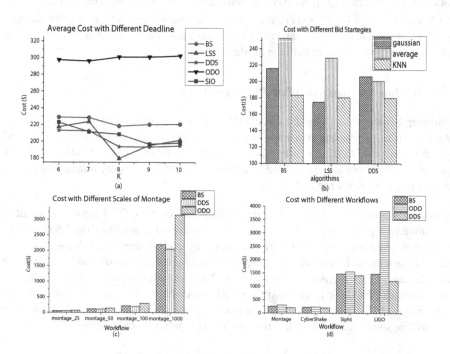

Fig. 3. Execution cost in different conditions

(3) KNN with 100 task nodes of workflow Montage and test three algorithms that uses spot instance with out-of-bid possibility (BS, LSS, DDS). Figure 3(b) shows the monetary cost under k = 10.

Different Workflows. Since the type and scale of workflows affects the cost of execution, we test different workflows: various numbers of task nodes and various kinds of science workflows. In order to show the effectiveness, we use ODO and BS as contrast algorithms. Figure 3(c) shows the average cost with different scales of Montage. Figure 3(d) shows the average cost in different structure of scientific workflows.

5.4   Results Analyses

As listed in Table 2, we can see that with relaxed deadline requirements, the scheduling success rate approaches 100 %. DDS algorithm shows higher success rate compared with other algorithms. Through Fig. 3(a), we can get that scheduling with spot instance can almost save 40 % money compared with on-demand only. Figure 3(b) demonstrates that different bidding strategies have their own advantages. Since KNN is based on history prices, able to predict periodical price waves, KNN has better performance. Figure 3(c) shows that the more nodes a workflow has, the more money is saved using DDS algorithm. In Fig. 3(d), since Montage has many serial structures and the competition of instances is not so fierce like Cybershake and Sipht that are highly parallel, BS can save 18 % monetary cost in Montage and 2 %, 5 % respectively in Cybershake and Sipht. And DDS have 32 %, 14 %, 10 % reduce in these three workflows for the global weight and group scheduling works in parallel structures. And since we only have 10 spot instances in our instance pool, if the parallel tasks is far more than 10, due to the constraint of the deadline, the algorithms have to use a large number of on-demand instances, so in Sipht and Cybershake, BS is almost the same as ODO, and DDS has better performance. In the complex workflows like LIGO, we can see that DDS saves 60 % than ODO and nearly 20 % monetary cost than BS.

6   Conclusion

Through the experiments, we can see that DDS algorithm shows stronger stability in different workflows and more than 20 % cut off than ODO. When the deadline requirement of workflows is relaxed a little, DDS may schedule successfully during most of the testing time. This demonstrates that DDS has advantage of reducing money while guaranteeing the customer's deadline requirement.

However, DDS algorithm is not perfect, still having space for improvement in the future. For example, in this paper, we did not consider the dynamic time of instances wasted in start or reboot operations. In addition, in the situation where customers use cloud storage services, such as Amazon S3, together with computing services, customers can save temporary results in S3, avoiding any recalculation when a spot instance fails. In the future, we plan to further improve the performance of DDS.

Acknowledgements. This work is partially supported by China National Science Foundation (Granted Number 61272438, 61472253), Research Funds of Science and Technology Commission of Shanghai Municipality (Granted Number 15411952502, 12511502704).

References

1. Case Studies of Amazon AWS service. http://aws.amazon.com/solutions/case-studies
2. Yu, J., Buyya, R., Tham, C.K.: A cost-based scheduling of scientific workflow applications on utility grids. In: Proceedings of the 1st IEEE International Conference on e-Science and Grid Computing, vol. 8 p. 147 (2005)
3. Duan, R., Prodan, R., Fahringer, T.: Performance and cost optimization for multiple large-scale grid workflow applications. In: SC 2007 Proceedings of the 2007 ACM/IEEE Conference on Supercomputing, 2007, p. 12. IEEE (2007)
4. Mao, M., Humphrey, M.: Auto-Scaling to minimize cost and meet application deadlines in cloud workflows. In: Proceedings of International Conference on High Performance Computing, Networking, Storage and Analysis (SC), pp. 49:1–49:12 (2011)
5. Byun, E.K., Kee, Y.S., Kim, J.S., et al.: Cost optimized provisioning of elastic resources for application workflows. Future Gener. Comput. Syst. **27**(8), 1011–1026 (2011)
6. Yi, S., Andrzejak, A., Kondo, D.: Monetary cost-aware checkpointing and migration on amazon cloud spot instances. IEEE Trans. Serv. Comput. **5**, 512–524 (2011)
7. Voorsluys, W., Buyya, R.: Reliable provisioning of spot instances for compute-intensive applications, pp. 542–549 (2012)
8. Zhong, H., Tao, K., Zhang, X.: An approach to optimized resource scheduling algorithm for open-source cloud systems. In: Fifth Annual China Grid Conference (2010)
9. Lin, C., Lu, S.: Scheduling scientific workflows elastically for cloud computing. In: IEEE 4th International Conference on Cloud Computing (2011)
10. Liu, H., Xu, D., Miao, H.: Ant colony optimization based service flow scheduling with various QoS requirements in cloud computing. In: 2011 First ACIS International Symposium on Software and Network Engineering (SSNE), pp. 53–58. IEEE (2011)
11. Yu, J., Buyya, R., Tham, C.K.: Cost-based scheduling of scientific workflow application on utility grids. In: First International Conference on e-Science and Grid Computing (2005)
12. Kllapi, H., Sitaridi, E., Tsangaris, M.M., Ioannidis, Y.: Schedule optimization for data processing flows on the cloud. In: Proceedings of the ACM SIGMOD International Conference on Management of Data (2011)
13. Wang, H., Jing, Q., Chen, R., He, B., Qian, Z., Zhou, L.: Distributed systems meet economics: pricing in the cloud. In: Proceedings of the Second USENIX Conference on Hot Topics in Cloud Computing (HotCloud), pp. 6–6 (2010)
14. Herodotou, H., Babu, S.: Profiling, what-if analysis, and cost-based optimization of mapreduce programs. Proc. VLDB Endowment **4**(11), 1111–1122 (2011)
15. Zhou, A.C., He, B., Liu, C.: Probabilistic scheduling of scientific workflows in dynamic cloud environments. In: CoRR (2013)
16. Juve, G., Chervenak, A., Deelman, E., et al.: Characterizing and profiling scientific workflows. Future Gener. Comput. Syst. **29**(3), 682–692 (2013)
17. Garey M.R., Johnson, D.S.: Computers and intractability: a guide to the theory of NP-completeness. W.H. Freeman & Co., (2003)
18. Kwok, Y.K., Ahmad, I.: Static scheduling algorithms for allocating directed task graphs to multiprocessors. ACM Comput. Surv. **31**(4), 406–471 (1999)

19. Sakellariou, R., Zhao, H.: A hybrid heuristic for DAG scheduling on heterogeneous systems. In: The 13th Heterogeneous Computing Workshop (HCW 2004), Santa Fe, New, Mexico, USA, 26 April 2004
20. Topcuoglu, H., Hariri, S., Wu, M.-Y.: Task scheduling algorithms for heterogeneous processors. In: 8th Proceedings of Heterogeneous Computing Workshop (1999)
21. Javadi, B., Thulasiramy, R.K., Buyya, R.: Statistical modeling of spot instance prices in public cloud environments. In: 2011 Fourth IEEE International Conference on Utility and Cloud Computing, pp. 219–228. IEEE Computer Society (2011)

GPU-based Static State Security Analysis in Power Systems

Yong Chen[1,2], Hai Jin[1(✉)], Han Jiang[2], Dechao Xu[2],
Ran Zheng[1], and Haocheng Liu[1]

[1] Services Computing Technology and System Lab,
School of Computer Science and Technology,
Huazhong University of Science and Technology, Wuhan 430074, China
hjin@hust.edu.cn
[2] China Electric Power Research Institute, Beijing 100192, China

Abstract. *Static State Security Analysis* (SSSA) is a key technology to ensure the stability of power systems. It is difficult to satisfy the computing requirement with traditional CPU-based concurrent methods, so that GPU is used to accelerate large amount of power flow calculations. The main issue of GPU-based SSSA is complex iterative operations in solving nonlinear equations. A GPU-based SSSA method is proposed for power systems, in which a novel algorithm is proposed for sparse matrix calculation and small partitioned matrices processing. GPU-based multifrontal algorithm is used to combine various small matrices into one matrix in multiplication for fast calculation. Compared with the execution on 4-cores CPU, the proposed method can decrease 40 % calculation time based on GPU to get a better performance.

Keywords: GPU computing · Static state security analysis · Power flow calculation · Power system

1 Introduction

Static State Security Analysis (SSSA) is the research on safety, feasibility, reliability and other characteristics of power system operations [1]. It can examine transmission equipment state of each device under disconnected or short-circuits circumstances to evaluate system state when an exception occurs. The essence of SSSA is to solve power flow calculation when devices are connected or disconnected for many times, and the essence of power flow calculation is to solve sparse nonlinear equations.

Solving nonlinear equations are data-intensive and computing-intensive operations, it will take a long time to complete the solving process on CPU. Some researches have been done on SSSA to improve the performance [2]. GPU is becoming an attractive accelerator for parallel computation [3], which is a many-core processor with high computation and data throughput. It is an interesting issue to use GPU for general-purposed computation because of its excellent computing ability, but there is no better GPU-based SSSA solution till now.

© Springer International Publishing Switzerland 2015
L. Yao et al. (Eds.): APSCC 2015, LNCS 9464, pp. 258–267, 2015.
DOI: 10.1007/978-3-319-26979-5_19

Many operations in SSSA can be parallelized on GPU to improve the performance. There are many complex iterative operations in solving nonlinear equations, but GPU is not good at iterative operations. A GPU-based static state security analysis method is proposed for power systems in the paper, in which an optimized algorithm is adopted for sparse matrix calculation and small partitioned matrices processing is proposed. At the same time, GPU-based multifrontal algorithm is used to solve large sparse matrices and combine various small matrices into one matrix in multiplication.

In this paper, the background and related work are given in Sect. 2. Section 3 describes the workflow of static state security analysis and its modules. Section 4 evaluates its performance. Section 5 draws conclusions and discusses possible future improvements.

2   Related Work

N-1 branch outage simulation is the necessary check for power grid safety [4]. DC (*Direct Current*) power method and sensitivity analysis method are two principal methods for static state security analysis.

DC power method [5] can convert nonlinear equation problem in power system into linear problem to reduce computational complexity to make calculation more efficient, but its weakness is poor precision. DC power method can only check overload situation in practical application. If voltage crosses the border, it will not be checked out. In sensitivity analysis method, the change of line on or off is regarded as a perturbation in normal circumstances [6]. It deduces sensitive matrix from the perspective of Taylor expansion simply and clearly, which can omit many intermediate steps of calculation, and significantly increase the effect of off-line analysis. This method can not only improve performance, but also obtain high calculation precision and speed, therefore it is a practical method.

Static state security analysis pays more attention on the calculation of linear algebra operation. Recently there are many research results on it based on GPU, in which the most used is CUBLAS. CUBLAS [10] is a linear algebra library on GPU, which implements almost every interface of BLAS, an application interface standard to regulate basic linear algebra operation of numerical library, based on CUDA (*Compute Unified Device Architecture*). Programming on CUBLAS is the same as BLAS without considering the details of threads model and storage model of CUDA programming.

There is still not a whole static state security analysis method based on GPU to finish N-1 branch outage simulation.

3   The Workflow and Modules of Static State Security Analysis

3.1   Overview of Static State Security Analysis

Sensitive analysis method is a highly parallel method, which is convenient to use GPU to accelerate the calculation of static state security analysis. Jacobi matrix

J is calculated by a full power flow calculation, and then sensitive matrix S_0 can be generated by substituting matrix J back. Using Taylor expansion [7], Formulas (1) and (2) can be figured out.

$$\triangle x = S_0 \triangle W_y \tag{1}$$

$$\triangle W_y = [I + L_0 S_0]^{-1} \cdot (-f'_y(X_0, Y_0)\triangle Y) \tag{2}$$

Here X_0 is status vector composed with node voltage and phase angle in normal situation, Y_0 is normal network parameter, $\triangle X$ is change amount of node voltage and phase angle, and $\triangle Y$ is change amount of network parameter. Take partial derivatives to them, we can calculate to get Formula (3).

$$f'_y(X_0, Y_0)\triangle Y = [0, ...0, P_{ij}, Q_{ij}, 0, ...0, P_{ji}, Q_{ji}, 0, ...0] \tag{3}$$

That is to say, there are only 16 non-zero elements in every line. We only need to focus on 4×4 matrix, which is a submatrix of sensitive matrix. Therefore the computing complexity is greatly simplified. Eventually we can only derivate the change of the grid W_y when the line is off with Formulas (4)–(7).

$$S = \begin{bmatrix} S_{ii}^{(1)} & S_{ii}^{(2)} & S_{ij}^{(1)} & S_{ij}^{(2)} \\ S_{ii}^{(3)} & S_{ii}^{(4)} & S_{ij}^{(3)} & S_{ij}^{(4)} \\ S_{ji}^{(1)} & S_{ji}^{(2)} & S_{jj}^{(1)} & S_{jj}^{(2)} \\ S_{ji}^{(3)} & S_{ji}^{(4)} & S_{jj}^{(3)} & S_{jj}^{(4)} \end{bmatrix} \tag{4}$$

$$L = \begin{bmatrix} -H_{ij} & 2P_{ij} - H_{ij} & H_{ij} & N_{ij} \\ -J_{ij} & 2Q_{ij} + L_{ij} & J_{ij} & L_{ij} \\ H_{ji} & N_{ji} & -H_{ji} & 2P_{ji} - N_{ji} \\ J_{ji} & L_{ji} & H_{ji} & 2Q_{ji} - L_{ji} \end{bmatrix} \tag{5}$$

$$H = I + S \cdot L \tag{6}$$

$$\triangle W_y = H^{-1} \begin{bmatrix} P_{ij} \\ Q_{ji} \\ P_{ij} \\ Q_{ji} \end{bmatrix} \tag{7}$$

The overall workflow of static state security analysis system is shown in Fig. 1. There are mainly 4 modules: I/O, preprocessing, power flow calculation and static state security analysis.

When original data is input, it will be preprocessed firstly for power flow calculation. Then, the system conducts a full power flow calculation to work out Jacobi matrix J, and prepares data from J for static state security analysis. At last, the system outputs the change parameter of power grid once a line is off.

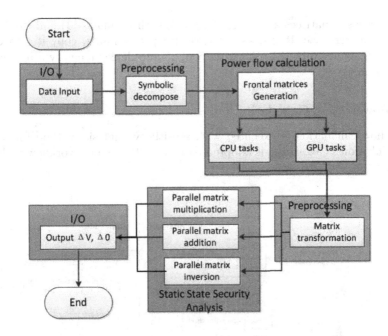

Fig. 1. The overall workflow of static state security analysis

3.2 Preprocessing

There are two preprocessings, whose purposes are to prepare data for power flow calculation and static state security analysis.

The first preprocessing is for power flow calculation. After raw data input, the data, a large sparse matrix, will be compressed in CSR format [8]. The subsequent nonzero elements in matrix rows are put in contiguous memory locations. Through matrix sorting, the rows and columns of the sparse matrix are exchanged as far as possible to avoid creating new filling elements in matrix decomposition, which will alleviate computational burden of subsequent matrix decomposition. At the same time, only nonzero elements are stored to save storage space. Elimination tree is built by symbolic decomposing sparse matrix to determine the position of injection elements in decomposition.

The second preprocessing is for static state security analysis, which is mainly to deal with the results of power flow calculation. Because three read operations are necessary when CSR-formatted data access each time, it will take much more costs for I/O operation. So matrix is decompressed before calculation. Assembling several matrices may take up a lot of space, and maybe memory space is not enough to store the matrices. Therefore, there are three operations in the second preprocessing:

1. Analyze the result of power flow calculation J, in which sensitive matrix is obtained by inverse S_0. Allocate memory to store assembled matrices from S_0.

2. Generate submatrices and vectors from S_0 with Formulas (4) and (5).
3. Check matrix size. If it does not exceed capacity limits, copy it into GPU device memory directly, otherwise, copy it into device memory after matrix partition.

3.3 Power Flow Calculation

Power flow calculation is executed with sensitivity analysis method for power grid. Multifrontal algorithm is adopted based on GPU, whose workflow is shown in Fig. 2.

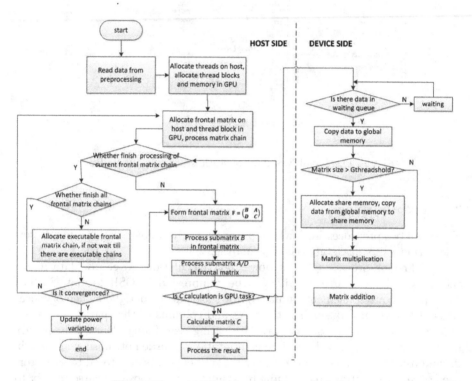

Fig. 2. The workflow of power flow calculation

Multifrontal algorithm [9] is an effective method to solve large sparse matrix operation. Frontal matrix is a small dense matrix. The term *multi* here refers to that there are multiple frontal matrices during the whole factorization. For any frontal matrix, before its factorization, other frontal matrices may still be active in the processing. Multifrontal method presented here is associated with the super-nodal implementation, such that multiple columns with the same nonzero patterns can be grouped together as a dense kernel for concurrent factorization.

First, the processing order of frontal matrices is determined by elimination tree from preprocessing. The order is expressed by frontal matrix chain, in which each frontal matrix is allocated as a leaf and processed by CPU thread. Frontal matrix F can be parted into four sub-matrices A, B, C and D with Formula (8).

$$\begin{pmatrix} B & A \\ D & C \end{pmatrix} = \begin{pmatrix} L_1 & 0 \\ L_2 & 1 \end{pmatrix} \begin{pmatrix} 1 & 0 \\ 0 & C - L_2 U_2 \end{pmatrix} \begin{pmatrix} U_1 & U_2 \\ 0 & 1 \end{pmatrix} \tag{8}$$

Threshold is defined to distinguish CPU or GPU tasks, derived from matrix calculation. If the amount of computations is larger than the threshold, main thread will allocate the task to GPU. Otherwise, small-scale matrix is computed on CPU. Nodes in frontal matrix chain are calculated one by one until all of them are finished. Then $\triangle X$ is computed through substitute decomposed matrices L and U backward to update X. If the absolute value of $\triangle X$ is less than 10^{-8}, which is regarded as converged, the calculation will be ended, else iteration will continue to be processed to get converged result.

3.4 Static State Security Analysis

In this module, sensitive method on GPU is used for analysis. There are mainly 4 steps:

Step 1: calculate $T = L \cdot S$ in Formula (6) by matrix multiplication.
Step 2: calculate $H = T + I$ in Formula (6) by matrix addition.
Step 3: calculate H^{-1} by matrix inversion from H.
Step 4: calculate Formula (7) by matrix multiplying vector.

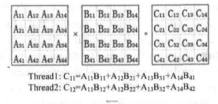

Thread1: $C_{11}=A_{11}B_{11}+A_{12}B_{21}+A_{13}B_{31}+A_{14}B_{41}$
Thread2: $C_{12}=A_{11}B_{12}+A_{12}B_{22}+A_{13}B_{32}+A_{14}B_{42}$
......
Thread16: $C_{44}=A_{41}B_{14}+A_{42}B_{24}+A_{43}B_{34}+A_{44}B_{44}$

Fig. 3. Submatrix multiplication with 16 threads

In matrix of 4×4 multiplication on GPU, 16 threads are executed for multiplying every element concurrently, shown in Fig. 3. 16 threads are completely independent and communicate with each other in their own block. But the code is actually executed in a warp with 32 threads. That is to say, whether a task is divided into 32 or 16 threads, it will launch a warp with same time costs. So if only one submatrix multiplication is in a block with only a warp, 16 threads will be launched and the other 16 threads will be idle.

The main idea to overcome it is to have more than one multiplication in a block. In fact, one block can contains 1024 threads at maximum, which can hold 64 submatrix multiplications. So there are 8×8 matrices in a block logically.

Host threads can combine them into a large matrix and copy them to share memory of this block. Because the instructions are consistent in the same block, it is necessary to coordinate transformation to operate correct data in each thread. The distribution of logical memory is shown in Fig. 4. Here we define that every 16 threads process a matrix multiplication and 64 matrix can be handled in a block.

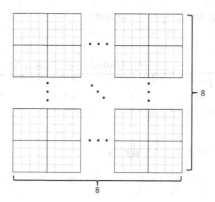

Fig. 4. The distribution of logical memory

Many elements are reused in matrix multiplication, and every element is read 4 times on average. For data input or output, share memory is much faster than global memory. Despite the space of share memory is only 48 KB per block, only $2 \times 64 \times 16 \times 4B = 8\,KB$ memory space is needed for 64 submatrices, so share memory of one block is big enough to store all data in computation. Before the optimization, it will take 4 operations in global memory, and after that, it will only take 1 operation on global memory and 3 operations in share memory.

In Step 2, there is an operation of matrix inversion. Generally there are two ways for common matrix inversion. One is adjoint matrix method with the calculation of $A^{-1} = A^*/|A|$. Adjoint matrix A^* of A must be calculated firstly. It is too complicated to deduct the expression of A^* when the matrix is higher than 3 order. The other is elementary transformation method, which is simpler than adjoint matrix method in high-order matrix, but it will be iterated much more times, which is inefficient on GPU. In order to be more efficient, Formula (9) is used to partition matrix inversion.

$$\begin{bmatrix} A & B \\ C & D \end{bmatrix}^{-1} = \begin{bmatrix} A^{-1} + A^{-1}B(D - CA^{-1}B)^{-1}CA^{-1} & -A^{-1}B(D - CA^{-1}B)^{-1} \\ D(D - CA^{-1}B)^{-1}CA^{-1} & (D - CA^{-1}B)^{-1} \end{bmatrix} \quad (9)$$

Only 4×4 matrix is concerned, which can be just divided into four 2×2 matrices. In this way, the task is assigned to each thread equally so that they can be finished at the same time. In Formula (9) there are only matrix multiplication, matrix addition and 2×2 matrix inversion. 2×2 matrix inversion can be expressed in Formula (10).

$$\begin{bmatrix} a & b \\ c & d \end{bmatrix}^{-1} = \begin{bmatrix} \frac{d}{ad-bc} & \frac{-b}{ad-bc} \\ \frac{-c}{ad-bc} & \frac{a}{ad-bc} \end{bmatrix} \tag{10}$$

In this way, four threads can be used with Formula (10) to calculate the inversion of A, B, C and D in Formula (9), then the results are substituted into Formula (9) to work out the inversion of 4×4 matrix.

4 Performance Evaluation

4.1 Dataset and Experimental Environment

The experiments are done on the server with Intel i7 950 (3.07 GHz) CPU, 16 GB memory, and NVIDIA GeForce GTX460. CentOS 5.9 Linux operation system with CUDA 4.0 is the software environment. 4 datasets are chosen from MATPOWER [11], a professional software in power calculation from real power grid or IEEE standard testing dataset, shown in Table 1.

Table 1. Testing dataset for SSSA experiments

Dataset Name	CA300	CA3012wp	CA3120sp	CA5472	CA5492wp	CA6024	CA6240
Number of Nodes	300	3012	3120	5472	5492	6024	6240
Number of Lines	411	3572	3693	9794	9824	10706	11069

The speedup ratio S is defined in Formula (11), in which T_{CPU_time} and T_{GPU_time} are the executing times on CPU and GPU respectively. The program executed on CPU is a multi-thread program with 4 cores CPU.

$$S = \frac{T_{CPU\_time}}{T_{GPU\_time}} \times 100\% \tag{11}$$

4.2 Evaluation of the Static State Security Analysis System

Respectively executed on GPU and CPU platform, the dataset in Table 1 is used for testing. The results are shown in Table 2.

From Table 2, we can find the execution on GPU is faster than CPU except the scale of 300 nodes. The reason is that GPU serves as a co-processor in the execution. The data is stored in host, so that it needs to be transferred from host to device before processing and transferred from device to host after processing. When the scale is 300 nodes, the transfer cost is much larger than the execution on GPU. On the other hand, only 300 sub-matrices are generated, in which every 2 sub-matrices are processed in a warp on average. But there are 336 cores

Table 2. Execution time comparison on CPU and GPU respectively

Scale	200	3000		5000		6000	
Dataset Name	CA300	CA3012wp	CA3120sp	CA5472	CA5492wp	CA6024	CA6240
CPU(ms)	10.0	446.6	476.7	2170.0	2186.7	2603.3	2793.3
GPU(ms)	19.7	269.3	269.6	1214.3	1213.0	1408.0	1483.7

on GTX 460, that means over half of cores are idle throughout the processing, which is a serious waste of GPU resource to hindle the whole performance.

Excluding the dataset with 300 nodes, the accelerating effect of GPU computing is reflected obviously, and the speedup is becoming more and more larger with the increasing of data scale. The increasing trend of GPU speedup is shown in Fig. 5.

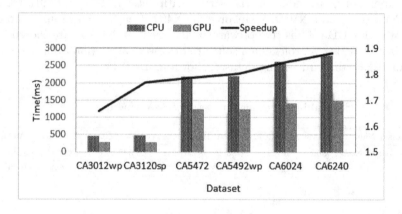

Fig. 5. Execution time comparison on CPU/GPU and GPU speedup

5   Conclusion and Future Work

In this paper, a static state security analysis based on GPU is proposed for power systems. GPU-based multifrontal method is implemented to solve large sparse matrix and sensitive method to deal with static state security analysis. GPU is used to accelerate the computation of matrix multiplication, addition and inversion. To make full use of GPU device, the system uses combining read of data in small-scale matrix multiplication and partition matrix method to invert matrix. In addition, the system is extended on GPU cluster.

Experimental results indicate that the proposed algorithm on GPU can improve system performance. Our results show up to 1.7–1.9 speedup with simulation case from 3000 to 6000 scale.

In future work, the way dealing with small-scale matrix can be used in scientific calculation in other fields, and the processing on special dimension matrix can be extended to all dimension matrices.

Acknowledgments. This work is supported by the National 973 Key Basic Research Plan of China (No. 2013CB2282036), Major Subject of State Grid Corporation of China (No. SGCC-MPLG001(001-031)-2012), the National 863 Basic Research Program of China (No. 2011AA05A118), the National Natural Science Foundation of China (No. 61133008) and the National Science and Technology Pillar Program (No. 2012BAH14F02).

References

1. Ao, L., Cheng, B., Li, F.: Research of power flow parallel computing based on MPI and P-Q decomposition method. In: Proceedings of the 2010 International Conference on Electrical and Control Engineering, pp. 2925–2928. IEEE (2010)
2. Green, R.C., Wang, L., Alam, M., Singh, C.: Intelligent and parallel state space pruning for power system reliability analysis using MPI on a multicore platform. In: Proceedings of 2011 IEEE PES Innovative Smart Grid Technologies, pp. 1–8. IEEE (2011)
3. Owens, J.D., Luebke, D., Govindaraju, N., Harris, M., Kruger, J., Lefohn, A.E., Purcell, T.J.: A Survey of general-purpose computation on graphics hardware. Comput. Graph. Forum **26**(1), 80–113 (2007)
4. Li, X., Guo, Z.: The transmission interface real power flow control based on N-1 static safety restriction. Electric Power **38**(3), 26–28 (2005)
5. Laurence, D.C., Silvio, C.B., Ednilton, B.F.: Stability analysis of power system including facts (TCSC) effects by direct method approach. Int. J. Electr. Power Energy Syst. **27**(4), 264–274 (2005)
6. Saha, T.K.: Determination of power system coherent bus groups by novel sensitivity-based method for voltage stability assessment. IEEE Trans. Power Syst. **18**(3), 1157–1164 (2003)
7. Wang, S., Zheng, Z., Wang, C.: Power system transient stability simulation under uncertainty based on Taylor model arithmetic. Front. Electr. Electron. Eng. China **4**(2), 220–226 (2009)
8. Greathouse, J.L., Daga, M.: Efficient sparse matrix-vector multiplication on GPUs using the CSR storage format. In: Proceedings of International Conference for High Performance Computing. Networking, Storage and Analysis, pp. 769–780. IEEE, New Orleans (2014)
9. Zheng, R., Wang, W., Jin, H., Wu, S., Chen, Y., Jiang, H.: GPU-based multi-frontal optimizing method in sparse Cholesky factorization. In: Proceedings of IEEE 26th International Conference on Application-Specific Systems, Architectures and Processors, pp. 90–97. IEEE, Toronto (2015)
10. Bientinesi, P., Gunnels, J.A., Myers, M.E., Quintana-Orti, E.S., Geijn, R.A.: The science of deriving dense linear algebra algorithms. ACM Trans. Math. Softw. **31**(1), 1–26 (2005)
11. Zimmerman, R.D., Murillo-Sanchez, C.E.: MATPOWER: steady-state operations, planning, and analysis tools for power systems research and education. IEEE Trans. Power Syst. **26**(1), 12–19 (2011)

Improved WSN Capabilities Through Efficient Duty-Cycle Mechanism

Zibouda Aliouat and Makhlouf Aliouat[✉]

Computer Science Department, Faculty of Sciences,
University Ferhat Abbas Sétif 1, Sétif, Algeria
{aliouat_zi,aliouat_m}@yahoo.fr

Abstract. A Wireless Sensor Network (WSN) is mission dependent network, deployed in an interesting area in order to collect data about a relevant observable environmental phenomenon and send them to end user through a base station. Due to their potential promising development, WSN increasingly attract research-er's attention in order to ensure them the expected maturity of widespread deploy-ment. However, many obstacles inherent to intrinsic sensor node characteristics may prevent achieving this goal. So, energy depleting is the most important hindering since node initial energy budget is poor. In this paper, we propose new hierarchical routing protocol sensitive to energy consumption and based on nodes duty-cycle management. This protocol improves WSN life time duration and data packets loss rate. The proposal was integrated to the well know LEACH protocol to enhance its performance. Simulation results via NS2 simulator showed that the proposal is convincing and outperforms the classical LEACH capabilities.

Keywords: WSN · Energy conservation · Life time duration · Data loss

1 Introduction

Due to their promising and rapid development, Wireless Sensor Networks (WSN) increasingly attracted attention of researcher's community in the domain. So, for more than a decade, WSNs spread out their beneficial influence over many areas such as those requiring control and monitoring environmental sizes. Although the development of WSN applications in several fields reaches important rate, efforts continue to be devoted in order to lower obstacle effects on WSNs allowing them achieving the expected maturity. Among the obstacles preventing WSNs to make great progress and widespread deployment, resource energy is the most constraining one. Thereby, since a sensor node is powered by a small battery, energy budget is limited. So, this poor energy amount has to be economically managed until the WSN successfully fulfills its mission.

There are many ways for saving node energy especially how and which level on the sensor node is concerned. To save energy, we have to identify the components and the node activities which are involved in speeding up depleting of that vital resource. So, sensing activity dissipates energy with small quantity, comes then the energy exhausted by computation at processor level. Communications involving node transceivers are the most energy consumer node activity since nodes usually operate as provider of sensed

© Springer International Publishing Switzerland 2015
L. Yao et al. (Eds.): APSCC 2015, LNCS 9464, pp. 268–277, 2015.
DOI: 10.1007/978-3-319-26979-5_20

data and also act as routers to forward collected data to their neighbors. The dissipated energy is determined by the data packet to be sent and the transmission distance between sender and receiver. So, when the sending signal has to range a large distance, the required transmission power must be higher [1]. This obviously induces higher energy consumption.

The used routing protocol also greatly impacts the energy consumption. Therefore, routing in WSNs is usually of multi-hop kind and routing packets from a given source to a destination is done through several intermediate nodes. Thus, a node consumes energy when transmitting its own data either for forwarding data of its neighbors.

In hierarchical networks, a clustering approach [2] is used. So, clustering divides a network into clusters and each cluster gathers a certain number of standard nodes among which one of them is elected as a Cluster-Head (CH). Only the CH is allowed to communicate with the Base Station (BS) or other CHs. So, all nodes in a Cluster send their data to their CH which in turn sends them to the BS directly or via nearest other CHs. A CH is then the node responsible for coordinating activities within the cluster. Among protocols using this technique, we cite the well known LEACH [3].

MAC layer also plays an important role in the coordination between nodes and minimization of the energy consumption. So, the main causes of energy consumption at this layer are: (a) packet retransmissions when collisions occur which cause additional energy consumption and information loss. (b) Idle listening of the channel for incoming packets can cause a significant loss of energy. To avoid this problem, nodes might be in sleeping mode as long as possible. (c) Overhearing occurs when a node receives packets that are not intended for it. Overhearing leads to a loss of extra energy because nodes are uselessly involved in data forwarding. (d) The overhead produced by control message exchanges such as signaling, connectivity, TDMA schedule, and collision avoidance increase the energy consumption. (e) Overemitting occurs when a node sends data to a recipient which is not ready to receive them. (f) The size of exchanged messages may have a negative effect on the energy consumption of sender/receiver nodes. Thus, the signal power is proportional to the data packet size to be transmitted and the distance separating the sender and receiver. The longer the distance is the higher the signal strength and the most energy dissipation will be. Since the radio is the most energy consumer component of a node; it must be turned off when a node is in idle state. This is what is known by the Duty-Cycling technique and constitutes the core of our present work in order to make WSNs resilient to energy lack problem.

The paper is organized as follows:

Section 2 presents some related works that impacted the duty cycle mechanism, while Sect. 3 describes our proposal and the performance evaluation of the latter is given is Sect. 4. We concluded our work with a conclusion and future work.

2 Related Work

Many mechanisms may be used to conserve energy of nodes in WSN. Among these mechanisms, the most significant ones are clustering nodes organization and nodes duty-cycle approach.

2.1   Nodes Clustering Organization

The clusters formation offers significant energy savings since only Cluster-Heads (CH) are involved in collecting, routing and aggregating data. So, as CHs consume more energy than standard nodes, thus CH role rotation has to be used in heterogeneous WSN. Several approaches can be used for clusters formation namely: CHs can be elected by their node members or designated in advance; they can be part of the set of nodes or be a separate set of nodes fitted with large resources; their number may remain fixed in time or change dynamically.

2.2   Duty Cycle Mechanism

An efficient way to make nodes saving energy is to alternate node radio from sleeping mode to transmitting/receiving one. Since even a node is in waiting state, its radio remains dissipating energy so, to avoid wasting this valuable resource, a node is forced to be in a sleeping state. Alternation between active and inactive period of the transceiver forms a Duty-Cycling. A duty-cycle is defined by the following equation: *Duty-Cycle = Pactive/(Pactive + Pinactive)*, where *Pactive* states for time period for which a node is in active state and *Pinactive* otherwise. Most of MAC protocols use the duty-cycle mechanism to minimize energy consumption in WSN.

So, for MAC protocols adopting Duty-Cycling, each node radio is turned off most of the time and node timer is set up for being waked up later. A synchronous duty cycle means that a set of neighboring nodes sleep together and weak up together. Unlike, asynchronous Duty-Cycle imposes each node to choose its own wake time without knowing the status of others. Synchronous protocols may have a lower delay compared to asynchronous ones, because the synchronous protocols know exact wake time of neighboring nodes. But in another hand, it is difficult to synchronize nodes of a large wireless sensor network. Asynchronous protocols are much easier to implement. An efficient MAC protocol for Duty-Cycle should also optimize the energy draining sources like: active listening, over-listened, collisions etc. [4]. The main factors for Duty-Cycle protocol design are scheduling [5] and routing [6].

Minimizing the energy consumed during the active listening implies to deploy some sensor nodes keeping dynamically in sleep/wake up states such that they will be awaked only when it should be necessary. This also may limit collisions that occur during retransmissions which also contribute to save energy.

In order to maintain WSN topology connected and to guarantee data packets delivery according to sleeping node schedules, MAC layer protocols must be developed carefully. The standard IEEE 802.15.4 [7] specifications, is the most known MAC protocol using Duty-Cycle. It is based on TDMA and CSMA/CA protocols offering so a fixed Duty-Cycling but does not allow traffic adaptation.

Ye et al. [8] proposed S-Mac protocol to coordinate and synchronize the sleep/wake cycles. To maintain synchronization, each node periodically broadcasts its time schedule in a message SYNC, so that the neighbors can update this information in their schedule tables. S-MAC may suffer from latency and integrity lack of SYNC exchange messages. The S-MAC requires that only cluster members be synchronized. The protocol works well only with fixed network topologies with infrequent changes.

In [9], T-MAC protocol has been proposed to deal with the problem of fixed Duty-Cycle of S-MAC. In T-MAC, nodes synchronize to put their transceiver active or asleep periodically as in S-MAC. However, T-MAC adapts a Duty-Cycle to different levels of network traffic. This Duty-Cycle is no longer fixed by application but varies according to the network traffic. If the traffic is important, nodes remain longer in active state in order to transmit more data. But, if the traffic is low, nodes remain briefly active in order to save energy. The drawback of T-MAC is over-listening since a node remains active during time-out period even it does not transmit.

In [10], MS-MAC protocol was designed to work in the two scenarios fixed and mobile sensor nodes. It can operate similarly to S-MAC with fixed nodes. In order to avoid long waiting periods to join a new cluster, each node detects mobility in its region from signal strength, periodically received by the SYNC messages. If a node notices a change in the received signal, it assumes that the neighbor or itself is in motion, and predicts the level of the movement speed. The message SYNC in MS-MAC also includes information on the estimated speed of its mobile neighbor. If there is more than one mobile neighbor, then the SYNC message includes only the maximum speed estimated among all the neighbors. This information of mobility is used to create an active area around a mobile node when it moves from one cluster to another, so that the mobile node can speed up the establishment of link with the new neighbors before it loss all its neighbors.

2.3 Cross-Layer Scheduling Algorithms for Power Efficiency

In [11], a scheduling inter-layer algorithm is proposed in order to conserve energy by disabling some nodes. The idea is to make nodes dynamically active/inactive such as nodes are awake only when it will be necessary. Scheduling and routing operates in a separate way in two different phases: Configuration/Reconfiguration and Stable phase.

Configuration and reconfiguration phase: This is the initialization phase of the WSN to update the routes and the queries. This phase is relatively short; its aim is to put in place the schedules that will be used during the second phase. The set up and reconfiguration algorithm is independent of the routing algorithm. So, most of the available algorithms for routing in ad-hoc networks and WSNs can be used.
The stable phase: Similar to transmission phase, it uses a scheduling established in the previous phase to transmit data to BS. The schedule of idle/active state is calculated according to the packets to be sent out.

In [12], the authors proposed a random algorithm to provide robustness to variations in the network connectivity. The algorithm does not require nodes to keep information of the state of their neighbors. Each node decides independently when it should sleep and when it must be awaken, based on local observations. The main constraints considered are latency time and the capacity to achieve a certain charge. The algorithm is not based on the time intervals or coordination with the neighbors. If the activity is estimated too low to satisfy the latency constraint, the node decides to wake up more often. Conversely, if the activity is greater than that required, the node decides to sleep longer. However, the algorithm may suffer from data packets loss.

In [13], authors proposed a protocol fitting better dynamic topologies and non-uniform energy consumption. It allows each node to set its own listening mode as a function of its local state. Initially, the nodes do not know their neighbors. First of all, each sensor periodically sends update messages of routes to declare its presence and state. A graph of routing is formed; the data streams are circulated through BS to find out how many descendants are in the graph. Depending on the number of descendant nodes, each node calculates its Duty-Cycle. If a node has too many children, it exhausts its energy source and announces its children to choose another parent to decrease the load.

A node in [14] decides to sleep according to its residual energy and its relative location. A node enters a sleep state after having known that its sensing area is fully covered by its neighbors. The node that decided to sleep sends a REQ packet to its neighbors who answer by acknowledgement. However, simultaneous sleep requests and ACK of neighboring nodes may cause problem of schedule sleeping/wake up.

The authors [15] focused their idea on the problem of maximizing the life time by using WSN as a grid with the emphasis on the problem of nodes sleeping scheduling. Nodes are able to buffer sensed data and there is a mobile BS visiting periodically the monitored area for collecting sensed data. This technique eliminates the need for multi-hop routing. The basic idea is to schedule nodes as active and inactive nodes dynamically, according to the schedule of mobile Base station movements.

A communication multi hops is used in each cluster to transmit the data to the CHs. The BS visits only CHs instead of visiting all nodes. Scheduling of Duty-Cycle is determined by the position of the mobile base station. So, when a BS arrives at a CH location (or any other node playing the role of data collector regarding the BS), the CH has to be in a wakeup state in order to send out collected data to BS.

3   Proposed Protocol

Our proposal consists of a new routing protocol based on a hierarchical sensor nodes organization with duty cycling scheme application. The aim of this combination is to minimize nodes energy consumption and prolonging WSN life time duration and also to minimize the data packets loss rate.

3.1   Assumptions

In this proposal, we consider a set of sensor nodes randomly distributed in a hierarchical topology in the form of grid of size (2 * 2) regions. The sensor nodes are supposed to be static after their deployment in the sensing area. Each region contains a cluster-head. The election of the latter will be done according to the remaining residual energy amount of nodes. Each cluster is split into two levels, such that the level 1 represents nodes close to the CH and the level 2 represents nodes the more distant.

All nodes of level 2 must send, during slot period, their data to the level 1 in which nodes must be in the active state to achieve this reception. All nodes of level 1 must send, during their slot, their data to the CH, and the latter will transmit them (after appropriate aggregation) to the BS. Every node has a next hop and a list of node predecessors. We therefore propose a new routing protocol using Duty-Cycle approach to reduce the loss of data and the energy consumption in WSN.

3.2   Description of the Proposed Protocol

3.2.1   Phase 1: Initialization and Grid Formation

1. Random Distribution of nodes over a well defined area
2. The grid formation by fixing the number of regions to use (2 * 2) in each round
3. Nodes of each region elect their cluster-head (Node with highest energy is chosen).
4. The Cluster-Head divides the cluster in two levels (depending on the distance between it and its member nodes, level1 for the nodes close to CH and level 2 for the distant nodes).
5. The CH creates the TDMA Schedule starting with nodes of level 2 and then nodes of level 1.
6. Nodes of level 2 send their data to the next hop selected (a list item of level1).
7. The Next Hope node must be in an Active state during the slot of its predecessor.
8. Nodes of level 1 send out their data to the selected cluster-head.
9. The cluster-heads send out received data (after aggregation) to the base station

The initialization phase is composed of 3 sub-phases: announcement, organization of grids and finally scheduling, which will be detailed below.

– The announcement sub-phase or advertisement:
   The aim of this sub-phase is to determine the elected CHs in each round executed by the protocol, taking into account the residual energy amount of each node.
– Sub-Phase of grids organization:
   The node elected as CH has to inform other nodes of its new role. For this, it sends an advertisement message ADV containing its identity. This broadcast allows ensuring that all nodes have received the message. Each node received ADV message, will choose its own cluster-head (according to distance sender/receiver).
– Scheduling Sub-Phase:
   After initialization, each CH has to coordinate data transmissions within its cluster. It splits the cluster into two sub levels (level1 and Level2) to create a TDMA Schedule starting by farther nodes and at the same time assigns to each member node a time slot during which it can transmit. All slots assigned to node members represent a frame. The duration of each frame differs depending on the number of members of the cluster. Furthermore, in order to minimize interference between transmissions in adjacent clusters, each CH chooses randomly a code in a CDMA code list. It transmits it subsequently to its members in order to use it during their transmissions.

3.2.2   Phase 2: Synchronization of Duty-Cycle

In LEACH, upon reception of TDMA schedule, each node knows its emission slot as well it switches in sleep mode until the arrival of its transmission time slot.

To achieve the Duty-Cycle in LEACH, we have to synchronize the time of waking up and sleeping for each node. Each node of level 1 knows the slots of waking up of its predecessors, and during these slots, it switches to listen mode to receive data from level 2. So, during this transmission, the CH remains asleep until the arrival of the emission slot of one of its predecessors of the level 1. Thus, a CH is in active state (Listening) only during the emissions of the nodes of the level 1.

4   Performance Evaluation

We have to evaluate the performance of our proposal integrated to the well Known hierarchical routing protocol LEACH. The comparison metrics we used are: the energy consumed by nodes, the network life time duration and the amount of data received by the base station compared with data send out by sensor nodes. So, from the latter, we deduce the rate of lost data. Note that when an acknowledgment is used to confirm to a sender of data that these data are received by the receiver, the higher the data loss rate the higher the amount of the energy consumption (Table 1).

Table 1. Simulation parameters

Parameters	Values
Network size	100 m * 100 m
Number of sensor nodes	100
Simulation duration	3600 s
Initial energy budget	3 J

4.1   Metrics of Comparison

The *WSN Life Time Duration* (WLTD) is the period of time during which the network can maintain enough connectivity and cover the entire sensing area. The WLTD is therefore tightly related to nodes life. The nodal life time depends essentially on two factors: the energy consumed as a function of time and the amount of residual energy.

The *Energy Consumed* (EC) is a critical parameter impacting the success or unsuccess of the WSN mission. So, energy consumption optimization remains the main factor to be considered in protocols design for WSNs.

The Received Data Amount (RDA) is the data amount effectively received by BS regarding to the total data amount sent to BS from WSN nodes.

4.2   Simulation Results

– Energy consumed

The curves of the Fig. 1 represent the energy consumed by sensor nodes of a WSN in relation to the time, during the simulation for the classical LEACH protocol and for the proposed Protocol named Duty-LEACH. The improvement of the latter is convincing.

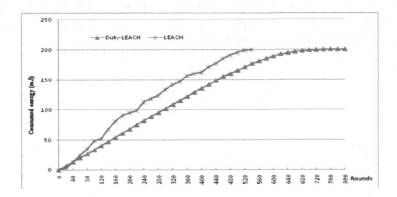

Fig. 1. LEACH vs Duty-LEACH regarding consumed energy

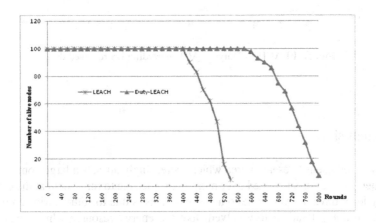

Fig. 2. LEACH vs Duty-LEACH regarding WSN life time

The Fig. 2 represents the curves giving the network life time duration as a function of the simulation time. We notice a significant improvement of this performance factor in case of Duty-Cycle integrated to standard LEACH protocol. Nodes start dying at the 400th round but with the proposal Duty-LEACH nodes begin to die at round 640, so we get an improvement of ≈60 %. The main reason is that when CHs remain in active mode during the entire round, they consume more energy and then reduce their life time duration quickly. At the same time, communication intra-cluster minimizes the energy consumption in the transmission of data to reach the cluster-head; this is due to the reduced distance which separates them.

– Data amount received by the base station.

Due to the appropriate use of time synchronization allowing a suited sleeping/
wakeup scheduling of nodes, there is almost no loss of data. This means that the used
duty-cycle mechanism operates in good conditions, because each sender node has one
receiver in active state to forward the data it receives (Fig. 3).

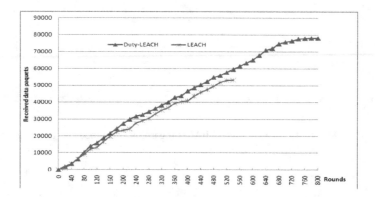

Fig. 3. LEACH vs Duty-LEACH regarding BS received data

5   Conclusion

The WSNs represent a research area which increasingly attracts a large community of
researchers. The efforts invested over more than a decade to raise them at the expected
maturity level have not yet come to a success. This is due to some intrinsic problems
that remain not yet satisfactory solved, like the energy resource which constitute a
serious obstacle to their advancement. In this paper, we are dealing with duty cycling
mechanism making nodes to switch alternately and advisedly from active mode to
sleeping one. This switchable operation mode, combined with adequate nodes organization, enables saving a substantial amount of energy which may be crucial to guarantee
a WSN achieving successfully its mission.

Our proposal has been integrated in a well know LEACH routing protocol for which
many improvements have been added such as good CHs distribution, a multi-hop intra-
cluster communication and an adequate communication slots scheduling correlated with
active/sleeping mode. The simulation results (via NS2 simulator) exhibited by the
proposal Duty-LEACH in terms of WSN life time duration, energy consumption and
data loss are convincing and outperform those provided by original LEACH. Our future
work will be the integration of robustness in duty- cycling approach which so far has
not been yet considered.

References

1. Biettner, M., Yee, G.V., Anderson, E., Han, R.: X-MAC: a short preamble MAC protocol for duty cycled WSNs. In: Proceedings of the 4th International Conference on Embedded Networked Sensor Systems. ACM (2006)
2. Liu, X.: A survey on clustering routing protocols in WSNs. Sensors **12**(8), 11113–11153 (2012)
3. Heinzelman, W., Chandrakasan, A., Balakrishnan, H.: Energy-efficient communication protocol for wireless microsensor networks. In: Proceedings of the 33rd Hawaii International Conference on System Sciences. IEEE (2000)
4. Maelin, C.J., Heinzelman, W.B.: Duty cycle control for low power listening MAC protocols. In: Proceedings of MASS 2008 (2008)
5. Zhou, H., Zhao, H., Chen, J., Liu, C.H., Fan, J.: Adaptive working schedule for duty-cycle opportunistic mobile networks. IEEE Trans. Veh. Technol. **63**, 4694–4703 (2014)
6. Ghadimi, E., Landsiedel, O., Soldati, P., Duquenoy, S., Johanss, M.: Opportunistic routing in low duty-cycle in WSNs. ACM Trans. Sens. Netw. **10**, 67 (2014)
7. Zheng, J., Lee, M.J.: A comprehensive performance study of IEEE 802.15.4. Department of Electrical Engineering, City College, University of New York, USA (2005)
8. Ye, W., Heidemann, J., Estrin, D.: An energy-efficient MAC protocol for wireless sensor networks. In: Proceedings of IEEE INFOCOM, June 2002
9. Dam, T.V., Langendoen, K.: An adaptive energy-efficient MAC protocol for WSNs. In: ACM SenSys 2003, November 2003
10. Pham, H., Jha, S.: An adaptive mobility-aware MAC protocol for sensor networks (MS-MAC). In: IEEE International Conference on Mobile Ad-hoc and Sensor Systems (2004)
11. Sichitiu, M.L.: Cross-layer scheduling for power efficiency in wireless sensor networks. In: INFOCOM (2004)
12. Greunen, J., Petrovic, D., Bonivento, A., Rabaey, J., Ramchandran, K., Sangiovanni-Vincentelli, A.: Adaptive sleep discipline for energy conservation and robustness in dense sensor networks. In: IEEE International Conference on ICC 2004 (2004)
13. Jurdak, R., Baldi, P., Lopes, C.V.: Energy-aware adaptive low power listening for sensor networks. In: Proceedings of 2nd International Workshop on Networked Sensing Systems, San Diego (2005)
14. Liu, M., Hsin, C.F.: Network coverage using low duty-cycled sensors: random and coordinated sleep algorithms. In: Proceedings of the 3rd International Symposium on Information Processing in Sensor Networks, pp. 433–442. ACM (2004)
15. Abawajy, J.H., Nahavandi, S., Al-Neyadi, F.: Sensor node activity scheduling approach. In: IEEE International Conference on MUE 2007 (2007)

Mining Multiple Periods in Event Time Sequence

Bing Xu, Zhijun Ding[(✉)], and Hongzhong Chen

The Key Laboratory of Embedded System and Service Computing,
Ministry of Education, Tongji University, Shanghai, China
{1336282,dingzj}@tongji.edu.cn

Abstract. The research of life pattern is a hot topic in the field of LBSN (Location Based Social Network). Periodic behavior is also a life pattern. In view of multiple periodic behaviors existed in time series, an algorithm which can mine all periods in time series is proposed in this paper. In view of periodic behaviors always occurred at the same time interval and the random access of matrix's characteristic, the algorithm creates a suspected periodic matrix which can store all suspected periods. By judging the validity of a suspected period in the matrix, the true periods can be mined accurately. Updating the suspected periodic matrix dynamically can reduce executing time.

Keywords: Data mining · Life pattern · Time sequence · Multiple periods

1 Introduction

LBSN (Location Based Social Network) is a new social network which combines the social network and the user's location data. Based on the location of a user we can mine this user's life pattern. Life pattern is a pattern that can describe the daily behavior of the user. LBSN give us the data from the user's daily behavior such as attendance data, GPS (Global Position System) trajectory and so on. We can mine the user's frequent mobile trajectory, the location which the user will most likely reach in different time of a day based on these data. In general some formal expression mined from the user's daily location data which can describe the user's daily life are called a user's life pattern. Periodic behavior is a kind of life pattern which can describe the user's periodicity to a certain place. In our daily life some people go to a place once a week. We call this life pattern a periodic behavior pattern. We found that a user may not only go to a place every Monday but also every two weeks on Tuesday and the classical algorithm can't mine all periods in the time sequence. So we propose an algorithm which can mine multi periods in a time sequence in this paper.

Peridogram is a classical algorithm for mining periods in the field of time sequence. It calculates the density of power spectrum for every frequency based on DFT (Discrete Fourier Transform). There is an explanation of DFT that the fourier coefficient is a relativity between its sine curve and original time sequence as shown in Eq. 1 and the period is the value of *len(S)/frequency* (*len(S)* means the length of the time sequence). When there are multiple periods in the time sequence and the *len(S)* is not the integer multiple of the period this algorithm can not mine all periods in the time sequence.

© Springer International Publishing Switzerland 2015
L. Yao et al. (Eds.): APSCC 2015, LNCS 9464, pp. 278–288, 2015.
DOI: 10.1007/978-3-319-26979-5_21

For the other period's timestamps still in the sequence it cannot avoid the interference of other periodic behaviors. For spectral leakage this algorithm can't mine periods automated.

$$F(\omega) = \mathcal{F}\left[f(t)\right] = \int_{-\infty}^{\infty} f(t)\, e^{-i\omega t} dt \tag{1}$$

As shown in Fig. 1 it is a result of the time sequence '00110101100'. We can know that there are two periods in this time sequence such as 4 and 3 directly. But as shown in Fig. 1 we can't get any integer periods directly because the period is not an integer. In the field of periodic pattern we should not only mine periods but also mine the time stamp the behavior happened in a period for predicting the user's movement in the future. For these reasons the classical algorithm can't meet our requirements.

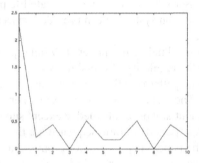

Fig. 1. The result of the time sequence '00110101100'.

We proposed an algorithm which can not only mine multiple periods but also mine the timestamp the target periodic behavior happened in a period in this paper. The rest of the paper is organized as follow. Section 2 introduces some related works. Section 3 formally states the problem definition. Section 4 introduces how to discover multiple periods in time sequence and some basic thought of this algorithm. We report our experimental results in Sect. 5, and conclude our research in Sect. 6.

2 Related Works

In the field of LBSN (Location Based Social Network), the research of recommendation systems has attracted many researchers' attention and produced a lot of research results. Based on locations and social relations, there are four different types of recommendations i.e. Friend, Location, Activity and Event recommendation. In the field of friend recommendation, Symeonidis et al. proposed Friendlink algorithm [1] which provides recommendations by traversing path of a user's social graph. Location recommendation is the hottest research topic in the field of LBSN. Ye et al. proposed a unified POI recommendation framework [2], which fuses user's preference to a POI with social and geographical influence. Activity are often recommended with a location that a user usually go to a location for a special activity. In the paper [3] Bellotti et al. proposed the

Magitti leisure guide based on other contextual information which can provides activity recommendations automatically. Event recommendation is a special case of activity recommendation, which recommends to a target user several events taking place either in the digital world or in a specific city. The main research topic is event detecting. In the paper [4], the authors detect the event based on the geo-tagged information posted by a large number of users.

Recently mining users' life pattern has attracted many researchers' attention and produced a lot of research results. In the paper [5], Zheng mined frequent traveling sequence from users' GPS logs and proposed a prediction model which can predict a user's next destination accurately based on this user's current place. In the paper [6], Rekimoto et al. propose a WiFi-based location detection technology for location logging and mine life patterns based on these logs. In the paper [7], the authors proposed the LP-Mine framework to effectively retrieves life patterns from raw individual GPS data. Life pattern can summarize a user's daily life. If we integrate life pattern with other applications such as recommendation system, it can provide services to comply with users' life habits more properly.

Periodic behavior [8] is a kind of life patterns. Mining periodic behaviors in time sequence can summarize the regularity of users' daily life in time correctly. Acquiring periods and mining periodic patterns in time sequence are the main research directions of the periodic behavior. The former is acquiring correct periods of time sequence by correlation algorithm without any prior knowledge except support threshold. The classical algorithm of acquiring periods in time sequence are the periodogram [9] and the autocorrelation [10]. In the paper [11], the author proposed a periodic behavior based application and mine the eagle's migration cycle successfully. In this paper we propose an algorithm which can mine multiple periods in the sequence.

3 Problem Definition

3.1 Event and States

For ease of description we adopt the definition of event and states. A time sequence is generated by some states of an event happened in different time stamp and ordered by the occurrence time of these states. An event contains multiple states.

Definition 1. Given an event E and its states $\{s_1, s_2, s_3 \ldots s_m\}$ (m is the count of states), a event time sequence is shown in Eq. 2.

$$S = s_1 s_2 s_1 \ldots s_n \tag{2}$$

3.2 Multiple Periods

For example a teacher in school who not only come to a campus every Monday, but also come to this campus on Friday every two weeks. The formal definition of multiple periods is shown in Definition 2.

Definition 2. Given an event sequence S. The states of this event are $\{s_1, s_2, s_3 \ldots\}$. There are multiple periods $\{p_1, p_2, p_3 \ldots\}$ of a state existed in this time sequence probably.

3.3 Support and Support Threshold

The support of a periodic behavior is the proportion of actual occurrence times and its theoretical occurrence times. As shown in Eq. 3, *theortical_count(p)* means the theoretical occurrence times of this period and *actual_count(p)* means the actual occurrence times of this period in this time sequence. *support (p)* means the support of this period. In this algorithm we should give a support threshold firstly. A periodic behavior is true when its support is more than the support threshold.

$$support\,(p) = \frac{actual\_count(p)}{theortical\_count(p)} \tag{3}$$

3.4 Basic Thoughts of Multiple Periods Acquisition Algorithm

We found that all true periods exist in the set which contains time intervals between any two timestamps the target state happened. We store all time intervals in a matrix. The columns of this matrix is mapping to the time stamp the target state happened. Creating the matrix at the first step can filter wrong periods preliminarily. The random access feature of matrix can improve the execution efficiency. In term of the characteristic of periodic behavior, we change the matrix dynamically after acquiring a true period to avoid judging the same periodic behavior many times.

4 Mining Multiple Periods

This chapter mainly introduces the basic process of the multiple periods acquiring algorithm with an example.

4.1 Preliminary Screening and Storage of Suspected Periods

In a time sequence all periods of a target state are in $[1, len(S)/2]$. As discussed in Sect. 2 we can create a matrix for storing these timestamps and time intervals. It can filter wrong periods preliminarily. The matrix is shown in Example 1.

Example 1. Given a time sequence $S =' 0001100101010011'$, the time stamps the state '1' happened are $\{3, 4, 7, 9, 11, 14, 15\}$. The matrix M is shown in Table 1. In M, $M[r][c] = c - r$ (if $M[r][c] > len(S)/2$ or $M[r][c] \leq 0$, then $M[r][c] = 0$).

Table 1. The matrix storing suspected periods

Time stamps	3	4	7	9	11	14	15
3	0	1	4	6	8	0	0
4	0	0	3	5	7	0	0
7	0	0	0	2	4	7	8
9	0	0	0	0	2	5	6
11	0	0	0	0	0	3	4
14	0	0	0	0	0	0	1
15	0	0	0	0	0	0	0

In nature column number and row number of matrix are increased from 0, we design a mapping function for mapping the column number and row number of the matrix to the time stamps the target state happened. So we can access the matrix randomly according to the timestamps.

4.2 Acquiring True Periods and Changing Matrix Dynamically

We judge all items stored in the matrix. The value of the item is a suspected period. The row number of this item is the first time this suspected periodic behavior happened. The column number is the second time this periodic behavior happened. When we judge a suspected period we generate the list of time stamps according to this item firstly. The time interval between any time stamps in this list which is equal to the period is stored in this matrix too. All these items are belong to the same period behavior. So if we judge a true period we should change these items' value to be 0 for avoiding judge the same period behavior many times. If the current item is not a true period we shouldn't change other related items discussed as above in the matrix. The process is shown in Example 2.

Example 2. Using the event time sequence shown in Example 1 and setting the support threshold of 80%, the result of matrix is show in Table 2 after judging $M[3][4]$ *and* $M[3][7]$. In the matrix shown in Table 2 the first row and the first column are all time stamps the target state happened.

While judging the item $M[3][4]$ the algorithm generates the actual happened time stamps list with the period of 1. In the event time sequence this item's all actual happened time stamps which started at the time stamp 3 are $\{3, 4, 7, 9, 11, 14, 15\}$. The theoretical happened time is 13 calculated by the formula $Count = (len(S) - begin)/period + 1$. *'begin'* means the periodic behavior's begin time. In this situation *'begin'* is equal to 3. Because the support of this suspected is $7/13 < 80\%$, this suspected period is not a true and the algorithm don't change the matrix.

Table 2. The result of matrix

Time stamps	3	4	7	9	11	14	15
3	0	0	0	6	8	0	0
4	0	0	3	5	7	0	0
7	0	0	0	2	0	7	8
9	0	0	0	0	2	5	6
11	0	0	0	0	0	3	0
14	0	0	0	0	0	0	1
15	0	0	0	0	0	0	0

While judging the item $M[3][7]$ the algorithm generates the actual happened time stamps list with the period of 4 and the start time of 3. The list of time stamps this period happened is $\{3, 7, 11, 15\}$. Because the actual happened time is 4 and the theoretical happened time is 4 too, the support of this periodic behavior is 100 % which is more than *80 %*. So this item is a true period. Then as discussed above the algorithm changes the matrix. In this situation the algorithm change the value of $M[3][7]$, $M[7][11]$ to be 0. It can reduce the counts of judging same periodic behavior.

5   Experiments

The whole detail of multiple periods acquisition algorithm proposed in this paper is described in Sect. 4. In this section we will mine periodic behaviors in 152 employees' attendance data. The attendance data is collected by an attendance machine for half a year. The attendance machine records the attendance information of every employee. Every employee should sign on this machine in the use of his fingerprint every working day. We will create a binary event time sequence for every employee firstly and then mine periodic behaviors.

5.1   Determining the Support Threshold

Before mining all employees' periodic behaviors we should give a good support threshold. In our experiments we get a good support threshold by testing the threshold from *10 %* to *100 %* step by *10 %*. When the support threshold is too small such as *10 %* the algorithm can mine some local periodic behavior as shown in Table 3. The periods of 2 and 5 is a local periodic behavior shown in the column of time stamps. In this experiment we find it can avoid mining local periodic behavior much properly when the support threshold is set as 80 %. The result is shown in Table 4.

Table 3. The result of 10 %

Periods	Time stamps
2	25,27,39,43,53,55,57,67,71,81,85,95,109,113,123
5	20,25,55,85,95

Table 4. The result of 80 %

Periods	Time Stamps
35	8,43,78,113
7	9,16,23,37,44,51,58,65,72,79,86,93,100,107,114,128,
14	10,24,38,52,66,80,94,108,122,

5.2   Analysis of All Employees' Attendance Data

This data was collected by the Attendance machine of an educational institution. In this company many employees are part-time workers. Some there is periodic behavior among them. Firstly we need to construct the time sequence according to the employees' attendance data in the time granularity of day. As shown in Example 3.

Example 3. It is a segment of an employee's attendance checking data shown in Table 5. The time sequence of this segment is shown in Table 6. The symbol of '0' means absence and the symbol '1' means attendance.

Table 5. A segment of a user's attendance data

User ID	Arrival time	Leave time
1	2013/9/27 7:50	
1	2013/10/9 12:49	2013/10/9 23:49

Table 6. A user's time sequence

User Id	Time sequence
1	...10010000000011...

Secondly we mine all users' periodic behaviors. According to the results of the experiment we found that some employees' behavior are not periodic and some employees have the periodic behavior of 1 week, 2 weeks, 3 weeks. Specific experimental results are shown in Figs. 2 and 3.

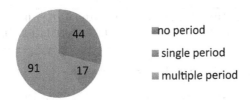

Fig. 2. Statistical table of periodic behavior

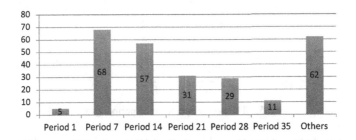

Fig. 3. The number of employees for a special period

As shown in Fig. 2, 44 employees have no periodic behavior, 17 employees just have single periodic behavior and 91 employees have multiple periodic behavior. In truth this company held a meeting weekly and every two weeks.

As shown in Fig. 3 we can find that most employees' periodic behavior are based on the week or weeks. There are five employees who has the periodic behavior of one day. After interviewing them we found these five employees' home are far from the company so they choose live in their company. From the period of 7 to 35 the number of people decreased, it is shown that the larger period is more hard to follow. The number of the other periods is the number of some users' specific periodic behavior which shows that some employees plan their working time for their own work.

5.3 Compared Experiment

In this section we mined periodic behaviors by using periodogram and the algorithm proposed in this paper. The time sequence is shown in Fig. 4. The result of our proposed algorithm is shown in Fig. 5. We can find that this user has two periodic behaviors directly. The employee come to this company every week on Friday and every two weeks on Tuesday.

> 0,1,0,0,1,0,0,0,0,0,1,1,1,0,1,0,0,0,0,0,0,0,0,0,1,0,0,0,1,0,0,1,0,0,0,0,0,
> 0,1,0,0,0,1,0,0,1,0,0,0,0,0,0,1,0,0,0,0,0,0,1,0,0,0,0,0,0,1,0,0,0,1,0,0,1,0,0,

Fig. 4. The time sequence

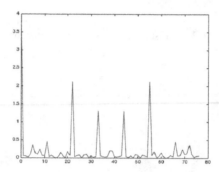

Fig. 5. The power spectrum of periodogram

Firstly we mine multiple periodic behaviors by using the proposed algorithm in this paper. The result can be seen in Table 7. We can find that the result of this experiment is correct.

Table 7. The result of the proposed algorithm

Periods	Time Stamps
7	2,16,30,44,72,
14	5,12,26,33,40,47,54,61,68,75,

Then we mine the periodic behaviors by using the algorithm of periodogram. The density of power spectrum is shown in Fig. 5. The X axis is the frequency and the Y axis is the density of power spectrum. We take the frequency of the power spectrum in the top as the true frequency. According to the Eq. 4 we can calculate the period. Because the period is an integer we get the period of 7. In Eq. 4 *len(s)* means the length of the time sequence.

$$period = \frac{len\,(s)}{frequency} \tag{4}$$

In this experiment we found that the periodogram can't mine all true periods because of spectral leakage. the period of 14 can't be divided by 77 (the length of this time sequence). So when we mine periodic behavior by periodogram we can't get the period of 14 directly. And the periodogram can't tell us when the behavior happen in a period. Based on this period we can't predict the user's behavior in future. The proposed algorithm solved these problem.

5.4 Time and Space Complexity of This Algorithm

The time complexity of this algorithm is $o(n^3)$. But when the algorithm mined a true period it can change the matrix dynamically to reduce the number of judging same periodic behavior. As a result the execution time of this algorithm is reduced. The space complexity of this algorithm is $o(n^2)$. n is depending on the occurrence time of the target state which we concerned.

6 Conclusion

In this paper a period acquiring algorithm is proposed. When there are multiple periodic behavior in a time sequence it can mine them without any pro-knowledge successfully. In this algorithm it avoid the inference of other periodic behavior. In real life periodic behavior can predict a user's future movement and help us understand the user's daily life more directly. In the future we focus on researching how to combine the periodic behavior with location based service for developing the service quality of location based service.

Acknowledgement. This work is partially supported by the National Natural Science Funds of P.R. China under Grants No. 61173042 and No. 61472004, Hongkong, Macao and Taiwan Science and Technology Cooperation Program of China under Grant No. 2013DFM10100, and Special Fund Project of Shanghai Economic and Information Committee under Grant CXY-2013-40.

References

1. Papadimitriou, A., Symeonidis, P., Manolopoulos, Y.: Friendlink: link prediction in social networks via bounded local path traversal. In: 2011 International Conference on Computational Aspects of Social Networks (CASoN). IEEE, pp. 66–71 (2011)
2. Ye, M., Yin, P., Lee, W.C., et al.: Exploiting geographical influence for collaborative point-of-interest recommendation. In: Proceedings of the 34th International ACM SIGIR Conference on Research and Development in Information Retrieval. ACM, pp. 325–334 (2011)
3. Bellotti, V., Begole, B., Chi, E.H, et al.: Activity-based serendipitous recommendations with the Magitti mobile leisure guide. In: Proceedings of the SIGCHI Conference on Human Factors in Computing Systems. ACM, pp. 1157–1166 (2008)
4. Lee, R., Wakamiya, S., Sumiya, K.: Discovery of unusual regional social activities using geo-tagged microblogs. World Wide Web **14**(4), 321–349 (2011)
5. Zheng, Y., Xie, X.: Learning location correlation from GPS trajectories. In: 2010 Eleventh International Conference on Mobile Data Management (MDM). IEEE, pp. 27–32 (2010)
6. Rekimoto, J., Miyaki, T., Ishizawa, T.: LifeTag: WiFi-based continuous location logging for life pattern analysis. In: Hightower, J., Schiele, B., Strang, T. (eds.) LoCA 2007. LNCS, vol. 4718, pp. 35–49. Springer, Heidelberg (2007)
7. Ye, Y., Zheng, Y., Chen, Y., et al.: Mining individual life pattern based on location history. In: Tenth International Conference on Mobile Data Management: Systems, Services and Middleware, 2009. MDM 2009. IEEE, pp. 1–10 (2009)

8. Sirisha, G., Shashi, M., Raju, G.V.P.: Periodic pattern mining–algorithms and applications. Global J. Comput. Sci. Technol. **13**(13), 19 (2014)
9. Bartlett, M.S.: Periodogram analysis and continuous spectra. Biometrika **37**, 1–16 (1950)
10. Wei, W.W.S.: Time Series Analysis. Addison-Wesley publ, Reading (1994)
11. Li, Z., Ding, B., Han, J., et al.: Mining periodic behaviors for moving objects. In: Proceedings of the 16th ACM SIGKDD International Conference on Knowledge Discovery and Data Mining. ACM, pp. 1099–1108 (2010)

Social Aware Mobile Payment Service Popularity Analysis: The Case of WeChat Payment in China

Yue Qu[1,2], Wenge Rong[1,2](✉), Yuanxin Ouyang[1,2],
Hui Chen[1,2], and Zhang Xiong[1,2]

[1] State Key Laboratory of Software Development Environment,
Beihang University, Beijing, China
[2] School of Computer Science and Engineering, Beihang University, Beijing, China
{quyuescse,w.rong,oyyx,chenhui,xiongz}@buaa.edu.cn

Abstract. Since its release in 2013, WeChat payment service has gradually become one of the most popular mobile payment services in China. Different from other mobile payment platforms, WeChat payment bundles with the most popular social network service in China, WeChat. It is then becoming interesting to investigate the reason beneath its popularity by combination of social network and mobile payment. In this research, we applied the technology acceptance model to predict the acceptability of WeChat payment and to identify the variables which attribute to the popularity of WeChat payment. Besides the primary explanatory variables of TAM, the proposed framework is further extended to include the constructs of Social Interaction, Trust, Perceived Enjoyment and Use Context. Online survey has been collected by respondents chosen randomly among users of WeChat payment. The results have shown that the proposed model is able to explain the variance in user's behaviour intention to use WeChat payment services. We hope this study can provide insights to understand the adoption behaviour of social aware mobile payment and service and help further improve their services.

Keywords: Mobile payment · WeChat · Technology acceptance model · Social interaction · Enjoyment · Trust

1 Introduction

With the development of information technology, online services have contributed much to the development of information systems [4]. Other than the rapidly increasing number of online services, most of them disappeared from the market before its widely adoption [9]. The qualitative research about the users acceptance of online services could find some useful hints to explain and suggest the development of service generalizability [3]. Technology Acceptance Model (TAM) is one of the leading approaches been used to analyse users behaviour

© Springer International Publishing Switzerland 2015
L. Yao et al. (Eds.): APSCC 2015, LNCS 9464, pp. 289–299, 2015.
DOI: 10.1007/978-3-319-26979-5_22

pattern [5]. During the last decades, researchers have successfully applied the TAM and/or its extended models to explain the users acceptance of a lot of information technology based systems [10].

In China's E-commerce services market, WeChat payment is a typical winner by its explosive user growth and continuous development [15]. Tencent released WeChat payment in 2013 and bundled it with WeChat 5.0 [16]. Moreover, through the promotion called "Red Envelope", WeChat payment attracted more than 8 million users giving away 3 billion RMB on the Chinese New Year's Eve in 2014 [16]. The success of WeChat payment draws people's attention. Compared with other online payment services, WeChat payment is obviously different due to its social factor. WeChat is currently the most popular mobile social network service application in China [15]. Due to the fact that online payment and social network apps are two of the most popular applications used daily on smartphones, it is deserved to study in depth the reasons beneath WeChat payment's wide usage. In this research, we propose an extended TAM model with several additional variables, such as social interaction, trust and others, to enhance the understanding of people's intention of using WeChat payment.

The remainder of this paper is organised as follow. In Sect. 2 we will introduce the background about TAM and social mobile payment. Section 3 will present the proposed extended TAM model and list the objectives and hypotheses. In Sect. 4 we will present the data processing and discuss the experimental results. At last Sect. 5 concludes the paper and points possible future work.

2   Theoretical Foundations and Related Work

2.1   Technology Acceptance Model

In the area of information systems, there is a need for researchers to understand the reasons behind the users' actual usage of IT system. Among the major approaches, Technology Acceptance Model (TAM) has become one of the most popularly used techniques to elaborate the rationality of user behaviour. During the last decades, TAM has been successfully transferred to different structures to applied to lots of research domains and related applications such as enterprise resource planning systems [10], education [11], finance [17] and E-commerce [3]. In this TAM model, the Behaviour Intention, which is influenced by Perceived Usefulness and Perceived Ease of Use, is the major determinant for a user to accept or reject a certain system. Furthermore, Perceived Usefulness and Perceived Ease of Use will be affected by several external stimulus in different research situations [6]. Due to that TAM has evolved into a leading model in predicting and explaining an information systems acceptance, it is believed the TAM is also appropriate to analyse the WeChat payment's popularity.

2.2   Social Network Based Mobile Payment Analysis

With the development of E-commerce and smartphones, mobile payment has been rapidly developed and has become an important part of E-commerce since it's convenience [14]. Moreover, with the privacy of mobile phone, people started to trust the mobile payment services more. From the studies by Mallat, they concluded the advantage of mobile payment was affected by the mobile technology, the using circumstance and the situational factors [12]. Apart from mobile payment, we also take focus on mobile social network services, such as Facebook, Twitter, WeChat and so on. A lot of researches on how people use social network have been conducted. For example, Klein summarised that the personal activity status of key members was highly correlated with their structural centrality measures, as such friends might always have the similar choices and practice [8].

If we put mobile payment services and social network services together, what will happen? The WeChat Payment is a good answer about this idea and has reached an initial success. In fact, socialism is very useful when generalising services including E-commerce services. Liébana-Cabanillas supplied a modification of the classical technological acceptance models (TRA and TAM) to analyse users' acceptance of mobile payment in VSN. They discovered that the acceptance process of online payment had big influence on the diffusion of mobile payment systems on VSN [9].

3 Hypotheses

This paper proposed an extended TAM model with some other factors to explain the relationship between social network with online payment. The basic TAM model includes factors as perceived ease of use (PEOU), perceived usefulness (PU), and behaviour intention (BI) [6]. Trust(TR) is the foundation of financial transactions, as ensuring the financial security is very important to people's choices about online payment [3]. Different from other online payment services, WeChat payment applies some interesting activities to improve the users' enthusiasm. Therefore perceived enjoyment (PE) is a new and important factor to polish the model. As a mobile service, use context (UC) is also a necessary factor which should be added in the model to show the affection of using situation and technology support [1]. Furthermore, considering the advantage of WeChat platform in social activity, we add the social interaction (SI) to the proposed model. The conceptual framework is depicted as in Fig. 1 and all factors and related hypotheses will be described in detail in following subsections.

3.1 TAM

Davis argued that the core constructs of TAM model is perceived usefulness (PU) and perceived ease of use (PEOU) [6]. In our research, PU is defined as the degree to which a technology is perceived as providing benefits in performing certain activities [6]. Based on the function of WeChat payment, PU can be defined various innovations because it reflects the influence on our life and habit by the service. PEOU is defined as the extent to which a technology is perceived as being easy to understand and use [6]. With the use experience, in this experiment, PEOU stands for easy-learning and easy-operating. Consequently, the following hypotheses is proposed:

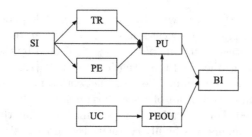

Fig. 1. Path model

H1. Perceived usefulness (PU) positively influences behaviour intention (BI) to use WeChat payment.
H2. Perceived ease of use (PEOU) positively influences behaviour intention (BI) on WeChat payment.
H3. Perceived ease of use (PEOU) positively influences perceived usefulness (PU) of using WeChat payment.

3.2   Trust

Trust is one of the most important foundations for financial transactions, especially in E-commerce [3]. In E-commerce applications, the trust degree mainly depends on the reliability of the system technology [18]. Because of privacy and account security, information confidentiality is very important. Furthermore, trust also depends on the reliability of the users' financial partners. Since trust may change user' awareness of a mobile payment service and judging on the usefulness, we made following hypotheses:

H4. Trust (TR) positively influences perceived usefulness (PU) of using social network mobile payment.

3.3   Perceived Enjoyment

Perceived enjoyment (PE) is a new consideration for E-payment, while it has principal influence on social network services. If the users can get enjoyment from one service, their attitude towards adoption will be positive [11]. In this research, PE reflects the pleasant experience of the service, specially with some activities. For example, from 2014, WeChat payment released an activity to offer the function that people could send "Red Envelope" to friends online with some wishing words by using WeChat payment. This activity spread quickly among Chinese people and attracted many users to WeChat payment because it fitted the Chinese traditional habits and provided entertainment experience. Based on the success of "Red Envelope", we made the following hypotheses:

H5. Perceived enjoyment (PE) positively influences perceived usefulness (PU) of using social network mobile payment.

3.4   Use Context

Use Context (UC) means the environment of using the technology and can be further divided into hardware environment and software situation [13]. In this research, the hardware environment can be considered as the using environment and technology of mobile phones. The software situation includes mobility, compatibility and relates to users' lifestyles [1]. As many contextual situations may reflect the users' experience on social network mobile payment, our study focused on some situations such with PEOU. For example, sometime users need to deal with a commerce transaction urgently but without cash and ATM machine. They can use WeChat payment to solve this problem without the limitations of the external devices. This situation may improve the user's convenience experience of the service. Hence, the hypothesis is posited as follows:

H6. Use context (UC) positively influences perceived ease of use (PEOU) of using social network mobile payment.

3.5   Social Interaction

Social interaction(SI) is very common in our daily life. In this research, we did the experiment on the case of WeChat payment, which offered a wide social network platform to traditional E-commerce. We proposed that SI was a necessary factor in the extended TAM model. Firstly, the social connections between a payer and a payee may determine the trust of financial transactions [19]. Moreover, most contractors in WeChat payment are friends in users' real life, so they trust each other more [18]. Consequently, we assume that SI is connected with TR. If someone takes part in some social activities on WeChat payment with his friends, the experience can influence his basic awareness of the service such as perceived enjoyment (PE) and perceived usefulness (PU). For instance, sharing coupon is an activity where users can share their online coupon in WeChat Friends Circle (this is a platform like Facebook to share information to friends). This activity increase the using frequency of WeChat payment and improve the connections in users' life. So we assume that SI is connected with PE and PU. For the above reasons, we propose the following hypotheses:

H7. Social interaction (SI) positively influences trust (TR) of social network mobile payment.
H8. Social interaction (SI) positively influences perceived enjoyment (PE) of social network mobile payment.
H9. Social interaction (SI) positively influences perceived usefulness (PU) of using social network mobile payment.

4   Analysis and Result

4.1   Data Collection

Our experiment questionnaire consists of two parts. The first one is the survey of basic personal information covering the gender, age and the basic user situation of mobile phone, WeChat, WeChat payment. There are 26 questions in the

Table 1. Questionnaire

Items	Questions
SI1	Many of my friends use WeChat payment
SI2	I like to join some online financial activities with my friends on Wechat payment
SI3	I enjoy sharing online coupon information with friends through Wechat payment
SI4	Sharing online coupon through Wechat payment is advantageous to me and others
TR1	Wechat payment always provides reliable financial services
TR2	The risk of abuse of usage information is low when using Wechat payment
TR3	The risk of abuse of billing information is low when using Wechat payment
PE1	I think that using WeChat payment is enjoyable
PE2	I have fun interacting with WeChat payment
PE3	The activities in Wechat payment such as red Envelope is interesting
UC1	I use Wechat payment because it can make the deal more easier
UC2	I think WeChat payment is easily accessible and portable
UC3	I believe WeChat payment is compatible with other mobile services
PEOU1	It is easy for me to perform the actions required to use Wechat payment
PEOU2	Skillfully using WeChat payment is easy for me
PEOU3	I expect that my interaction with WeChat payment would be clear and easy
PEOU4	I expect that learning how to use WeChat payment would be easy for me
PU1	WeChat payment improves my living and working efficiency
PU2	Using WeChat payment makes it easier for me to conduct financial transactions
PU3	In some special cases, WeChat payment is more useful than the traditional payment ways
BI1	I expect my use of Wechat payment to continue in the future
BI2	Now I always pay for purchases by using WeChat payment
BI3	I recommend Wechat payment to others who intend to use mobile payment

second part about users' experience on the seven hypothetical factors. In the 26 questions there are 3 reverse questions for judging whether an exam attitude seriously enough. Each question was measured on a 7-point Likert scale from "strongly agree (7)" to "strongly disagree (1)". Finally we have 23 questions for analysis, as shown in Table 1.

Initially we obtained 420 groups of data. To improve the data's validity, we eliminated (1) the responses of the informants who never used WeChat payment service; (2) insincere responses through data filtering on the three reverse questions; (3) insincere responses which looks like "Straight-Line" answers or "Wave" answers. Finally, 320 groups of data are filtered to conduct the experiment.

4.2   Exploratory Factor Analysis

It is necessary to conduct exploratory factor analysis to valid if the sample data is acceptable as the experimental basis. In this research, we firstly analyse

Table 2. Results of factor analysis

Dimensions	Items	Mean	Std.D	Loading	V.E	AVE	Cronbach's alpha	CR
SI		5.302				0.5	0.722	0.797
	SI1	4.819	1.921	0.612	0.628			
	SI2	5.322	1.403	0.702	0.618			
	SI3	5.422	1.494	0.696	0.668			
	SI4	5.644	1.429	0.797	0.685			
TR		4.465				0.58	0.734	0.807
	TR1	4.466	1.527	0.767	0.690			
	TR2	4.363	1.492	0.801	0.718			
	TR3	4.566	1.598	0.721	0.684			
vPE		5.608				0.54	0.702	0.779
	PE1	5.041	1.654	0.702	0.612			
	PE2	5.831	1.356	0.793	0.726			
	PE3	5.953	1.339	0.709	0.612			
UC		5.646				0.51	0.757	0.750
	UC1	5.738	1.279	0.742	0.693			
	UC2	5.916	1.151	0.793	0.758			
	UC3	5.284	1.450	0.578	0.561			
PEOU		6.177				0.52	0.732	0.810
	PEOU1	6.238	2.528	0.708	0.601			
	PEOU2	6.188	1.210	0.805	0.712			
	PEOU3	6.378	1.049	0.733	0.578			
	PEOU4	5.903	1.339	0.620	0.523			
PU		5.739				0.51	0.732	0.753
	PU1	5.697	1.350	0.705	0.601			
	PU2	5.906	1.345	0.683	0.645			
	PU3	5.613	1.486	0.742	0.654			
BI		4.585				0.52	0.717	0.761
	BI1	4.078	1.849	0.646	0.576			
	BI2	4.678	1.655	0.788	0.693			
	BI3	5.000	1.689	0.714	0.627			

the reliability of data to see whether the data can be fit for exploratory factor analysis. The analysis result is shown in Table 2.

Firstly, we use Cronbach's alpha coefficient to test the internal consistency of data. The results is listed in Table 2, where the total Cronbach's alpha coefficient is 0.872 and all the coefficients of each factor are more than 0.700, which means the collected data meet the general requirement that the total Cronbach's alpha coefficient should be greater than 0.8 and the coefficient of each element should be greater than 0.7. We also test the factor loading of each item and the AVE of each factor. The results are all exceed 0.5, which indicates the convergent validity. All the composite reliability (CR) values are over 0.7, which means the reliability of each factor is satisfied. From the exploratory factor analysis, it is concluded that the collected data is satisfied to conducting further analysis [7].

Secondly, we need to conduct validity analysis for the construct validity of the assessment. Construct validity includes two categories, i.e., convergent validity and differentiate validity. We used Bartlett's test of sphericity and the Kaiser-Meyer-Olkin (KMO) measure of sampling adequacy to measure the collected data. Based on the commonly used KMO measures by Kaiser, the value over 0.7 will be acceptable [13]. Our KMO measure = 0.860, which means the collected data has valid convergent validity. $X^2 = 2366.528$ (Sig. < .000) means the data has distinction validity [13].

Thirdly, we also need to judge if the distinction validity of the data is reasonable or not. In Table 3, the results shown at diagonal are the squared root of the average variance extracted (AVE) value for each factor, which are greater than the correlation between different factors.

Table 3. Intercorrelations between factors

	SI	TR	PE	UC	PEOU	PU	BI
SI	.707						
TR	.342	.762					
PE	.374	.264	.735				
UC	.403	.393	.459	.714			
PEOU	.231	.241	.232	.364	.721		
PU	.251	.337	.377	.451	.399	.714	
BI	.283	.349	.303	.479	.173	.503	.721

4.3 Confirmatory Factor Analysis

Confirmatory factor analysis, also known as structural equation modelling (SEM), is statistical method to analyse the relationship between the variables of covariance matrix. After conducting exploratory factor analysis, we need to use confirmatory factor analysis method to test the proposed hypothesis model. Confirmatory factor analysis results are normally divided into two parts, i.e., path analysis and fit test of model.

Path analysis is used to estimate the relationship between the variables in the proposed model. We used the statistical tools to add the data to the hypothesis model and calculate the weight coefficients of the hypothetical path. The result is shown as Fig. 2, where we can judge whether the path is accepted by the result of the weight coefficients. Moreover, the coefficients also reflect the strength of the causal relationships.

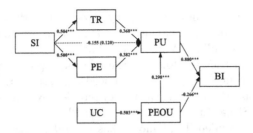

Fig. 2. Path verification

Table 4. Fit indices for the measurement

Indices	X^2	GFI	AGFI	RMSEA	CFI	NFI
Results	1.068	0.947	0.926	0.015	0.994	0.913
Criteria	<5.0	>0.85	>0.80	≤0.06	>0.90	>0.85

The result of model fit indices is shown at Table 4. Since all the fit indexes meet the requirements, it is concluded that the proposed model's fitness is satisified.

4.4 Discussion

This study developed a theoretical framework based on TAM model and discussed the structural equation modelling analysis of social aware mobile payment service adoption. The findings in this study provided empirical support for the proposed model. The results clarified our understanding of people's attitudes and intentions towards using social aware mobile payment services, and also helped reveal implications for the successful implementation of WeChat payment in China.

This study shows that PU has a strong effect on BI. It implies that usefulness-oriented technologies will be paid much attention by the markets. However, the relationship between PEOU and BI is weaken and negative. Because of the steady trend of PEOU data, we think PEOU isn't a typical character of WeChat payment and excessively simple steps can make people to doubt the safety of the software [1]. Some information checking steps for security are necessary in E-commerce services.

From the experimental study, it is found that SI has strong influences on PE and TR. Profited from the activities like "Red Envelop", WeChat payment provides a new platform for users to communicate with each other and then closes the relationship among them. This result corroborates previous researches on SI and PE that the interactions between friends can greatly enhance the entertainment of the using experience [2]. Similarly, SI also shows significant role on TR. Trust heavily depends on the social interaction of the partners in the transactions [18]. So the relationship of friends devotes to strong the user's trust attitude to WeChat payment. In addition, it is shown that SI doesn't have direct effect on PU, while SI has indirect effect by PE and TR. Previous studies have suggested that the PU of online payment was supported by the use experience of dealing financial trade online. While in WeChat payment, the online trade will not be limited within friends. However, the relationship of friends can increase the using entertainment of trade directly indeed, and strong trust attitude both of which can increase the using frequency of mobile payment services. It could be concluded that SI has indirectly influence on PU.

This proposed extended TAM model has several practical and theoretical implications for researchers and engineers to develop popular mobile social payment services. This study provided some in-depth analysis of popularity of WeChat payment in China and then can be applied into development of E-commerce industry. It is argued that SI can bring more chances to the popularity of mobile services, which also be verified by other applications, such as games and E-learning [1,11]. The mobile services should provide excellent social interaction experience to encourage users to share their fun.

5   Conclusion and Future Work

In this research, we use WeChat payment as a case to study the determinants beneath this popular applications. To this end, technology acceptance model is employed and some amending predictors are integrated from social and mobile's perspective. It is believed this research provide insights relevant for Web services to understand the adoption behaviour and further improve their service computing.

Similar to other research, there are several limitations in this study. The major issue is related to the users of WeChat payment. All responses of the questionnaire are from Mainland China. Thus, the results of this study may be difficult to generalise to other context. It would be interesting to extend this work to an international context and perhaps consider other social networks. Furthermore, the proposed model and questions of the investigation mainly focus on the usage intention of using WeChat payment service, and the result revealed the factors and their relationship for people to use WeChat payment or not. However, usage continuation is also important for an application to survive in the market competition. It would be meaningful to excavate the factors about continuity in following research.

Acknowledgements. This work was partially supported by the State Key Laboratory of Software Development Environment of China (No. SKLSDE-2015ZX-23), the National Natural Science Foundation of China (No. 61472021), the National High Technology Research and Development Program of China (No. 2013AA01A601), and the Fundamental Research Funds for the Central Universities.

References

1. Liu, Y., Li, H.: Exploring the impact of use context on mobile hedonic services adoption: an empirical study on mobile gaming in china. Comput. Hum. Behav. **27**(2), 890–898 (2011)
2. Hausman, D.A., Soares, A.M., Pinho, J.C.: Advertising in online social networks: the role of perceived enjoyment and social influence. J. Res. Interact. Mark. **8**(3), 245–263 (2014)
3. Carminati, B., Ferrari, E., Tran, N.H.: Enforcing trust preferences in mobile person-to-person payments. In: International Conference on Social Computing, pp. 429–434 (2013)
4. Ciganek, A.P., Haines, M.N., Haseman, W.D.: Challenges of adopting web services: experiences from the financial industry. In: Proceedings of Annual Hawaii International Conference on System Sciences, pp. 168b–168b (2005)
5. Davis, F.: Perceived usefulness, perceived ease of use, and user acceptance of information technology. MIS Q. **13**(3), 319–340 (1989)
6. Davis, F., Bagozzi, R., Warshaw, P.: User acceptance of computer technology: a comparison on two theoretical models. Manage. Sci. **35**(8), 729–736 (1989)
7. Hair, J.F., Black,W.C., Babin, B.J., Anderson, R.E., Tatham, R.L.: Multivariate data analysis, 6th Edition. Pearson Prentice Hall Upper Saddle River, NJ (2006)
8. Klein, A., Ahlf, H., Sharma, V.: Social activity and structural centrality in online social networks. Telematics Inform. **32**(2), 321–332 (2015)
9. Liébana-Cabanillas, F., Sánchez-Fernández, J., Muñoz-Leiva, F.: The moderating effect of experience in the adoption of mobile payment tools in virtual social networks. Int. J. Inf. Manage. **34**(2), 151–166 (2014)
10. Lin, J.C., Lu, H.: Towards an understanding of the behavioural intention to use a web site. Int. J. Inf. Manage. **20**(3), 197–208 (2000)
11. Mackey, T.P., Ho, J.: Exploring the relationships between web usability and students' perceived learning in web-based multimedia tutorials. Comput. Educ. **50**(1), 386–409 (2008)
12. Mallat, N.: Exploring consumer adoption of mobile payments-a qualitative study. J. Strateg. Inf. Syst. **16**(4), 413–432 (2007)
13. Martin, F., Ertzberger, J.: Here and now mobile learning: an experimental study on the use of mobile technology. Comput. Educ. **68**, 76–85 (2013)
14. Mohamadi, M., Ranjbaran, T.: Effective factors on the success or failure of the online payment systems, focusing on human factors. In: e-Commerce in Developing Countries (2013)
15. Team, C.: Insights of china mobile payment (wechat, alipay) users (2014). http://www.chinainternetwatch.com/10561/mobile-payment-users-insight/
16. Team, C.: echat, weibo or alipay? who won hongbao war in 2015? (2015). http://www.chinainternetwatch.com/12424/hongbao-war-2015/
17. Verma, N., Singh, J.: Improved web mining for e-commerce website restructuring. In: Computational Intelligence & Communication Technology (CICT), pp. 155–160 (2015)
18. Yang, Q., Pang, C., Liu, L., Yen, D.C., Tarn, J.M.: Exploring consumer perceived risk and trust for online payments: an empirical study in chinas younger generation. Comput. Hum. Behav. **50**, 9–24 (2015)
19. Zhou, T.: An empirical examination of continuance intention of mobile payment services. Decis. Support Syst. **54**(2), 1085–1091 (2013)

Towards Truly Elastic Distributed Graph Computing in the Cloud

Lu Lu, Xuanhua Shi, and Hai Jin[✉]

Services Computing Technology and System Lab, Cluster and Grid Computing Lab,
School of Computer Science and Technology,
Huazhong University of Science and Technology, Wuhan 430074, China
{llu,xhshi,hjin}@hust.edu.cn

Abstract. Elasticity is very important to the scale-out distributed systems running on today's large-scale multi-tenant clouds, regardless public or private. An elastic distributed data processing system must have the capability of: (1) dynamically balancing the computing load among workers due to their performance heterogeneity and dynamicity; (2) fast recovering the lost memory state of failure workers with acceptable overheads during the regular execution.

Unfortunately, we found that the design of the state-of-the-art distributed graph computing system only works well in small sized dedicated clusters. We implement a distributed graph computing prototype, *X-Graph*, and demonstrate the capabilities of being elastic in three ways. First, we present *menger*, a novel two-level graph partition framework, which further splits one worker-level partition into several sub-partitions as the basic migration units, and each has the "migration affinity" with one of the other workers. Second, we implement a dynamical load balancer based on *menger*, which prefers the worker that has the affinity of the sub-partition to be migrated as the destination, and completely avoids the costly sophistical graph re-partitioning algorithms. Third, we implement a differentiated replication frame-work, which supports parallel recovery for lost partitions just like general-purpose dataflow systems.

1 Introduction

Graph mining is important for many "Big Data" applications such as webpage ranking and link prediction for social networks. Recently several parallel graph computing frameworks have been proposed for both the single multi-core node [13, 14, 17, 19] and the cluster [1, 8, 15]. These frameworks follow the Google Pregel's think-as-a-vertex programming model [16] in which each vertex updates its own data based on the data of neighboring vertices, and perform the vertex update function on all vertices iteratively until the computation converges.

The state-of-the-art parallel graph computing frame-work project, GraphLab [2], introduces several classical optimization methods into the system engine to accelerate the computation convergence rate, such as asynchronous computing, dynamic priority ordering and delta caching. PowerGraph (GraphLab 2.1) [8] uses a novel vertex-cut partitioning framework to evenly distributed load among workers statically before the job execution, and also minimizes the cross-node communication during the iterative

© Springer International Publishing Switzerland 2015
L. Yao et al. (Eds.): APSCC 2015, LNCS 9464, pp. 300–309, 2015.
DOI: 10.1007/978-3-319-26979-5_23

computation. It also uses the classical periodical full checkpointing to stable storage for fault tolerance, which has to re-computing the full state of application from the last checkpoint even lost only one partition on the failure worker.

PowerGraph works well in small sized dedicated clusters which typically includes tens to hundreds of homogeneous machines, and failures are much less common at such scale. Unfortunately, today's large-scale cloud data centers bring many new challenges and need for **graph-aware** dynamic load balancing and fault tolerance for elastic distributed graph computing:

Hardware Heterogeneity. Thanks to the rapid development of power-efficient processor micro architecture such as ARM and Atom, the system administrators are planning toward a mix of high-performance and low-power nodes of heterogeneous CPU architectures for cloud data centers, which could exploits the application diversity to provide differentiated price-performance and to optimize power, real-estate foot print, and cost. Even heterogeneity is not introduced by purchase planning, the practice of staged addition of nodes to the cluster over time implies that the nodes are often some hardware generations apart resulting in performance variation. The popular cluster scheduling model used in cloud data centers that treats the heterogeneous CPU cores as homogeneous "task slots" can significantly impact the performance of distributed applications [16]. A possible solution is dividing the heterogeneous physical CPU cores into several homogeneous virtual CPU cores statically. But the performance difference between high-performance and low-power micro architectures dynamically varies for different running applications [4].

Resource Management of Multi-tenant Clouds. Both of the public cloud and the private cloud are platforms sharing by diverse applications and different users. The resources of these platforms are usually managed by the recently proposed two-level cluster schedulers such as Mesos [10], YARN [12] and Omega [21]. These kinds of cluster schedulers usually could dynamically allocate hardware resources to distributed applications during the job execution. Tasks are put into a logical containers such as virtual machines or cgroups, each occupies a portion of the resource of a server. Google also founds the current isolation methods have few defenses against performance interference in shared components such as CPU caches and memory buses when several tasks of different applications are consolidated onto a same machine, so distributed applications can experience unpredictable performance caused by other tasks behavior [25]. Therefore, an elastic graph computing framework must have the ability to migrate tasks to newly arrived or light-loaded nodes after the static graph partition.

Node Failures and Task Preemption. In large-scale data centers that full of commodity hardware, component failures and straggler workers are the norm rather than the exception [7]. The regular running batched tasks could also be killed by the resource manager according to internal scheduling policies. YARN resource manager supports task preemption scheduling and may kill tasks of overcommitting jobs to guarantee the fair sharing among different users [12]. Googles central cluster scheduler periodically collects the CPI sample data from worker nodes to detect performance anomalies, and may kill these outlier tasks to guarantee the QoS of the longrunning latency-sensitive services [25]. These situations requite the distributed graph computing frameworks support low-overhead fault tolerance as well as fast recovery after worker failures and stragglers occur.

Our goal is to develop a distributed graph processing system that fulfills all of these requirements and supports the user-friendly vertex-centric programming models, while achieving the better performance in the largescale cloud data centers. This paper makes the following contributions: (1) a novel two-level graph partitioning framework called menger that supports differentiated processing of outer vertices (that have replicas on remote workers) and inner vertices (that have only local replicas). It prefers to groups the outer vertices that have remote replicas on the same worker into the same sub-partition; (2) a graph aware dynamic load balancing based on menger partitioning; (3) a graph aware fault tolerance based on menger which supports low-overhead fault tolerance and fast parallel recovery.

2   Background and Model

Most parallel graph computing systems, including PowerGraph, provides an abstraction of core graph computation logic to programmers as a user-defined vertex update function F, which is executed in parallel on each vertex $v \in V$ of the input graph $G = \{V,E\}$. The userdefined vertex-update function $F(v)$, which can only read and modify the values of the current updating vertex v and its adjacent edges. The vertex-update function is executed iteratively on all the vertices of the input graph until a convergent condition is satisfied.

Most existing parallel graph computing systems implement the Bulk-Synchronous Parallel (BSP) model executing computation in lock-step, which defines that user-defined functions can only observe values from the previous iteration. BSP is widely adopted in parallel and distributed computing as it is simple to implement and allows maximum degree of parallelism during the computation. PowerGraph also provides an asynchronous model of computation which makes an vertexupdate function be able to access the most recent updated values of the edges and the vertices of the current iteration round. PowerGraph adapts this model to not synchronize the computing threads by lock-step barriers at all to tolerate the straggler workers. Because the avoidance of synchronize barriers has very limited effects [6], our X-Graph framework supports the synchronous model and the asynchronous model, both perform computation in lock-step to simplify runtime load balancing and fault tolerance.

3   System Architecture

We have implemented the PowerGraph system. Figure 1 presents the architecture of our system, which is formed by two main components: (1) the *computing worker*, an extended PowerGraph execution engine with menger partitioning, dynamic load balancing and differentiated replication. (2) the *coordinator*, a central service which coordinates iterative computation and makes the plans for migration and replication.

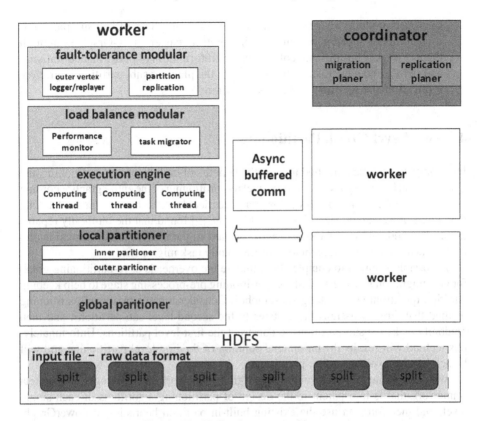

Fig. 1. The X-Graph system architecture

Computing Worker: In the X-Graph system, the input graph blocks are loaded from the underlying distributed file system such as HDFS, then partitioned and processed by each computing worker in a distributed fashion.

(1) *Menger partitioner:* The global partitioner uses the built-in vertex-cut partitioner of PowerGraph to partition all the vertices of input graph into the first-level partition for each worker. The local partitioner further assigns the vertices of local partitions into second-level sub-partitions.

(2) *Execution Engine:* A modified version of Power-Graph's execution engine. It launches several computing threads to perform the update-function computation on each local sub-partition one by one. It also buffers and exchanges the updated values of outer vertices for remote replicas.

(3) *Load balance modular:* It continually collects the runtime performance statistics during the task execution and sends them to the coordinator for planning. Once the coordinator triggers a sub-partition migration, the load balance modular on the sender will push the sub-partition to the receiver.

(4) *Fault tolerance modular:* It logs the values of all the outer vertices to the local disks every lock-steps and replicates each full sub-partition to remote workers every I iterations. Once the coordinator triggers a recovery action, the fault

tolerance modular on a recovering node uses the replicas of lost sub-partitions to recompute them. Coordinator: the central master of X-Graph system. It detects worker failures and collects useful information of performance statistics and vertex and edge distributions to make the plans of sub–partition migration, replication and recovery of the lost partitions.

4   Two-Level Graph Partitioning

Intuitively, the simplest method to keep the benefits of both the graph partitioning and dynamic load balancing is using vertex level re-partitioning of the entire graph according to the migration plan whenever the load balancer decides to re-balance the load among workers. Unfortunately, Hendrickson and Devine [9] found that the on-the-fly repartitioning methods are either too expensive or hard to parallelize, the computation and communication cost is unacceptable for frequently task migrations.

A possible solution to completely eliminate the overheads of re-partitioning graph for task migration is using a two-level partitioning pre-processing stage to help keeping partition quality after migrating sub-graphes. Consider a naive two-level partitioning method that directly assigns the vertices to the second-level sub-partitions, and then randomly picks the sub-partitions to construct the first-level partitions. Unfortunately, this method will results the un-optimal solution of the first-level partitioning and will be helpless for avoidance of re-partitioning.

We design a novel two-level graph partition framework called menger. It is a meta framework that is transparent to the design of partitioning algorithms of each single level, and therefore can use the existing built-in partition heuristics of PowerGraph. A general workflow of a menger two-level partitioning is a first-level global graph partitioning followed by an extra local reassignment phase to generate sub-partitions from the node-level partition on each worker. As shown in Fig. 2, the left side illustrates the global view of execution flow of graph partitioning using the menger framework, and right side demonstrates the corresponding view for the input graph in each partitioning phase. First, the worker on each machine loads raw graph data splits from underlying distributed file systems such as HDFS in parallel, and dispatches them using the single-level partitioning algorithms chosen by the users. The edges and their owning vertices are assigned to the machines according to some heuristic strategies, usually based on the current statistics of edge number and vertex degree distribution, and one vertex replica will be chosen as the master replica for this vertex for computation. Then all the n workers start to perform a second-level node-locally partitioning phase, each worker will further split its first-level partition to $n-1$ sub-partitions, each sub-partition will be suitable for migrating to one of the other $n-1$ workers exclusively. To get finer migration granularity, it can also generate $m \times (n-1)$ sub-partitions for each node-level partition. During the locally partitioning, each worker firstly counts the replica number of the vertices in its local partition to distinguish the outer vertices that have remote replicas from the inner vertices just owned by this worker only. After that, all the outer vertex are reassigned before the inner vertices to sub-partitions according to the rules: if one of the adjacent vertices of an edge is an outer vertex that has a replica on node k, and another is an inner vertex, this edge will be forcibly assigned to the sub-partition k; if both of the two adjacent vertices of one edge have remote replicas on node k, this edge will be also forcibly assigned to the sub-partition k; if an edge has two outer adjacent vertices and their remote replica sets are disjoint, to break the ties, it will be randomly assigned to a sub-partition according to the set joint by the two remote replica sets. Finally, each worker assigns the inner vertices of its local partition to sub-partitions according to the user chosen single-level partitioning algorithm again.

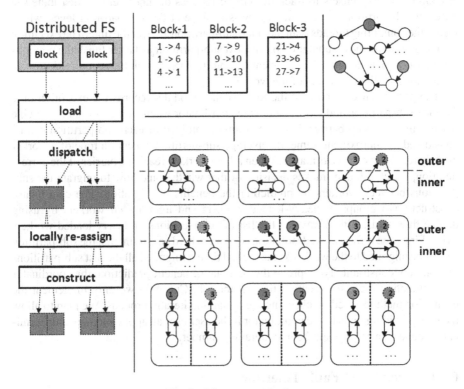

Fig. 2. Menger partitioning

According to the above rules, the sub-partition k will own mostly all the outer vertices that have a replica on worker k. Because the messages sent to a vertex along its edges will be combined locally on each worker, the remote data exchanges are only used for synchronized the value between replicas of each outer vertex. All the data exchange channels of outer vertices in sub-partition k will be "absorbed" after migrating it to the worker k.

5   Execution Engine and Dynamic Load Balancing

X-Graph does not need to maintain a full global vertex node mapping after graph partitioning neither on the computing workers nor on the coordinator. Because the execution engine of each worker performs computation just on its own sub-graph, it only needs to access the mapping information of the replicas of local outer vertices. Just like Power-Graph, X-Graph allocates a bitset to record the replica-worker mapping for each outer vertex. X-Graph also uses the data structure similar to the PowerGraph partition for menger sub-partitions. It includes (1) a vertex value array sorted by vertex id; (2) an adjacent edges list sorted by the source vertex id first and then by the destination vertex id; (3) a corresponding edge value array with the same ordering of the adjacent list. X-Graph also uses bitsets to track the virtual replicas of inner vertices distributed in different sub-partitions. The concurrency is achieve at the sub-partition level, which means the computing threads scan the same sub-partition all over the time. The execution engine only starts the computation of the next sub-partition just after finish the previous sub-partition. The number of concurrent computing threads will be set to the same with the number of allocated cores of the worker.

The performance monitor of the load balance modular records the execution time for each sub-partition and send them to the coordinator whenever the execution engine reaches the lock-step barrier. Because computation load of each sub-partition can be considered as approximate same, the average sub-partition execution time of a worker directly represents its performance. When all the workers reach the barrier of the current iteration, the coordinator makes decision of that whether and how to migrate the sub-partitions between workers by predicting the benefits of migration. The coordinator constructs two groups of heavy-loaded workers and light-loaded workers by using double thresholds. It then traverse the list of sub-partitions on heavy-loaded workers twice. First time it choose the destinations under the hard constraint that the sub-partition k must be migrated to the worker k; the second time it migrated all the rest sub-partitions without any constraint. To improve the accurate of benefit prediction, the coordinator also estimation the rest execution time of the job. Currently we only implement the estimation method based on checking the remaining iteration rounds. We plan to allow the user providing hints to the coordinator if the user uses an application-specific termination condition rather than the max iteration number.

6   Differentiated Fault Tolerance

The existing distributed graph computing frameworks uses classical checkpointing/recovery mechanism for fault-tolerance. The shared stable storage devices for checkpointing are rarely deployed in cloud data centers due to the poor scalability. Even checkpointing the full memory state into a distributed file system is viable, it is also very important to choose the right number of checkpointing interval due to the significant I/O costs.

We implement a fault tolerance mechanism in X-Graph which does not checkpoint but just maintain sub-partition replicas on all the workers. More importantly, it differentiated the replication process of inner vertices and outer vertices in each sub-partitions. Because the outer vertices have already been replicated actually, X-Graph dose not explicitly replicate them again and only replicate inner vertices every I iteration rounds. The value of I can be set relative big because the lost sub-partitions can be re-computed in parallel. The outer vertices logger in fault tolerance modular writes their values on disks because that the recovery computation needs to replay these values of every iterations since the last inner vertices replication. The logging files also can be garbage-collected every I iterations. Each worker try to replicates the sub-partition k to worker k. Therefore the recovery computation can be performed by all the alive workers whenever one of the workers crashes.

7   Preliminary Evaluation and Discussions

We compare our X-Graph with the original PowerGraph on a 65 node cluster, using a single application that computes the pagerank value of the input graph. The machines are running 64-bit CentOS 6, are equipped with two Octa-core processors and 32 GB of memory, and have one 1 TB SAS disk each. One of the node runs the master only and the rest 64 nodes run computing workers. We use graph generation tools to generate a typical power-law graph and a regular graph both have 40 million vertices and 1.5 billion edges.

Firstly, we evaluate the replication factor of vertices of three different static partitioning algorithms before job execution: the default PowerGraph partitioner, simple random two-level partitioner and the menger partitioner, because replicate factor directly dominates the runtime communication overheads. The replicate factors are 15.39, 17.43 and 15.39 perspectively for the power-law graph, 14.90, 16.02 and 14.90 for the regular graph. Menger performs exactly as well as the default Power-Graph partitioner because it does not affect the first-level global partitioning at all.

Then we compare the pagerank computing performance of X-Graph with Power-Graph We launches background CPU-intensive threads to simulate slow workers and a 10× slower straggler. The results show that X-Graph increase the job performance more than 40 % than PowerGraph for the regular graph but only increase less than 15 % for the power-law graph. We found that X-Graph works much worse on the power-law graph because the partitioning of this kind of graphs generate a large number of hard-to-assign edges in the second level local partitioning of menger. It significantly reduces the number of absorbed exchange channel after the sub-partition migration. Currently, we are working on designing more sophistic heuristics of edge assigning for outer vertices to address this problem. We also notice that it is important to make the coordinator making right decisions of migration and replication planning in the real environment. We are also working on a comprehensive performance study for dynamic load balancing and differentiated fault tolerance by using public-available real traces collected from production clouds such as Google data centers. We plan to design adaptive planing algorithms based on this study in the future.

8   Related Work

In the past few years, a number of large scale data processing systems have been proposed. In this section, we mainly examine the design aspects of existing work of scalable distributed computing systems for achieving dynamic load balance and fault tolerance.

The Google Pregel and its clones such as the open source project Giraph [1] use the hash-based and range-based partitioning schemes to simplify the vertex location tracking and dynamic load balancing. The partition number is often multiplies of the worker number. They randomly groups partitions into a worker, and blindly pick the partition on heavy-loaded workers to migrate to light-loaded workers. The distributed Graphlab (GraphLab 2.0) [15] uses very similar approaches. The performance of this design suffer from the poor partition quality significantly [22].

Mizan [11] follows the BSP and message passing model of Pregel, but it uses a graph-unaware dynamic load balancing at the granularity of vertex level. It is not designed for the resource heterogeneity and dynamicity but for the instinct load imbalance during the execution for contain types of applications. GPS [20] uses runtime re-partitioning to further improve the partition quality after the static partitioning, but does not support dynamic load balancing. Surfer [5] designs a adaptation version of static METIS partitioning algorithm for the network heterogeneity in clouds. All of the popular distributed dataflow frameworks [3, 7, 23, 24] can easily perform the parallel recovery to re-compute the lost results of complete tasks on failure workers. This simplicity is achieved through immutable data abstraction borrowed from functional programming.

All the computation is transforming an input dataset to generate a new output dataset. This method can hardly be used by a distributed system maintains large updatable states in main memory. Piccolo [18] allows the users' code performing fine-grained updating on the in-memory state and adopts a full asynchronous computation model which forces it using an expensive distributed snapshot protocol for fault tolerance.

9 Conclusion

In this paper, we present X-Graph, an elastic distributed graph computing system that supports dynamically balancing the computing load among workers and fast recovering the lost memory state of failure workers with acceptable overheads. X-graph uses a novel two-level graph partition framework which further splits one worker-level partition into several sub-partitions, and achieves load balance by migrating the sub-partitions. It implement a differentiated replication frame-work that supports parallel recovery for lost partitions. Our evaluation results show that X-Graph is up to 1.4× faster than the original GraphLab system.

Acknowledgement. This paper is partly supported by the NSFC under grant No. 61433019 and No. 61370104, International Science and Technology Cooperation Program of China under grant No. 2015DFE12860, MOE- Intel Special Research Fund of Information Technology under grant MOE-INTEL-2012-01, and Chinese Universities Scientific Fund under grant No. 2014TS008.

References

1. Giraph. https://giraph.apache.org/
2. Graphlab. http://graphlab.org/
3. Hadoop. http://hadoop.apache.org/

4. Ahmad, F., Chakradhar, S., Raghunathan, A., Vijaykumar, T.N.: Tarazu: optimizing mapreduce on heterogeneous clusters. In: ASPLOS 2012, pp. 61–74 (2012)
5. Chen, R., Weng, X., He, B., Yang, M., Choi, B., Li, X.: Improving large graph processing on partitioned graphs in the cloud. In: SoCC 2012, p. 3 (2012)
6. Cipar, J., Ho, Q., Kim, J.K., Lee, S., Ganger, G.R., Gibson, G.: Solving the straggler problem with bounded staleness. In: HotOS 2013 (2013)
7. Dean, J., Ghemawat, S.: MapReduce: simplified data processing on large clusters. Commun. ACM **51**, 107–113 (2008)
8. Gonzalez, J.E., Low, Y., Gu, H., Bickson, D., Guestrin, C.: PowerGraph: distributed graph-parallel computation on natural graphs. In: OSDI 2012, p. 2 (2012)
9. Hendrickson, B., Devine, K.: Dynamic load balancing in computational mechanics. Comput. Methods Appl. Mech. Eng. **184**, 485–500 (2000)
10. Hindman, B., Konwinski, A., Zaharia, M., Ghodsi, A., Joseph, A.D., Katz, R., Shenker, S., Stoica, I.: Mesos: a platform for fine-grained resource sharing in the data center. In: NSDI 2011, p. 22 (2011)
11. Khayyat, Z., Awara, K., Alonazi, A.: Mizan: a system for dynamic load balancing in large-scale graph processing. In: EuroSys 2013, pp. 169–182 (2013)
12. Kumar, V., Vavilapallih, Murihyh, A.C., Douglasm, C., Agarwali, S., Konarh, M., Evansy, R., Gravesy, T., Lowey, J., Shahh, H., Sethh, S., Sahah, B., Curinom, C., Omaleyh, O., Radiah, S.: Apache hadoop YARN: yet another resource negotiator. In: SoCC 2013, p. 5 (2013)
13. Kyrola, A., Blelloch, G., Guestrin, C.: GraphChi: large-scale graph computation on just a pc. In: OSDI 2012, pp. 31–46 (2012)
14. Low, Y., Gonzalez, J., Kyrola, A., Bickon, D., Guestrin, C., Hellersten, J.M.: GraphLab: a new framework for parallel machine learning. In: UAI 2010 (2010)
15. Low, Y., Gonzalez, J., Kyrola, A., Bickon, D., Guestrin, C., Hellersten, J.M.: Distributed GraphLab: a framework for machine learning in the cloud. In: PVLDB 2012, pp. 716–727 (2012)
16. Malewicz, G., Austern, M.H., L., Hundt, R.: Whare-Map: heterogeneity in homogeneous warehouse-scale computers. In: ISCA 2013, pp. 619–630 (2013)
17. Nguyen, D., Lenharth, A., Pingali, K.: A lightweight infrastructure for graph analytics. In: SOSP 2013, pp. 456–471 (2013)
18. Power, R., Li, J.: Piccolo: building fast, distributed programs with partitioned tables. In: OSDI 2010, pp. 1–14 (2010)
19. Roy, A., Mihailovic, I., Zwaenpoel, W.: X-Stream: edge-centric graph processing using streaming partitions. In: SOSP 2013, pp. 472–488 (2013)
20. Salihoglu, S., Widom, J.: GPS: a graph processing system. In: SSDBM 2013, p. 8 (2013)
21. Schwarzkopf, M., Konwinski, A., Abdelmalek, M., Wilkes, J.: Omega: flexible, scalable schedulers for large compute clusters. In: EuroSys 2013, pp. 351–364 (2013)
22. Stanton, I., Kliot, G.: Streaming graph partitioning for large distributed graphs. In: KDD 2012, pp. 1222–1230 (2012)
23. Yu, Y., Isard, M., Fetterly, D., Budiu, M., Lfarer-Lingsson, Kumar, P., Currey, G.J.: DryadLINQ: a system for general-purpose distributed data-parallel computing using a high-level language. In: OSDI 2008, pp. 1–14 (2008)
24. Zaharia, M., Chowdhury, M., Das, T., Dave, A., Ma, J., Mccauley, M., Frankin, M.J., Shenker, S., Stoca, I.: Resilient distributed datasets: a fault-tolerant abstraction for in-memory cluster computing. In: NSDI 2012 (2012)
25. Zhang, X., Tune, E., Hagmann, R., Jnagal, R., Gokhale, V., Wilkes, J.: CPI2: CPU performance isolation for shared compute cluster. In: EuroSys 2013, pp. 379–391 (2013)

Author Index

Printed in the United States
By Bookmasters